BIOENERGY

Opportunities and Challenges

BIOENERGY
Opportunities and Challenges

Edited by
R. Navanietha Krishnaraj
Jong-Sung Yu

Apple Academic Press Inc.	Apple Academic Press Inc.
3333 Mistwell Crescent	9 Spinnaker Way
Oakville, ON L6L 0A2	Waretown, NJ 08758
Canada	USA

© 2016 by Apple Academic Press, Inc.

First issued in paperback 2021

Exclusive worldwide distribution by CRC Press, a member of Taylor & Francis Group

No claim to original U.S. Government works

ISBN 13: 978-1-77463-383-0 (pbk)
ISBN 13: 978-1-77188-109-8 (hbk)

Library and Archives Canada Cataloguing in Publication

Bioenergy : opportunities and challenges / edited by R. Navanietha Krishnaraj, Jong-Sung Yu.

Includes bibliographical references and index.
ISBN 978-1-77188-109-8 (bound)
1. Biomass energy. I. Krishnaraj, R. Navanietha, author, editor II. Yu, Jong-Sung author, editor

| TP360.B45 2015 | 662'.88 | C2015-903924-X |

Library of Congress Cataloging-in-Publication Data

Bioenergy: opportunities and challenges / [edited by] R. Navanietha Krishnaraj, and Jong-Sung Yu.

pages cm
Includes bibliographical references and index.
ISBN 978-1-77188-109-8 (alk. paper)
1. Biomass energy. 2. Microbial fuel cells. 3. Biodiesel fuels. 4. Lignocellulose.
I. Navanietha Krishnaraj, R. II. Yu, Jong-Sung.

| TP339.B526 2015 | 662'.88--dc23 | 2015020359 |

Apple Academic Press also publishes its books in a variety of electronic formats. Some content that appears in print may not be available in electronic format. For information about Apple Academic Press products, visit our website at **www.appleacademicpress.com** and the CRC Press website at **www.crc-press.com**

ABOUT THE EDITORS

R. Navanietha Krishnaraj

R. Navanietha Krishnaraj, PhD, is currently working as the Technical Officer at the National Institute of Technology, Durgapur. Before joining NIT Durgapur, he was working at CSIR-Central Electrochemical Research Institute. He holds a BTech in biotechnology and PhD in the field of microbial fuel cells. His areas of interests include bioelectrocatalysis and nanobiotechnology. He has authored several research articles in reputed journals. He is a life member of several professional bodies.

Prof. Jong-Sung Yu

Prof. Jong-Sung Yu is one of the most renowned scientists in the field of electrochemical science. He is currently a Professor in the Department of Advanced Materials Chemistry at Korea University in Korea. He has a BS from Sogang University in Korea and a PhD in chemistry from the University of Houston in USA. He has several years of rich postdoctoral experience at Ohio State University, University of Houston, Pennsylvania State University, and Northwestern University. He is currently on the editorial boards of the *Journal of Experimental Nanoscience* (Taylor and Francis), *Frontiers in Fuel Cells*, and *Carbon Letter* (Korean Carbon Society). He was awarded an Excellent Research Award in the Inorganic Chemistry Division in 2011 and Material Chemistry Division in 2004 from the Korean Chemistry Society, and also was given a Korea Strategic Research Grant (2010–2015) in Energy Application. He has authored about 190 peer-reviewed articles and 60 patents. His research interests include nanostructured materials for storage, delivery, and conversion, with particular attention toward electrochemical applications such as fuel cell, battery, capacitor, solar cell, and sensors.

CONTENTS

LIST OF CONTRIBUTORS

Rintu Banerjee
Microbial Biotechnology and DSP Laboratory, Agricultural and Food Engineering Department, IIT Kharagpur, Kharagpur, West Bengal – 721302

Aditya Bhalla
Department of Chemical and Biological Engineering, South Dakota School of Mines and Technology, 501 East St. Joseph Street, Rapid City, SD 57701–3995 USA

Mohit Bibra
Department of Chemical and Biological Engineering, South Dakota School of Mines and Technology, 501 East St. Joseph Street, Rapid City, SD 57701–3995 USA

Nandita Dasgupta
School of Bio Sciences and Technology, VIT University, Vellore, Tamil Nadu, India

Vijay Kumar Garlapati
Microbial Biotechnology and DSP Laboratory, Agricultural and Food Engineering Department, IIT Kharagpur, Kharagpur, West Bengal – 721302

Sannapaneni Janardan
Department of Chemistry, School of Advanced Sciences, VIT University, Vellore

J. Jayapriya
Department of Chemical Engineering, A.C. Tech, Anna University, Chennai – 600025, India

R. Navanietha Krishnaraj
Department of Biotechnology, National Institute of Technology Durgapur, Durgapur, West Bengal, India–713209

B. Ashok Kumar
Department of Genetic Engineering, School of Biotechnology, Madurai Kamaraj University, Madurai, Tamil Nadu

S. Vasanth Kumar
Department of Chemistry, Karunya University, Coimbatore – 641114, Tamilnadu, India

Sudhir Kumar
Department of Chemical and Biological Engineering, South Dakota School of Mines and Technology, 501 East St. Joseph Street, Rapid City, SD 57701–3995 USA

M. J. Angelaa Lincy
Department of Molecular Microbiology, School of Biotechnology, Madurai Kamaraj University, Madurai, Tamil Nadu

D. Malini
Professor, Department of Marine Biotechnology, Bharathidasan University, Tiruchirappalli–620024, Tamilnadu

P. Malliga
Professor, Department of Marine Biotechnology, Bharathidasan University, Tiruchirappalli–620024, Tamilnadu

Vinita Mishra
Cell Biology Lab, School of Biological Sciences and Biotechnology, Indian Institute of Advanced Research, Koba Institutional Area, Gandhinagar, 382 007, Gujarat, India; E-mail: vinita.mbif@gmail.com

Kayla Morisette
Department of Chemical and Biological Engineering, South Dakota School of Mines and Technology, 501 East St. Joseph Street, Rapid City, SD 57701–3995 USA

A. Obadiah
Department of Chemistry, Karunya University, Coimbatore – 641114, Tamilnadu, India

B. B. Pavankumar
Department of Chemistry, School of Advanced Sciences, VIT University, Vellore

Subramanian M. Raj
Department of Chemical and Biological Engineering, South Dakota School of Mines and Technology, 501 East St. Joseph Street, Rapid City, SD 57701–3995 USA

Chidambaram Ramalingam
School of Bio Sciences and Technology, VIT University, Vellore, Tamil Nadu, India

V. Ramamurthy
Department of Biotechnology, PSG College of Technology, Coimbatore – 641004, India

Shivendu Ranjan
School of Bio Sciences and Technology, VIT University, Vellore, Tamil Nadu, India

Lakshmi Shri Roy
Microbial Biotechnology and DSP Laboratory, Agricultural and Food Engineering Department, IIT Kharagpur, Kharagpur, West Bengal – 721302

R. Sabitha
Professor, Department of Marine Biotechnology, Bharathidasan University, Tiruchirappalli–620024, Tamilnadu

Deeksha Sachdeva
Department of Biochemistry, Punjab Agricultural University, Ludhiana, Punjab, India–141004; E-mail: sachdeeksha@gmail.com

David Salem
Department of Chemical and Biological Engineering, South Dakota School of Mines and Technology, 501 East St. Joseph Street, Rapid City, SD 57701–3995 USA

Melvin Samuel
School of Bio Sciences and Technology, VIT University, Vellore, Tamil Nadu, India

Rajesh K. Sani
Department of Chemical and Biological Engineering, South Dakota School of Mines and Technology, 501 East St. Joseph Street, Rapid City, SD 57701–3995 USA

Asha Sinhda
Department of Mycology and Plant Pathology, Institute of Agricultural Sciences, Banaras Hindu University, Varanasi – 221 005, U.P., India

Akella Sivaramakrishna
Department of Chemistry, School of Advanced Sciences, VIT University, Vellore

Isha Srivastava
Department of Biotechnology, Delhi Technological University (Formerly Delhi College of Engineering), Shahbad Daulatpur, Main Bawana Road, Delhi-110 042, India

Seweta Srivastava
Department of Mycology and Plant Pathology, Institute of Agricultural Sciences, Banaras Hindu University, Varanasi – 221 005, U.P., India; E-mail: shalu.bhu2008@gmail.com

P. Varalakshmi
Department of Molecular Microbiology, School of Biotechnology, Madurai Kamaraj University, Madurai, Tamil Nadu

V. S. Vasantha
Department of Natural Products, School of Chemistry, Madurai Kamaraj University, Madurai, Tamil Nadu

Kari Vijayakrishna
Department of Chemistry, School of Advanced Sciences, VIT University, Vellore

V. Viswajith
Professor, Department of Marine Biotechnology, Bharathidasan University, Tiruchirappalli–620024, Tamilnadu

Jia Wang
Department of Chemical and Biological Engineering, South Dakota School of Mines and Technology, 501 East St. Joseph Street, Rapid City, SD 57701–3995 USA

Jong-Sung Yu
Department of Advanced Materials Chemistry, Korea University, 2511 Sejong-ro, Sejong, 339-700, Korea

LIST OF ABBREVIATIONS

AC	activated charcoal
AEM	anion exchange membrane
AFC	alkaline fuel cell
AGCS	amber glass bottle with closed space
AGCSN	amber glass bottle with closed space containing nitrogen
AGOS	amber glass bottle with open space
ATP	adenosine triphosphate
ATSDR	Agency for Toxic Substances and Disease Registry
BEAMR	bioelectrochemically assisted microbial reactor
BES	bromoethanesulfonic
BioH2	biohydrogen
BPV	bio-photovoltaic cell
BRA	biotechnology risk assessment
BSES	biological simultaneous enzyme production and saccharification
BTL	biomass-to-liquid
CBH	contaminated barley hulls
CBP	consolidated bioprocess
CMCEs	chemically modified carbon electrodes
CN	cetane number
CNT	carbon nanotubes
COD	chemical oxygen demand
CoTMPP	cobalt tetramethylphenylporphyrin
CSTR	continuous stirred tank reactor
CV	cyclic voltammetry
CVD	chemical vapor deposition
CWP	cheese whey powder
DDG	dried distiller's grain
DG	diglycerides

DME	dimethyl ethanol
DMRB	dissimilatory metal reducing bacteria
DNSA	dinitrosalicylic acid
EAB	electrochemically active bacterial
FAME	fatty acid methyl ester
FFA	free fatty acids
FT	Fischer–Tropsch diesel
GC	gas chromatography
GECE	graphite/epoxy composite electrode
GHG	greenhouse gas
GM	genetic modification
GTZ	Gesellschaft für Technische Zusammenarbeit
GVL	g-valerolactone
HMF	hydroxymethylfurfural
HRT	hydraulic retention time
IP	induction point
JBD	Jatropha curcus biodiesel
KV	kinematic viscosity
ldh	lactate dehydrogenase gene
MB	methylene blue
MBR	membrane bioreactors
MCC	microcrystalline cellulose
MEC	microbial electrolysis cell
MFC	microbial fuel cell
MG	monoglycerides
MSW	municipal solid wastes
NAD	nicotinamide adenine dinucleotide
NR	neutral red
OGCS	ordinary glass bottle with closed space
OGCSN	ordinary glass bottle with closed space containing nitrogen
OGOS	ordinary glass bottle with open space
OSI	oil stability index
PBD	Pongamia (karanja) biodiesel
PEC	photoelectrochemical
PEM	proton exchange membrane

PFR	plug flow reactor
PNS	purple-nonsulfur
PSI	photosystem I
PSII	photosystem II
PY	pyrogallol
RTD	residence time distribution
RVC	reticulated vitreous carbon
RW	refinery wastewater
SDPF	sequential dark-photo fermentations
SEA	separator electrode assembly
SHF	separate hydrolysis and fermentation
SNG	synthetic natural gas
SPA	widely spaced electrodes
SSF	simultaneous saccharification and fermentation
SWG	switchgrass
TG	triglycerides
UF	ultra filtration
VSS	volatile suspended solids
WCO	waste cooking oil

PREFACE

Energy is one of the prime needs in the modern world. Energy demands are exponentially increasing in the recent years owing to rapid advancements in industrialization and population explosion. The conventional fossil fuels are depleting at rapid rates. The use of conventional sources such as coal or nuclear sources also causes several hazards to the environment.

Bioenergy sources such as bio-hydrogen, biodiesel, bioethanol, bio-fuel cells seem to be an ideal option for fulfilling the increasing energy demands. Several researchers all over the world are working in this sector, and it has led to drastic improvements in bioenergy research. However, the technology gap between research and commercialization still exist.

The objective of the book, *Bioenergy: Opportunities and Challenges*, is to offer engineers and technologists from different disciplines information to gain knowledge on the breadth and depth of this multifaceted field. The field of bioenergy is interdisciplinary. It requires the knowledge of biologists, chemists, physicists, and engineers. There is an urgent need to explore and investigate the current shortcomings and challenges of the current bioenergy technologies. The book, in short, will explore and convey the key concepts on factors hindering the real-time application of these bioenergy sources.

Bioenergy: Opportunities and Challenges provides valuable information for engineers and technologists working in this cutting-edge field. The book will also serve as a text for BTech and MSc courses in biotechnology and other allied disciplines. The book will also be useful to management students for bioenergy management and related courses.

PART 1

BIOHYDROGEN PRODUCTION

CHAPTER 1

THERMOPHILIC BIOHYDROGEN PRODUCTION: CHALLENGES AT THE INDUSTRIAL SCALE

SUDHIR KUMAR[1,2], ADITYA BHALLA[1], MOHIT BIBRA[1], JIA WANG[1], KAYLA MORISETTE[1], SUBRAMANIAN M. RAJ[1], DAVID SALEM[1], and RAJESH K. SANI[1]

[1]*Department of Chemical and Biological Engineering, South Dakota School of Mines and Technology, 501 East St. Joseph Street, Rapid City, SD 57701–3995 USA*

[2]*Department of Biotechnology and Bioinformatics, Jaypee University of Information Technology, Waknaghat, Solan-173234, Himachal Pradesh, India*

CONTENTS

ABSTRACT

Biohydrogen (BioH$_2$) production using lignocellulosic biomass has been extensively studied at laboratory scale. Localized availability of suitable feedstocks, their pretreatment costs, and lower rates of BioH$_2$ production make the industrial operation uneconomical. Several thermophilic and hyperthermophilic microorganisms can produce elevated levels of H$_2$ (~4 mol H$_2$/mol glucose) through the dark fermentation process. Decentralized thermophilic BioH$_2$ production in a single-step, consolidated process is desirable for successful industrial operation. In addition to lignocellulosic biomass, food wastes, and industrial effluents have been projected as economical substrates for H$_2$ production. Using these inexpensive and easily available substrates, the concept of 'waste to energy' can be achieved. There is, however, a lack of successful case studies of long-term industrial operations producing BioH$_2$. This chapter focuses on the need for BioH$_2$ production using thermophiles at realistic, commercial scales. Issues related to studying scale up requirements are discussed, together with goals for successful, industrial-scale production of BioH$_2$.

1.1   INTRODUCTION

Non-renewability and depletion of fossil fuels coupled with enhanced greenhouse gas emissions by their combustion warrants global search for alternative fuels. A variety of alternative fuels including biobutanol, biodiesel, bioethanol, biogas, biohydrogen (BioH$_2$) and other biomass-derived products are currently being examined. Among these, H$_2$ due to its high-energy efficiency (calorific value of 143 MJ/kg) and its ability to be converted into electrical energy in fuel cells, has gained much interest as a sustainable biofuel in the past few years. At present, most of the current H$_2$ supply (around 96%) comes from fossil fuels but with high production cost and negative environmental impact. To lower the production cost of H$_2$, microbe-mediated H$_2$ (BioH$_2$) can be an option, produced using readily available, carbon neutral, inexpensive, and renewable feedstocks such as lignocellulosic biomass. Though lignocellulosic biomass is inexpensive (US $2–4/GJ at a cost of US $39–60/dry ton biomass), its pretreatment cost (US \$15–25/GJ) exceeds that of fossil fuel (US\$ 5–15/GJ), which makes the lignocellulosic $BioH_2$ production uneconomical [1]. In this context, a consolidated bioprocess (CBP) using lignocellulosics to biofuels in a single step could provide an efficient and economical process [2, 3]. Therefore, microorganisms that have the capability of enzymatic hydrolysis of lignocellulose and sugar fermentation to produce H_2 in a single reactor are desirable.

In the past few years, extensive research has been carried out on $BioH_2$ production under dark fermentation conditions using thermophilic and hyperthermophilic microorganisms due to their inherent advantage of improved reaction kinetics and stability at high temperatures [4–6]. Thermophilic microorganisms such as *Caldicellulosiruptor saccharolyticus*, *Thermococcus kodakaraensis*, and *Clostridium thermocellum* have been reported to produce cellulosic $BioH_2$ under dark fermentation processes [7–9]. $BioH_2$ production via dark fermentation is widely preferred over photo-fermentation and biophotolysis. The major advantages of dark fermentation are: (i) simplicity of reactor design; (ii) more efficient process operation; (iii) a wide variety of feedstocks utilization; and (iv) higher H_2 production rates compared to other biological methods of H_2 production [10–13].

Caldicellulosiruptor spp. are emerging as potential H_2-producers through CBP. Recently it has been reported that *C. saccharolyticus* DSM8903 produced H_2 (11.2 mmol/g switchgrass) directly from switchgrass without any physicochemical or biological treatment making it a promizing candidate for CBP of lignocellulosic biomass [4]. *C.bescii* produces a combination of lactate, acetate, H_2, and CO_2, and pyruvate acts as a major branch point with carbon routed to acetyl-CoA or lactate. Deletion of lactate dehydrogenase gene (ldh) from the chromosome of *C. bescii* has also been reported. This deletion resulted in production of acetate and H_2 through redirection of electron flow from lactate. These modified strains have been reported to grow on biomass without any conventional treatment and produced 3–4 mol H_2/mol glucose [14]. There are few reports on CBP of lignocellulosic feedstocks for H_2 production using thermophiles. It has been suggested that thermophilic CBP has advantages over mesophilic processes; for example, achieving theoretical maximum yield of 4 mol H_2/mol glucose, utilization of a wide range of biomasses as substrates, increased substrate availability due to high temperatures, and reduced risk of microbial contamination [4, 7, 15–20].

BioH$_2$ is projected as a dominant biofuel, as a result of readily available lignocellulosic feedstocks, increasing knowledge of H$_2$-related metabolic pathways of H$_2$-producing native heterotrophs (dark fermentation) and phototrophs (photofermentation), and genetically engineered microbes. BioH$_2$ production using lignocellulosic biomass is well documented and has been proven at the laboratory scale [2, 4, 21, 23–26]. Globally available inexpensive raw materials in the form of food waste and industrial effluents have also been gaining interest for BioH$_2$ production [27–31]. Toyota and Hyundai have also been working to promote H$_2$ fuel vehicles in the USA and UK, and therefore investment opportunities are increasing in the H$_2$ sector. However, economical industrial production of this biofuel is a challenge for a number of reasons including high processing costs of raw materials, lower levels of substrate conversion efficiency, lack of appropriate microorganisms as well as bioreactor design, and low stoichiometric H$_2$ yield. This review briefly discusses issues related to scale up of BioH$_2$ production and demonstrates that this biofuel is a promising alternative fuel, especially when produced with thermophiles.

1.2   FEEDSTOCKS FOR H$_2$ PRODUCTION

Many different processes (e.g., electrolysis of H$_2$O, thermocatalytic reformation, solar thermochemical gasification, supercritical conversion, direct and indirect biophotolysis, photofermentation, and dark fermentation) have been demonstrated for H$_2$ production from a variety of feedstocks [9, 32–43]. However, local availability and cost of feedstocks and conversion technologies to produce H$_2$ are key factors determining the choice of feedstock and method of H$_2$ production. Since these factors vary from region to region, it is not necessary to develop a universal technology for H$_2$ production; one of several processes and feedstocks can be selected for H$_2$ production according to its cost-effectiveness in a given nation or region.

Production of H$_2$ from easily fermentable substrates such as glucose, xylose, sucrose, and starch has been achieved with high yields (in some cases the maximum of 4 mol/mol glucose equivalents) as well as high rates, for example, 94 mmol/L/h [16, 32, 35–38]. The use of these easily fermentable substrates for the production of H_2 is not economically viable due to their high costs. For example, the cost of glucose is US $400–600/Metric Ton (MT). In theory, roughly 44 kg of H_2 can be obtained from one MT of glucose (assuming that 4 mol of H_2 can be produced from one mole of glucose under dark fermentative conditions), which can be sold for $31–$135 depending on the mode (on-site gas or off-site liquid) of H_2 production [44]. This means the cost of raw material is more than 3 folds higher than the price of product H_2, illustrating the need to select a cost-effective feedstock for the sustainable production of H_2. Table 1.1 summarizes various types of feedstocks (e.g., energy crops lignocellulosic residues, wastewater, complex solid wastes, and microalgae) for BioH$_2$ production with their comparative advantages and disadvantages.

TABLE 1.1   Feedstocks for BioH$_2$ Production

Feedstocks	Advantages	Disadvantages	References
Energy crops (e.g., sugarcane, sugar beet, sorghum, corn, wheat)	• Carbon neutral • High sugar and low lignin contents • Renewable • Short incubation time for H_2 production	• Environmental problems due to high dose of fertilizers and pesticides • Expensive • Increase in food price • Scarcity of arable land	[6, 22, 54]
Lignocellulosic residues (e.g., rice, wheat straw and bran, corn stover, sugarcane and sweet sorghum bagasse, wood waste, wood fibers, tree and shrub residues)	• Carbon neutral • Globally abundant, sustainable • Inexpensive • No competition for arable land, waste, land management • Renewable	• Depletion of soil organic content • High cost of pretreatment and saccharification • Inhibitory compounds • Long incubation time of H_2 production • Soil erosion hazard	[1, 3, 16, 111–113]

TABLE 1.1   (*Continued*)

Feedstocks	Advantages	Disadvantages	References
Wastewater (e.g., paper mill, distillery and molasses based, starch hydrolysate, cattle, cheese whey, olive mill, dairy, food industry waste water)	• Carbon neutral, renewable • High sugar/starch content • Inexpensive • Pollution control and wastewater treatment • Reduction of COD	• Complex chemical composition • Inhibition due to high organic load • Low reproducibility of the waste • Selection and separation of waste components	[35, 114–117]
Complex solid waste (e.g., kitchen food, municipal, mixed wastes)	• Carbon neutral • High sugar/starch content • Low conc. of inhibitors • Low cost and abundant • Reduction of BOD • Renewable	• Additional cost of separation of fermentable fraction of the waste • Low H_2 conversion efficiency due to high contents of proteins and fats	[6, 50, 118]
Microalgae	• Absence of lignin • Carbon neutral • CO_2 sequestration • Growth on wastewater • High carbohydrate and lipid contents • No need of arable land • Potency of genetic manipulation to increase productivity • Rapid growth rate • Short harvesting time	• Cross contamination of algal monocultures in open ponds • High cost of photobioreactors and dewatering process • Susceptibility of cross contamination of algae from bacteria, fungi and predators like protozoans, rotifers and crustaceans • Low H_2 production potential	[52, 53, 119, 120]

Lignocellulosic biomass may be the ideal feedstock for H_2 production because it is renewable, sustainable, and available abundantly. Crop residue is inexpensive (\$39–60/MT) and produced worldwide at a level of about 2802×10^6 Mg/yr for cereal crops and 3758×10^6 Mg/year for 27 food crops [45]. Although, lignocellulosic biomass fits well in the overall concept of renewable fuel generation, the cost of pretreatment of lignocellulosic feedstocks makes the process industrially impractical as the pretreatment process alone adds the cost of US \$15–25/GJ. Once pretreated, the efficiency of H_2 production largely depends upon the type of sugars and inhibitors released. The other globally available feedstocks such as municipal solid wastes (MSW), algal biomass, and industrial effluents also have good potential to produce H_2 [31, 35, 46–50]. Replacement of expensive carbon feedstocks with inexpensive wastes, for example, industrial effluents, could be one of the ways to overcome the economic barriers of BioH$_2$ production. Also, the use of inexpensive, abundantly available wastewaters as feedstock is environmentally advantageous [6]. However, due to the complex dynamic nature of wastewater (e.g., organic load, pH, salinity, sterility, and presence of biodegradable organics), extensive research is needed to understand the overall feasibility of this feedstock for BioH$_2$ production.

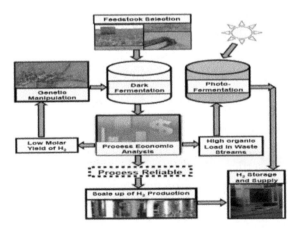

FIGURE 1.1 Global network for industrialization of H_2 production.

Recently, algal biomass has gained attention for production of various biofuels including H_2 [51–53], due to rich fermentable content. Rapid growth of algae in seawater, a short harvesting cycle, and the CO_2 sequestration property can make algae the best renewable source for biofuel production. Compared to lignocellulosics, microalgae require only a mild pretreatment (enzymatic saccharification) due to absence of lignin and hemicelluloses [22, 47, 54–56]. However, difficulties in propagating a monoculture of algae in open ponds need to be overcome for developing a cost-effective H_2 production system (Table 1.1).

1.3   ROUTES OF BIOH$_2$ PRODUCTION, MICROORGANISMS, AND H$_2$ YIELDS

1.3.1   BIOH$_2$ PRODUCTION FROM READILY FERMENTABLE SUBSTRATES

Biophotolysis of water using cyanobacteria and algae, photofermentation using photosynthetic bacteria, and dark fermentation using anaerobic fermentative bacteria are the routes of BioH$_2$ production. During the process of biophotolysis, microalgae or cyanobacteria use solar energy and splitting of water occurs in microbial photosynthesis. In photosystem I of microalgae, reduction of ferrodoxin takes place-producing H_2 by a hydrogenase enzyme (for details refer to Ref. [57]). While in photofermentation, ATP-requiring nitrogenase enzyme catalyzes H_2 production using various substrates in the absence of oxygen and nitrogen. Also in photofermentation, due to presence of reverse hydrogenase enzyme, H_2 oxidizes back to protons. So the net gain is the difference between the evolved H_2 by nitrogenase and H_2 consumed by reversed hydrogenase. H_2 uptake hydrogenase-free mutants are required to see the full potential of the photofermentation process. Hydrogenase encoding genes (hup) can be modified by insertion of resistance genes or by deletion of hup genes. Apart from this low efficiency of light conversion, high cost of photobioreactors and oxygen sensitivity of hydrogenase enzyme are the major limitations of photofermentation and biophotolysis.

During dark fermentation process, anaerobic bacteria use carbohydrates rich substrates under mesophilic (25–40 °C), thermophilic (40–65 °C) and hyperthermophilic (>80 °C) conditions [58]. Oxygen sensitivity to hydrogenase enzyme is not an issue in dark fermentation and there is no need for solar energy. Thermophilic microbial dark fermentation, mainly due to high H_2 production rates compared to mesophilic dark fermentation, photofermentation, and biophotolysis, is a promising approach for BioH$_2$ production. Through metabolic engineering, the modified mesophilic strains of *Escherichia coli* are limited to produce 2 mol of H_2/mol of glucose [59, 60]. In contrast native or genetically modified thermophilic strains of *Clostridium* sp. can produce 4 mol H_2/mol glucose [61]. Elevated levels of H_2 production (i.e., ≥3.6 mol/mol of glucose) have been reported from various pure or mixed thermophilic and hyperthermophilic cultures (Table 1.2). Hyperthermophilic microorganisms such as *Thermotoga neapolitana* DSM 4359 and *Pyrococcus furiosus* DSM 3638 produced 3.8 mol H_2/mol of glucose consumed [62–64].

The thermophilic BioH$_2$ production process poses some inherent benefits at elevated temperature, such as lower risk of contamination from mesophiles, no need for reactor cooling, better mixing and lower viscosity, and removal of volatile products, which could minimize the chances of product inhibition [5, 65]. Another advantage of using thermophiles and hyperthermophiles (e.g., *Thermoanaerobacter tengcongensis* and *Pyrococcus furiosus*) is the presence of membrane bound NADPH-dependent hydrogenases, which can reduce transport energy input, and thus can make the process thermodynamically more favorable [66– 70]. Enhanced BioH$_2$ production has been observed with increase in reactor temperature. For example, Gadow et al. [71] showed higher H_2 yield (19.01 mmol H_2/g sugar consumed) by using a mixed culture in a continuous stirred tank reactor (CSTR) at 80 °C as compared to the temperatures 55 °C (15.2 mmol H_2/g sugar consumed) and 37 °C (0.6 mmol H_2/g sugar consumed). In another report, Kargi et al. [72] showed higher BioH$_2$ production from cheese whey powder at 55 °C (0.81 mol/mol glucose) as compared to the fermentation carried out at 35 °C (0.47 mol/mol glucose).

H_2 production by mixed cultures may be desirable when compared to pure culture fermentation due to the capability of producing greater yields [73]. Zeidan et al. [74] reported higher H_2 yield (3.7 mol/mol hexose sugar) by co-culture of *Caldicellulosiruptor* sp. as compared to monoculture (3.5 mol/mol hexose). Cell-free supernatant of *C. saccharolyticus* also enhanced the growth of *C. kristjanssonii* in batch culture leading to increased biomass concentration, which proves a positive interspecies interaction. Table 1.2 summarizes H_2 production from readily fermentable sugars as first generation feedstock by pure cultures, co-cultures, and mixed cultures of thermophiles and hyperthermophiles.

TABLE 1.2 Overview of H_2-Producing Thermophiles and Hyperthermophiles Utilizing Readily Fermentable Substrates

Organism, Growth Temperature (°C), and Culturing Type	Substrate	End Products	H_2 Yield (mol/mol)[a]	Reference
Thermoanaerobacter mathranii A3N, 70 °C, and Batch	Glucose	Acetate, Butyrate, Lactate, Ethanol	2.64	[121]
Thermotoga neapolitana, 75 °C, and Batch	Xylose	Acetate, Lactate	2.22	[122]
Thermoanarobacterium thermosaccharolyticum, 60 °C, and Chemostat	Glucose	Acetate, Butyrate, Ethanol	2.07	[123]
Thermotoga neapolitana DSM 4359, 80 °C, and Batch	Glucose	Acetate, Lactate	3.8	[62]
Thermotoga neapolitana DSM 4359, 77 °C, and Batch	Glucose	Acetate, Butyrate	3.2	[76]
Thermoanaerobacterium sp., 50 °C, and Batch	Glucose	Acetate, Butyrate, Lactate, Ethanol	1.59	[18]
Thermoanaerobacterium saccharolyticum YS485, 55 °C, and Batch	Cellobiose	Acetate, Lactate, Ethanol	0.87	[124]
Thermotoga neapolitana DSM 4359, 85 °C, and Batch	Glucose	Acetate, Lactate	3.8	[63]
Pyrococcus furiosus DSM 3638, 90 °C, and Chemostat	Cellobiose	Acetate, Alanine, Ethanol	3.8	[67]
Cladicellulosiruptor saccharolyticus, 72 °C, and Batch	Glucose	Acetate, Lactate	3.3	[39]

TABLE 1.2   (*Continued*)

Organism, Growth Temperature (°C), and Culturing Type	Substrate	End Products	H$_2$ Yield (mol/mol)[a]	Reference
Thermococcus kodak-araensis TSF100, 85 °C, and Chemostat	Starch	Acetate, Alanine	3.3	[9]
Thermoanaerobac-ter tengcongensis JCM11007, 75 °C, and Batch	Starch	Acetate, Ethanol	2.8	[68]
Thermotoga elfii DSM 9442, 65 °C, and Batch	Glucose	Acetate	3.3	[43]
Caldicellulosiruptor sac-charolyticus, 70 °C, and Controlled batch DSM 8903	Sucrose	Acetate, Lactate	3.3	[43]
Mixed Or Co-Cultures				
Thermoanaerobacterium thermosaccharolyticum, *Klebsiella* sp., *Bacillus coagulans* and *Sporolac-tobacillusnakayamae*, 70 °C, and Continuous	Arabinose or Glucose	Ethanol, Lactate, Acetate	1.10 or 0.75	[125]
Dominant species of *Thermoanarobacterium* and *Thermoanaerobac-ter*, 70 °C, and Batch	Glucose	Acetate, Butyrate, Ethanol, Propio-nate	2.38	[126]
Bacillus sp., *Anoxybacil-lus* sp., *Caloramator* sp., *Clostridium* sp., *Thermo-anaerobacterium* sp. and *Caldicellulosiruptor* sp., 60 °C, and Batch	Xylose or Glucose	Acetate, Butyr-ate, Propionate, Ethanol, Lactate, Butanol	1.58 or 1.71	[127]
Bacillus coagulans and Thermoanaero-bacterium, 60 °C, and Chemostat	Glucose	Lactate, Acetate, Formate, Butyr-ate, Ethanol	0.85	[128]
Co-culture (*Caldicellu-losiruptor kristjanssonii* and *C. saccharolyticus*), 70 °C, and Batch	Glucose-xylose	Acetate	3.6	[37]

TABLE 1.2 (*Continued*)

Organism, Growth Temperature (°C), and Culturing Type	Substrate	End Products	H_2 Yield (mol/mol)[a]	Reference
Thermobrachiumcelere, Thermoanaerobacterium aotearoense, Clostridium thermopalmarium, 70 °C, and Batch	Xylose	Acetate, Ethanol	1.84	[129]
Mixed culture, 70 °C, and Batch	Xylose	Acetate, Ethanol, Lactate	1.62	[130]
Compost microflora and *Caldicellulosiruptor saccharolyticus,* 70 °C, and Batch	Glucose-Xylose	Acetate, Lactate	2.3	[88]

[a]mol of H_2 produced/mol of substrate consumed.

1.3.2 BIOH$_2$ PRODUCTION FROM COMPLEX SUBSTRATES

As discussed above, thermophilic and hyperthermophilic microbes can ferment mono- and disaccharides (e.g., glucose, xylose, arabinose, and cellobiose, Table 1.2) efficiently. However, the use of monosaccharides for biofuels competes with food use, and also the cost of monosaccharides (e.g., glucose) is more than 3-fold higher than the price of BioH$_2$ produced. Inexpensive feedstocks including lignocellulosic materials have been used as substrates for H$_2$ production by pure or mixed cultures of thermophiles and hyperthermophiles (Table 1.3). For example, rice straw [75, 76], cornstalk waste [77, 78], wheat straw [79, 80], corn stover [81], sweet sorghum [82], carrot pulp hydrolysate [83], vegetable kitchen waste [84], maize leaves and sugarcane baggase [7], shredded paper waste [31], and delignified wood fibers [20] have been reported for H$_2$ production by thermophiles.

Among reported thermophiles, *Caldicellulosiruptor saccharolyticus* has been identified as a suitable candidate for H$_2$ production. *C. saccharolyticus* produces thermostable cellulolytic and xylanolytic enzymes, and grows at ≥65 °C on a wide variety of lignocellulosic biomasses including switchgrass, sweet sorghum, sugarcane bagasse, wheat straw, maize

TABLE 1.3   Overview of H_2-Producing Thermophiles and Hyperthermophiles Utilizing Complex Substrates

Organism, Growth temperature (°C), and Culturing Type	Substrate	End Products	H_2 Yield	Reference
Clostridium pasteurianum, *Clostridium stercorarium* and *Thermoanaerobacterium saccharolyticum*, 55 °C, and Batch	Rice straw	Acetate, Butyrate	1.11 mmol H_2/g TS	[75]
Anaerobic mixed microflora, 55 °C, and CSTR	Cellulose	Acetate, Butyrate	15.16 mmol H_2/g Cellulose	[71]
Anaerobic mixed microflora, 60 °C, and CSTR	Cellulose	Acetate, Butyrate	19.01 mmol H_2/g Cellulose	[71]
Co-culture (*Clostridium thermocellum* and *Clostridium thermosaccharolyticum*), 55 °C, and Batch	Cornstalk waste	Ethanol, Acetate, Butyrate	3.04 mmol H_2/g Cornstalk	[77]
Thermoanaerobacterium saccharolyticum, *T. thermosulfurigenes*, *T.* sp., *Bacillus* sp., *Anoxybacillus* sp., *B. lentus*, and *Geobacillus* sp., 60 °C, and Batch	Sago starch	Acetate, Ethanol	19.72 mmol H_2/g Starch	[131]
Mixed anaerobic culture from rotted wood crumb, 50 °C, and Batch	Lime treated cornstalk waste	Not Reported	6.93 mmol H_2/g-TVS	[78]
Thermoanaerobacterium sp., *Thermoanaerobacter* sp., *Caloramater* sp. and *Anoxybacillus* sp., 60 °C, and Batch	Oil palm trunk hydrolysate	Acetate, Butyrate, Ethanol, Lactate, Butanol	13.42 mmol H_2/g sugar consumed	[127]
Thermoanaerobacterium saccharolyticum, *Thermoanaerobacterium thermosaccharolyticum*, *Anoxybacillus* sp., *Geobacillus* sp. and *Clostridium* sp., 60 °C, and Continuous	Cassava starch processing wastewater	Acetate, Ethanol	10.84 mmol H_2/g sugar consumed	[35]
Caldicellulosiruptor saccharolyticus, 60 °C, and Batch	Carrot pulp hydrolysate	Acetate, Lactate	2.8 mol H_2/mol Hexoses	[83]

TABLE 1.3 (*Continued*)

Organism, Growth temperature (°C), and Culturing Type	Substrate	End Products	H₂ Yield	Reference
Thermotoga neapolitana, 60 °C, and Batch	Carrot pulp hydrolysate	Acetate, Lactate	2.7 mol H_2/mol Hexoses	[83]
Thermoanaerobacterium thermosaccharolyticum, Clostridium sp., 55 °C, and CSTR	Vegetable kitchen waste	Butyrate, Lactate	1.7 mmol H_2/g COD	[84]
Thermotoga neapolitana, 75 °C and Batch	Potato stem peels	Acetate	2.6–3.8 mol H_2/mol Glucose	[36]
Caldicellulosiruptor saccharolyticus, 72 °C and Batch	Potato stem peels	Acetate	2.4–3.4 mol H_2/mol Glucose	[36]
Thermotoga neapolitana, 75 °C, and Batch	Rice straw	Not Reported	2.7 mmol H_2/g Rice Straw	[76]
Thermoanaerobacterium sp., 70 °C, and Batch	Sweet sorghum baggase	Acetate, Lactate	2.6 mol H_2/mol Hexoses	[82]
Thermoanaerobacterium thermosaccharolyticum W16, 60 °C, and Batch	Corn stover	Acetate, Butyrate, Ethanol	108.5 mmol H_2/L#	[81]
Caldicellulosiruptor saccharolyticus, 72 °C, and Batch	Miscanthus hydrolysate	Acetate, Lactate	3.4 mol H_2/mol Hexoses	[25]
Co-culture (*Clostridium thermocellum* and *Clostridium thermopalmarium*), 55 °C, and Batch	Cellulose	Acetate, Butyrate, Ethanol	1.36 mol H_2/mol Glucose Equivalent	[132]
Caldanaerobacter subteraneus, Thermoanaerobacter subteraneus, Thermoanarobacterium thermosaccharolyticum, 70 °C, and CSTR	Wheat straw hydrolysate	Acetate, Butyrate, Propionate, Ethanol, Lactate, Formate	9.46 mmolH_2/g Sugars	[79]
Thermotoga neapolitana, 80 °C, and Batch	Miscanthus hydrolysate	Acetate, Lactate	2.9 mol H_2/mol Hexoses	[25]

TABLE 1.3 (*Continued*)

Organism, Growth temperature (°C), and Culturing Type	Substrate	End Products	H$_2$ Yield	Reference
Caldicellulosiruptor saccharolyticus, 70 °C, and Batch	Maize leaves	Not Reported	1.87 mmol H$_2$/g of VS	[7]
Caldicellulosiruptor saccharolyticus, 70 °C, and Batch	Sugarcane baggase	Not Reported	0.87 mmol H$_2$/g of VS	[7]
Co-culture (*Clostridium thermocellum* JN4 and *Thermoanaerobacterium thermosaccharolyticum* GD17), 60 °C, and Batch	Cellulose	Acetate, Butyrate, Ethanol	1.8 mol H$_2$/mol Glucose	[8]
Clostridium thermocellum ATCC 27405, 60 °C, and Chemostat	Dried distiller's Grain (DDG)	Acetate, Lactate, Ethanol, Formate	1.7 mol H$_2$/mol Fermented Substrate	[133]
Thermoanaerobacterium sp., 60 °C, and Packed-bed reactor	Pulverized garbage and shredded paper waste	Not Reported	2.4 mol H$_2$/mol Hexoses	[31]
Clostridium thermocellum, 60 °C, and Batch	Delignified wood fibers	Acetate, Ethanol, Lactate, Formate	1.6 mol H$_2$/mol Fermented Substrate	[20]

#Concentrations.

leaves, and silphium [4, 7, 85]. *C. saccharolyticus* has a high number of ABC transporters for a wide variety of substrates, and can tolerate high partial pressures of H$_2$ [14, 86, 87]. Another thermophile, *C.owensensis* (ATCC 700167/DSM 13100) is capable of using a broad range of carbohydrates and producing H$_2$ as a metabolic by-product. Co-culturing of *C.owensensis* with the enriched compost microflora resulted in fast, simultaneous consumption of both glucose and xylose in the medium with a high specific H$_2$ production rate (3.6 mol H$_2$/mol sugar consumed, [88]). *C. Thermocellum* JN4 produces 0.8 mol H$_2$/mol glucose whereas, a two-fold increase (1.8 mol H$_2$/mol glucose) in H$_2$ production was observed from microcrystalline cellulose when JN4 was co-cultured with *Thermo-anaerobacterium thermosaccharolyticum* GD17 [8]. Overall, many thermophilic and hyperthermophilic microbes use complex substrates, and produce high H_2 yield near to 4 mol H_2/mol glucose, but with higher thermal energy consumption (at least 2.5 fold compared to mesophiles) for maintenance of their temperatures.

As we know, organic acids (acetate, butyrate, formate, lactate, and propionate) are by-products of dark fermentation. The highest theoretical yields of H_2 are associated with acetate as the only volatile fatty acid product. Even with 4 mol H_2/mol glucose ($C_6H_{12}O_6 + 2H_2O \rightarrow 2CH_3COOH + 2CO_2 + 4H_2$), we harness only 33% yield of H_2 because 12 mol of H_2 are theoretically available in one mole of glucose ($C_6H_{12}O_6 + 6H_2O \rightarrow 12H_2 + 6CO_2$). It is necessary to enhance $BioH_2$ production, and one such method is sequential dark-photo fermentations (SDPF). Organic acids produced during dark fermentation can be converted to H_2 and CO_2 by photofermentation [57, 89], and CO_2 produced can be used as a substrate during photofermentation. A microalgal-based SDPF system was shown to produce 6.5 mol H_2/mol hexose [90]. This SDPF system has advantages because it produces $BioH_2$ with low CO_2 emissions as well as high reduction in COD. The SDPF system has certain limitations, which poses a challenge for its successful implementation. The cost of photofermentation in the SDPF system is the main bottleneck in scaling up the process. Preliminary cost estimates in the literature indicate that photobiological H_2 could be produced in large-scale systems (100 ha and more) at a cost of 10–15 Euro/GJ (http://edepot.wur.nl/17056).

1.3.3 BIOH₂ PRODUCTION FROM UNTREATED COMPLEX SUBSTRATES

Thermophilic H_2 production is currently being carried out using pretreated (alkali and acid based, enzymatic hydrolysis-natural or synthetic mixture, hydrothermal, or ammonia) lignocellulosic substrates. For example, poplar leaves were first pretreated using acid and alkali as well as enzymes, and then fermented using anaerobic mixed culture of bacteria for H_2 production [91]. This suggests that conversion of lignocellulose from specific energy crops, agricultural residues, or forest products to H_2 will require: (i) various physiochemical pretreatments; (ii) enzymatic pretreatment (enzymes cocktails to hydrolyze cellulose and hemicellulose into to fermentable sugars usually under conditions of high temperature and extreme pH); (iii) removal of inhibitors (e.g., 5-hydroxymethylfurfural, furfural, acetic acid, and phenolic compounds) released during pretreatment; and (iv) robust fermentative microbes. The inclusion of the pretreatment and inhibitors removal steps reduces the overall cost-efficiency of the conversion process.

A viable option to lower the costs of feedstock and lignocellulolytic commercial enzymes is to screen for and identify potent H_2-producing microorganisms (saccharolytic fermentative extremophiles) capable of using cellulose and hemicellulose directly from the biomass without pretreatment. This would eliminate the need for separate steps of pretreatment, lignocellulolytic thermostable enzymes production, enzymatic hydrolysis of biomass, and improve the process economics. However, information on such microorganisms, which produce H_2 from untreated lignocellulosic biomass, is relatively scarce. *Caldicellulosiruptor bescii* sp. nov. DSM 6725 used various types of untreated plant biomasses (poplar, low-lignin grasses such as Napier and Bermuda grasses, and high-lignin containing grasses such as switchgrass), as well as cellulose and xylan [19]. A thermophilic mixed culture [75], a co-culture of *Clostridium thermocellum* and *C. thermosaccharolyticum* [77], and *Caldicellulosiruptor saccharolyticus* [7] have also been shown to use untreated lignocellulosic biomasses. Recently *Thermoanaerobacterium thermosaccharolyticum* M18 was shown to produce H_2 with a yield of 3.23 to 3.48 mmol/g corn cob, corn stalk, or wheat straw without any physicochemical or biological pretreatment [15]. Su et al. [92] reported that the presence of thermostable cellulases and hemicellulases supported the microbes to hydrolyze plant cell wall polysaccharides into simple sugars. Table 1.4 summarizes the recent studies of H_2 production from untreated lignocellulosic biomass using thermophiles.

TABLE 1.4 H_2 Production from Untreated Lignocellulosic Materials

Feedstock	Temp (°C)	Bacterial Strain	Pretreatment	H_2 Yield	Reference
Corn cob, Corn stalk, and Wheat straw	60	*Thermoan-aerobacterium thermosaccharo-lyticum* M18	None	3.23 to 3.48 mmol H_2/g substrate	[15]
Switchgrass (SWG) and microcrystalline cellulose (MCC)	65	*Caldicellulo-siruptor saccha-rolyticus* DSM 8903	None	11.2 mmol H_2/g SWG and 9.4 mmol H_2/g MCC	[4]
Rice straw	55	Heat treated sludge	Physical pulverization	1.11 mmol H_2/g TS	[75]
Lignocellulosic feedstock*	30	Enriched sludge	None	0.56 mmol H_2/g Raw Sample	[134]
Potato stem peels	75	*Thermotoga neapolitana*	None	3.8 mol/mol Glucose Equivalents	[36]
Potato stem peels	72	*Caldicellulo-siruptor saccha-rolyticus*	None	3.5 mol/mol Glucose Equivalents	[36]
Maize leaves	70	*Caldicellulo-siruptor saccha-rolyticus*	None	0.68 mmol H_2/g Dry Biomass	[7]
Poplar, and Switchgrass	70	*Anaerocellum thermophilum* DSM 6725	None	1.6 mmol/L for Poplar and 3.8 mmol/L for Switch-grass[#]	[19]
Corn husk, Ground nut shell, and Rice husk*	37	Mixed consortia from buffalo dung compost	None	2.93 mmol H_2/g TVS	[135]

TABLE 1.4 (*Continued*)

Feedstock	Temp (°C)	Bacterial Strain	Pretreatment	H₂ Yield	Reference
Dried distillers grain (DDG), Barley hulls (BH), and Fusarium head blight contaminated barley hulls (CBH)	60	*Clostridium thermocellum* ATCC 27405	None	1.07 mmol H_2/g DDG, 1.07 mmol H_2/g BH, and 1.06 mmol H_2/g CBH	[133]

*Mesophilic study.

#Concentrations.

1.4 INDUSTRIAL OPERATION OF H₂ PRODUCTION

The majority of the laboratory studies on $BioH_2$ production under dark fermentation conditions have been performed in a batch reactor system, whereas industrial scale H_2 production may need a continuous or at least semicontinuous reactor system for enhanced and uninterrupted $BioH_2$ production [93]. CSTR is one of the preferred choices for fermentative H_2 production from various substrates [94, 95]. A biofilm-based or granule-based reactor design to retain active microbes for $BioH_2$ production has been recommended [46]. Steady state operation of CSTRs has been successfully demonstrated up to a period of 3 months producing $BioH_2$ at a rate of 56 mL H_2/L/h at a biomass concentration of 1.2 g volatile suspended solids (VSS) /L [96, 97]. Lee et al. [98] increased biomass loadings (20 and 40 g VSS/L) resulting in improved rate of H_2 production (541 and 9300 mL H_2/L/h, respectively) using CSTR with modifications. For commercial feasibility of the bioprocess of H_2 production, a waste (condensed molasses) was used as substrate in laboratory scale CSTR with *Clostridium* spp. being the dominant bacteria. In this study at an organic loading rate of 320 g COD/L with a hydraulic retention time (HRT) of 3 h, 390 mmolH_2/L was produced [99]. Kim and Lee [94] used tofu processing waste and digester sludge for H_2 production with 2.3 molH_2/mol glucose equivalent in a batch, and also in a continuous mode using a thermophilic mixed culture.

Vatsala et al. [30] used mesophilic cocultures of *Citrobacter freundii*, *Enterobacter aerogenes* E10, and *Rhodopseudomonas palustris* P2 to produce H_2 at laboratory scale with glucose as carbon source. They also scaled up their process to 100 m^3 to evaluate the feasibility of H_2 production using distillery effluent from the sugarcane industry, and successfully produced 21.38 kg of H_2 in 40 h in a batch reactor with an average yield of 2.76 mol H_2/mol of glucose using mesophilic bacterial strains. Conversion of distillery effluents or wastewaters for H_2 production clearly well suits the overall concept of waste to energy. Use of wastewater or effluent streams for H_2 production poses dual advantages: production of energy and significant reduction of COD and BOD of wastewater.

H_2 production by dark fermentation is a strictly anaerobic process. Therefore O_2 elimination is required from the growth medium. In most of the laboratory studies, researchers use N_2 or Ar gas to remove O_2. At industrial scale, use of N_2 or Ar gas may not be feasible due to cost constraints. However, fabrication of continuous or semicontinuous reactors with recirculation of the purged gas may reduce the cost of operation. Also, a separate facility is required to recover and store $BioH_2$ produced during the industrial operation to shift the equilibrium towards H_2 production, otherwise increase in the partial pressure of H_2 can inhibit electron carriers (e.g., NADH and ferrodoxin). Thus, industrial scale-up of $BioH_2$ has multiple issues that need to be addressed adequately. These include selection and cost of the process, creating anaerobic conditions in a large reactor, purification and storage of $BioH_2$, and also safety parameters. Integration of the existing optimized technologies of dark and photofermentation may be a good strategy to produce $BioH_2$ at competitive rates.

Another challenge is overlapping features of $BioH_2$ production to methanogenic anaerobic digestion. By-products of $BioH_2$ production process such as acetate and CO_2 are quickly consumed by methanogens to form methane. To scale-up the process, H_2-consuming bacteria need to be replaced by a typical mixed culture involved in anaerobic digestion. Temperature, pH, and HRT are the most applicable methods to prevent methanogenesis in an industrial scale system of H_2 production. Extremethermophilic conditions (>70 °C) inhibit methanogens and thermodynamically favor H_2 production as end product [100]. Short HRT (< 3 days) would favor H_2 production as methanogens need longer HRT [101]. Low pH is also an effective parameter to limit H_2 consumers thus promoting H_2 producers [102]. Heat shock and addition of 2-bromoethanesulfonic (BES) acid have been reported to inhibit methanogens. Heat shock treatment (100 °C or above) before the experiments kill methanogens, but enrich H_2 producing sporulating bacteria such as *Clostridium* sp. [103]. Inducing low pH and addition of BES may not be suitable for large-scale industrial application of H_2 production. Chen et al. [97] achieved effective removal of methanogens at 12 h HRT and effectively produced $BioH_2$ from wastewater. However, pH control alone, even at 4.5, could not inhibit methanogens at long HRT (9 days) in semi-CSTR reactors [101].

Many times laboratory results that look promising may differ significantly at industrial scale. Thus, there is a need to strengthen the focus on projects leading from laboratory to scale-up processes for $BioH_2$ production. Also, discrepancies of H_2-production have been obtained under various experimental conditions by researchers making it difficult to compare their results. Uniformity in expressing the yields of H_2 obtained will simplify the comparative analyzes. Optimized pilot-scale demonstration projects are required to explore the potential of H_2 research at industrial scale. It was estimated that H_2 produced by a 1000L bioreactor connected with "5.0 kW" proton exchange membrane fuel cell would be sufficient to meet the electricity requirement of an average residential load [13]. On site production and utilization of $BioH_2$ from renewable resources is considered to be a cost-effective and viable industrial model [104]. This will not only eliminate the transportation cost of biomass, but also produce energy that can be used to partially support the parent industrial activity that is creating the feedstock as waste material or by-product. Utilization of inexpensive substrates and appropriate bioreactors are critical for sustainable $BioH_2$ production. Combining dark and photofermentation will further help to achieve higher yields of H_2. Bridging the gap between scientists who work on $BioH_2$ production and engineers who develop H_2 fuel cell technologies is necessary for success in this field. Fuel cell engineers need to know about the rate of H_2 production by biological systems and scientists should understand the H_2 requirement in the fuel cell [105].

1.5   DECENTRALIZED INDUSTRIAL SCALE H$_2$ PRODUCTION

Any suitable and abundant substrate for BioH$_2$ production in one geograph-ical location can be scarce at another location. Therefore, a decentralized system of BioH$_2$ production can be useful to materialize the scale up of this energy source [106]. Locally owned biorefineries based on regional bio-mass or wastes utilization may reduce treatment and handling cost with no need of arable land. In cities and towns, wastewater and kitchen waste (as a part of municipality waste) are the most appropriate substrates for BioH$_2$ production. In remote and less populated areas, however, locally available abundant feedstocks will likely be suitable substrates. In Midwestern US (Illinois, Indiana, Iowa, Minnesota, Missouri, Nebraska, North Dakota, South Dakota, and Wisconsin) corn stover, prairie cordgrass, switchgrass, and ponderosa pine are abundant feedstocks for BioH$_2$ generation. For example, approximately 750,000 tons of wood-waste from the Black Hills of South Dakota is already available as slash from previous forest thin-ning, and could be converted into BioH$_2$. On the other hand, Asia is rich in agricultural residues, especially rice straw, wheat straw, and sugarcane baggase, which can be used for BioH$_2$ production. Using abundantly avail-able, inexpensive, and renewable feedstock can be the way to lower the BioH$_2$ production cost for a given region.

The U.S. Department of Energy has been promoting a decentralized system for H$_2$ production [107]. It circumvents the need for a costly dis-tribution infrastructure. In a study of a decentralized system in south-west Georgia, H$_2$ was produced using pelletized peanut shells and pyrolysis of biomass yielded soluble and insoluble fractions of oil. Soluble bio-oil was converted to H$_2$ via catalytic steam reforming and shift conversion. The maximum yield achieved was 1.72 g H$_2$/100 g bio-oil in the over-all stoichiometry whereas the insoluble fraction was used as a source of adhesive formulations [108]. After cost analysis of an efficient pyrolysis process of bio-oil (soluble fraction), the selling price of H$_2$ was $6.05/GJ. This figure is promising, reflecting the abundant supply of local feedstock (peanut shells) and, coproduct value, and small-scale operation. In other research on H$_2$ production, Liu et al. [109] demonstrated cost-effective H$_2$ production through acetate in an electricity-mediated electrolysis process of an anaerobic microbial fuel cell. Acetate is a coproduct made during

H_2 production via dark fermentation. The energy generation from organic wastes such as acetate, if coupled with a dark fermentation process, has the potential to produce >4 mol H_2/mol glucose. However, in reality, existence of cost-effective scaled-up reactors with purified H_2 is scarce[104]. Integrating agricultural and industrial activities for $BioH_2$ production seem suitable for small-scale decentralized systems [110]. Developments in this direction will truly demonstrate the potential of H_2 energy as a major, global energy source.

1.6 SUMMARY AND CONCLUSIOINS

$BioH_2$ seems to be a lucrative alternate energy but it depends on how economical the $BioH_2$ process can be at industrial scale. To produce H_2 inexpensively a few of the prerequisites are abundant and economical biodegradable wastes—like agricultural residues, industrial effluents, or municipal waste, robust microbes, and suitable bioreactor configuration. Establishing H_2 production plants in the vicinity of a parent industry producing waste or near the site of biomass availability will surely reduce the $BioH_2$ production cost. This decentralized H_2 production strategy will reduce the cost of transport and storage. Most of the existing laboratory scale processes of $BioH_2$ production from complex wastes, such as lignocellulosic feedstocks, use several steps (e.g., pretreatment, hydrolysis, removal of inhibitors, and fermentation). The inclusion of several steps reduces the overall cost-efficiency of the process. An alternative to the use of various steps in H_2 production is the development of an efficient and cost-effective single step process using thermophiles (innovative thermophilic CBP), which can degrade and ferment untreated-lignocellulose. The use of elevated temperatures offer several potential advantages such as: (i) improved hydrolysis of cellulosic substrates; (ii) higher mass transfer rates leading to better substrate solubility; and (iii) lowered risk of microbial contamination, thus improving the overall economics of the process. For these reasons, thermophilic H_2 production in a consolidated mode (e.g., CBP) can be a preferred choice at the industrial scale; and additional advantages include increased H_2 yield, favorable thermodynamic conditions, and the ability to use a wide variety of substrates. The development of innovative thermophilic CBP will likely impact current multiple-step conversion processes of complex wastes to H_2 by providing a more efficient, sustainable, and economical process. However, there are also drawbacks of H_2 production at higher temperature, including: (a) thermophilic H_2 production under anaerobic fermentation process needs energy for heating and maintenance; and (b) low cell densities at higher temperatures reduce the H_2 yield. To overcome the first issue, energy from hot wastewater of industries such as distilleries, food and paper processing units could be recovered to maintain temperature of thermophilic bacteria. To address the second problem, cell immobilization technologies would be required to retain the cells in the reactors.

Regional industrial-academia partnership is also pertinent for successful production of H_2. A process that lends itself well to this type of cooperation is the coupling of dark fermentation and photofermentation processes of H_2 production along with localized energy production (see, for example, Fig. 1.1). Identified research groups could work in the area of substrate selection for dark fermentation, optimization of bioreactor conditions for dark fermentation or photofermentation, cost analysis, life cycle assessment studies, and issues of safety or storage of $BioH_2$, to evaluate feasibility of a sustainable industrial process. H_2 energy funding agencies, scientists, and engineers may then reliably predict the probability of successful scale-up based on thorough laboratory scale analysis of H_2 production. Finally we need to focus time and energy on building a global research group or network, working towards a common goal of industrialization of H_2 production.

1.7 ACKNOWLEDGEMENT

Aditya Bhalla, Jia Wang, David Salem, and Rajesh Sani gratefully acknowledge the financial support provided by National Science Foundation – Industry/University Cooperative Research Center. The support from the Department of Chemical and Biological Engineering at the South Dakota School of Mines and Technology is also acknowledged. Sudhir Kumar is thankful to Jaypee University of Information Technology, Solan, HP, India for granting a sabbatical leave for doing research work at the South Dakota.

KEYWORDS

- **Biohydrogen**
- **Extremophiles**
- **Hyperthermophiles**
- **Thermophiles**
- **Untreated Lignocellulosics**

REFERENCES

1. Lange, J. P. (2007). Lignocellulosic conversion: An introduction to chemistry, process and economics. *Biofuel. Bioprod. Bior. 1*, 39–48.
2. Bhalla, A., Bansal, N., Kumar, S., Bischoff, K. M., Sani, R. K. (2013). Improved lignocellulose conversion to biofuels with thermophilic bacteria and thermostable enzymes. *Bioresour. Technol. 128*, 751–759.
3. Lynd, L. R., van Zyl, W. H., McBride, J. E., Laser, M. (2005). Consolidated bioprocessing of cellulosic biomass: an update. *Curr. Opin. Biotechnol. 16*, 577–583.
4. Talluri, S., Raj, S. M., Christopher, L. P. (2013). Consolidated bioprocessing of untreated switchgrass to hydrogen by the extreme thermophile *Caldicellulosiruptor saccharolyticus* DSM 8903. *Bioresour Technol. 139*, 272–279.
5. Barnard, D., Casanueva, A., Tuffin, M., Cowan, D. (2010). Extremophiles in biofuel synthesis. *Environ. Technol. 31*, 871–888.
6. Ntaikou, I., Antonopoulou, G., Lyberatos, G. (2010). Biohydrogen production from biomass and wastes via dark fermentation: a review. *Waste Biomass Valor. 1*, 21–39.
7. Ivanova, G., Rakhely, G., Kovacs, K. L. (2009). Thermophilic biohydrogen production from energy plants by *Caldicellulosiruptor saccharolyticus* and comparison with related studies. *Int. J. Hydrogen Energy. 34*, 3659–3670.
8. Liu, Y., Yu, P., Song, X., Qu, Y. (2008). Hydrogen production from cellulose by co-culture of *Clostridium thermocellum* JN4 and *Thermoanaerobacterium thermosaccharolyticum* GD 17. *Int. J. Hydrogen Energy. 33*, 2927–2933.
9. Kanai, T., Imanaka, H., Nakajima, A., Uwamori, K., Omori, Y., Fukui, T., Atomi, H., Imanaka, T. (2005). Continuous hydrogen production by the hyperthermophilic archon, *Thermococcuskodakaraensis* KOD1. *J. Biotechnol. 116*, 271–282.
10. Saripan, A. F., Reungsang, A. (2013). Biohydrogen production by *Thermoanaerobacterium thermosaccharolyticum* KKU-ED1: culture conditions optimization using mixed xylose/arabinose as substrate. *Electronic J. Biotechnol. 16(1),* DOI: 10.2225/vol16-issue1-fulltext-1.
11. Oh, Y. K., Raj, S. M., Jung, G. Y., Park, S. (2011). Current status of the metabolic engineering of microorganisms for biohydrogen production. *Bioresour. Technol. 102*, 8357–8367.
12. Hallenbeck, P. C., Ghosh, D., Skonieczny, M. T., Yargeau, V. Microbiological and engineering aspects of biohydrogen production. *Indian J. Microbiol.* (2009). *49,* 48–59.

13. Levin, D. B., Pitt, L., Love, M. (2004). Biohydrogen production: prospects and limitations to practical application. *Int. J. Hydrogen Energy. 29*, 173–185.
14. Willquist, K., Zeidan, A. A., van Niel, E. W. J. (2010). Physiological characteristics of the extreme thermophile *Caldicellulosiruptor saccharolyticus*: an efficient hydrogen cell factory. *Microb. Cell Fact. 9*, 89.
15. Cao, G. L., Zhao, L., Wang, A. J., Wang, Z. Y., Ren, N. Q. (2014). Single-step bioconversion of lignocellulose to hydrogen using novel moderately thermophilic bacteria. *Biotechnol. Biofuels. 7*, 82.
16. Raj, S. M., Talluri, S., Christopher, L. P. (2012). Thermophilic Hydrogen production from renewable resources: current status and future perspectives. *BioEnerg. Res. 5*, 515–531.
17. Shaw, A. J., Hogsett, D. A., Lynd, L. R. Natural competence in *Thermoanaerobacter* and *Thermoanaerobacterium* species. *Appl. Environ. Microbiol.* (2010). *76*, 4713–4719.
18. Karadag, D., Mäkinen, A. E., Efimova, E., Puhakka, J. A. (2009). Thermophilic biohydrogen production by an anaerobic heat treated-hot spring culture. *Bioresour. Technol. 100*, 5790–5795.
19. Yang, S. J., Kataeva, I., Hamilton-Brehm, S. D., Engle, N. L., Tschaplinski, T. J., Doeppke, C., Davis, M., Westpheling, J., Adams, M. W. (2009). Efficient degradation of lignocellulosic plant biomass, without pretreatment, by the thermophilic anaerobe "*Anaerocellumthermophilum*" DSM 6725. *Appl. Environ. Microbiol. 75*, 4762–4769.
20. Levin, D. B., Sparling, R., Islam, R., Cicek, N. (2006). Hydrogen production by *Clostridium thermocellum* 27405 from cellulosic biomass substrates. *Int. J. Hydrogen Energy. 31*, 1496–1503.
21. Saratale, G. D., Saratale, R. G., Chang, J. S. (2013). Biohydrogen from renewable resources. In: Pandey A (ed.), *Biohydrogen*. Elsevier, San Diego. 185–221.
22. Cheng, C. L., Lo, Y. C., Lee, K. S., Lee, D. J., Lin, C. Y., Chang, J. S. (2011). Biohydrogen production from lignocellulosic feedstock. *Bioresour. Technol. 102*, 8514–8523.
23. Baghchehsaraee, B., Nakhla, G., Karamanev, D., Margaritis, A. (2010). Fermentative hydrogen production by diverse microflora. *Int. J. Hydrogen Energy. 35*, 5021–5027.
24. Bianchi, L., Mannelli, F., Viti, C., Adessi, A., Philippis, R. D. (2010). Hydrogen-producing purple nonsulfur bacteria isolated from the trophic lake Averno (Naples, Italy). *Int. J. Hydrogen Energy. 35*, 12216–12223.
25. de Vrije, T., Bakker, R. R., Budde, M. A., Lai, M. H., Mars, A. E., Claassen, P. A. M. (2009). Efficient hydrogen production from the lignocellulosic energy crop *Miscanthus* by the extreme thermophilic bacteria *Caldicellulosiruptor saccharolyticus* and *Thermotoganeapolitana. Biotechnol. Biofuels. 2*, 12.
26. Ananyev, G., Carrieri, D., Dismukes, G. C. (2008). Optimization of metabolic capacity and flux through environmental cues to maximize hydrogen production by the cyanobacterium "*Arthrospira (Spirulina) maxima*". *Appl. Environ. Microbiol. 74*, 6102–6113.
27. Gómez, X., Fernández, C., Fierro, J., Sánchez, M. E., Escapa, A., Morán, A. (2011). Hydrogen production: two stage processes for waste degradation. *Bioresour. Technol. 102*, 8621–8627.

28. Lakshmidevi, R., Muthukumar, K. (2010). Enzymatic saccharification and fermentation of paper and pulp industry effluent for biohydrogen production. *Int. J. Hydrogen Energy. 35,* 3389–3400.

29. Kim, J. K., Nhat, L., Chun, Y. N., Kim, S. W. (2008). Hydrogen production conditions from food waste by dark fermentation with *Clostridium beijerinckii* KCTC 1785. *Biotechnol. Bioprocess Eng. 13,* 499–504.

30. Vatsala, T. M., Raj, S. M., Manimaran, A. (2008). A pilot-scale study of biohydrogen production from distillery effluent using defined bacterial coculture. *Int. J. Hydrogen Energy. 33,* 5404–5415.

31. Ueno, Y., Fukui, H., Goto, M. (2007). Operation of a two-stage fermentation process producing hydrogen and methane from organic waste. *Environ. Sci. Technol. 41,* 1413–1419.

32. Masset, J., Calusinska, M., Hamilton, C., Hiligsmann, S., Joris, B., Wilmotte, A., Thonart, P. (2012). Fermentative hydrogen production from glucose and starch using pure strains and artificial cocultures of *Clostridium* spp. *Biotechnol. Biofuels. 5,* 35.

33. Pandu, K., Joseph, S. (2012). Comparisons and limitations of biohydrogen production processes: a review. *Int. J. Adv. Eng. Technol. 2,* 342–356.

34. Rittmann, S., Herwig, C. (2012). A comprehensive and quantitative review of dark fermentation biohydrogen production. *Microb. Cell Fact. 11,* 115.

35. O-Thong, S., Mamimin, C., Prasertsan, P. (2011). Effect of temperature and initial pH on biohydrogen production from palm oil mill effluents: long-term evaluation and microbial community analysis. *Electronic J. Biotechnol. 14(5),* DOI: 10.2225/vol14-issue5-fulltext-9.

36. Mars, A. E., Veuskens, T., Budde, M. A. W., van Doeveren, P. F. N. M., Lips, S. J., Bakker, R. R., de Vrije, G. J., Claassen, P. A. M. (2010). Biohydrogen production from untreated and hydrolyzed potato steam peels by the extreme thermophiles *Caldicellulosiruptor saccharolyticus* and *Thermotoganeapolitana. Int. J. Hydrogen Energy. 35,* 7730–7737.

37. Zeidan, A. A., Radstorm, P., van Niel, E. W. J. (2010). Stable coexistence of two *Caldicellulosiruptor* species in a de novo constructed hydrogen producing coculture. *Microb. Cell Fact. 9,* 102.

38. Chen, S. D., Lee, K. S., Lo, Y. C., Chen, W. M., Wu, J. F., Lin, C. Y., Chang, J. S. (2008). Batch and continuous biohydrogen production from starch hydrolysate by *Clostridium* species. *Int. J. Hydrogen Energy. 33,* 1803–1812.

39. de Vrije, T., Mars, A. E., Budde, M. A. W., Lai, M. H., Dijkeme, C., de Waard, P., Claassen, P. A. M. (2007). Glycolytic pathway and hydrogen yield studies of the extreme thermophile *Caldicellulosiruptor saccharolyticus. Appl. Microbiol. Biotechnol. 74,* 1358–1367.

40. Lin, C. Y., Hung, C. H., Chen, C. H., Chung, W. T., Chen, L. H. (2006). Effects of initial pH on fermentative hydrogen production from xylose using natural mixed cultures. *Process Biochem. 41,* 1383–1390.

41. Kádár, Z., deVrije, T., van Noorden, G. E., Budde, M. A. W., Szengyel, Z., Réczey, K., Claassen, P. A. M. (2004). Yields from glucose, xylose, and paper sludge hydrolysate during hydrogen production by the extreme thermophile *Caldicellulosiruptor saccharolyticus. Appl. Biochem. Biotechnol. 114,* 497–508.

42. Lin, C. Y., Jo, C. H. (2003). Hydrogen production from sucrose using an anaerobic sequencing batch reactor process. *J. Chem. Technol. Biotechnol. 78*, 678–684.

43. van Niel, E. W. J., Budde, M. A. W., de Haas, G. G., van der Wal, F. J., Claassen, P. A. M., Stams, A. J. M. (2002). Distinctive properties of high hydrogen producing extreme thermophiles, *Caldicellulosiruptor saccharolyticus* and *Thermotogaelffi. Int. J. Hydrogen Energy. 27*, 1391–1398.

44. Vehicle Technologies Program: Fact #205: February 25, 2002. Hydrogen Cost and Worldwide Production. eere.energy.gov. Retrieved 2009–09–19.

45. Lal, R. (2004). Soil carbon sequestration impacts on global climate change and food security. *Sci. 304*, 1623–1627.

46. Ren, N., Guo, W., Liu, B., Cao, G., Ding, J. (2011).Biological hydrogen production by dark fermentation: challenges and prospects towards scaled-up production. *Curr. Opin. Biotechnol. 22*, 365–370.

47. Yang, H. H., Guo, L. J., Liu, F. (2010). Enhanced bio-hydrogen production from corncob by a two-step process: dark- and photo-fermentation. *Bioresour. Technol. 101*, 2049–2052.

48. Jayalakshmi, S., Joseph, K., Sukumaran, V. (2009). Biohydrogen generation from kitchen waste in an inclined plug flow reactor. *Int. J. Hydrogen Energy. 34*, 8854–8858.

49. Karlsson, A., Vallin, L., Ejlertsson, J. (2008). Effects of temperature, hydraulic retention time and hydrogen extraction rate on hydrogen production from the fermentation of food industry residues and manure. *Int. J. Hydrogen Energy. 33*, 953–962.

50. Lay, J. J., Fan, K. S., Chang, J., Ku, C. H. (2003). Influence of chemical nature of organic wastes on their conversion to hydrogen by heat-shock digested sludge. *Int. J. Hydrogen Energy. 28*, 1361–1367.

51. Wijffels, R. H., Barbosa, M. J. An outlook on microalgal biofuels. *Sci.* (2010). *329*, 796–799.

52. Williams, P. L., Laurens, L. M. L. (2010). Microalgae as biodiesel and biomass feedstock: review and analysis of the biochemistry, energetics and economics. *Energ. Environ. Sci. 3*, 554–590.

53. Harun, R., Danquah, M. K., Forde, G. M. (2010). Microalgal biomass as a fermentation feedstock for bioethanol production. *J. Chem. Technol. Biotechnol. 85*, 199–203.

54. Balat, H., Kirtay, E. (2010). Hydrogen from biomass – resent scenario and future prospects. *Int. J. Hydrogen Energy. 35*, 7416–7426.

55. Demirbas, A. H. (2010). Biofuels for future transportation necessity. *Energy Educ. Sci. Technol. Part A. 26*, 13–23.

56. Singh, J., Gu, S. (2010). Commercialization potential of microalgae for biofuels production. *Renew. Sust. Energ. Rev. 14*, 2596–2610.

57. Hallenbeck, P. C. (2011). Microbial paths to renewable hydrogen production. *Biofuels. 2*, 285–302.

58. Sinha, P., Pandey, A. (2011). An evaluative report and challenges for fermentative biohydrogen production. *Int. J. Hydrogen Energy. 36*, 7460–7478.

59. Abo-Hashesh, M., Wang, R., Hallenbeck, P. C. (2011). Metabolic engineering in dark fermentative hydrogen production; theory and practice. *Bioresour. Technol. 102*, 8414–8422.

60. Hallenbeck, P. C., Ghosh, D., Abo-Hashesh, M., Wang, R. (2011). Metabolic engineering for enhanced biofuels production with emphasis on the biological production of hydrogen. *Adv. Chem. Res. 6*, 125–154.
61. Hallenbeck, P. C., Abo-Hashesh, M., Ghosh, D. (2012). Strategies for improving biological hydrogen production. *Bioresour. Technol. 110*, 1–9.
62. Eriksen, N. T., Riis, M. L., Holm, N. K., Iversen, N. (2011).Hydrogen synthesis from pentoses and biomass in *Thermotoga* spp. *Biotechnol. Lett. 33*, 293–300.
63. Munro, S. A., Zinder, S. H., Walker, L. P. (2009). The fermentation stoichiometry of *Thermotoganeapolitana* and influence of temperature, oxygen, and pH on hydrogen production. *Biotechnol. Progr. 25*, 1035–1042.
64. Chou, C. J., Shockley, K. R., Conners, S. B., Lewis, D. L., Comfort, D. A., Adams, M. W. W., Kelly, R. M. (2007). Impact of substrate glycoside linkage and elemental sulfur on bioenergetics of and hydrogen production by the hyperthermophilic archon *Pyrococcusfuriosus*. *Appl. Environ. Microbiol. 73*, 6842–6853.
65. Verhaart, M. R. A., Bielen, A. A. M., van der Oost, J., Stams, A. J. M., Kengen, S. W. M. (2010). Hydrogen production by hyperthermophilic and extremely thermophilic bacteria and archaea: mechanisms for reductant disposal. *Environ. Technol. 31*, 993–1003.
66. Atomi, H., Sato, T., Kanai, T. (2011).Application of hyper-thermophiles and their enzymes. *Curr.Opin. Biotechnol. 22*, 618–626.
67. Chou, C. J., Jenney, F. E. Jr., Adams, M. W. W., Kelly, R. M. (2008). Hydrogenesis in hyperthermophilic microorganisms: implications for biofuels. *Metab. Eng. 10*, 394–404.
68. Soboh, B., Linder, D., Hedderich, R. (2004). A multisubunit membrane-bound [NiFe] hydrogenase and an NADH-dependent Fe-only hydrogenase in the fermenting bacterium *Thermoanaerobacter tengcongensis*. *Microbiol. 15*, 2451–2463.
69. Smith, E. T., Odom, L. D., Awramko, J. A., Chiong, M., Blamey, J. (2001). Direct electrochemical characterization of hyperthermophilic *Thermococcuseler* metalloenzymes involved in hydrogen production from pyruvate. *J. Biol. Inorg. Chem. 6*, 227–231.
70. Silva, P. J., van den Ban, E. C., Wassink, H., Haaker, H., de Castro, B., Robb, F. T., Hagen, W. R. (2000). Enzymes of hydrogen metabolism in *Pyrococcusfuriosus*. *Eur. J. Biochem. 267*, 6541–6551.
71. Gadow, S. I., Li, Y. Y., Liu, Y. (2012). Effect of temperature on continuous hydrogen production of cellulose. *Int. J. Hydrogen Energy. 37*, 15465–15472.
72. Kargi, F., Eren, N. S., Ozmihci, S. (2012). Bio-hydrogen production from cheese whey powder (CWP) solution: comparison of thermophilic and mesophilic dark fermentations. *Int. J. Hydrogen Energy. 37*, 8338–8342.
73. Zuroff, T. R., Curtis, W. R. (2012). Developing symbiotic consortia for lignocellulosic biofuel production. *Appl. Microbiol. Biotechnol. 93*, 1423–1435.
74. Zeidan, A. A., van Niel, E. W. J. A quantitative analysis of hydrogen production efficiency of the extreme thermophile *Caldicellulosiruptor owensensis* OL^T. *Int. J. Hydrogen Energy*. (2010). *35*, 1128–1137.
75. Chen, C. C., Chuang, Y. S., Lin, Y. C., Lay, C. H., Sen, B. (2012). Thermophilic dark fermentation of untreated rice straw using mixed cultures for hydrogen production. *Int. J. Hydrogen Energy. 37*, 15540–15546.

76. Nguyen, T. A. D., Kim, K. R., Kim, M. S., Sim, S. J. (2010). Thermophilic hydrogen fermentation from Korean rice straw by *Thermotoganeapolitana*. *Int. J. Hydrogen Energy*. *35*, 13392–13398.

77. Li, Q., Liu, C. Z. (2012). Co-culture of *Clostridium thermocellum* and *Clostridium thermosaccharolyticum* for enhancing hydrogen production via thermophilic fermentation of cornstalk waste. *Int. J. Hydrogen Energy*. *37*, 10648–10654.

78. Cao, G. L., Guo, W. Q., Wang, A. J., Zhao, L., Xu, C. J., Zhao, Q. L., Ren, N. Q. (2012). Enhanced cellulosic hydrogen production from lime-treated cornstalk wastes using thermophilic anaerobic microflora. *Int. J. Hydrogen Energy*. *37*, 13161–13166.

79. Kongjan, P., Angelidaki, I. (2010). Extreme thermophilic biohydrogen production from wheat straw hydrolysate using mixed culture fermentation: Effect of reactor configuration. *Bioresour. Technol*. *101*, 7789–7796.

80. Kongjan, P., O-Thong, S., Kotay, M., Min, B., Angelidaki, I. Biohydrogen production from wheat straw hydrolysate by dark fermentation using extreme thermophilic mixed culture. *Biotechnol. Bioeng*. (2010). *105*, 899–908.

81. Ren, N. Q., Cao, G. L., Guo, W. Q., Wang, A. J., Zhu, Y. H., Liu, B. F., Xu, J. F. (2010). Biological hydrogen production from corn stover by moderately thermophile *Thermoanaerobacterium thermosaccharolyticum* W16. *Int. J. Hydrogen Energy*. *35*, 2708–2712.

82. Panagiotopoulos, I. A., Bakker, R. R., de Vrije, T., Koukios, E. G., Claassen, P. A. M. (2010). Pretreatment of sweet sorghum bagasse for hydrogen production by *Caldicellulosiruptor saccharolyticus*. *Int. J. Hydrogen Energy*. *35*, 7738–7747.

83. de Vrije, T., Budde, M. A. W., Lips, S. J., Bakker, R. R., Mars, A. E., Claassen, P. A. M. (2010). Hydrogen production from carrot pulp by the extreme thermophiles *Caldicellulosiruptor saccharolyticus* and *Thermotoganeapolitana*. *Int. J. Hydrogen Energy*. *35*, 13206–13213.

84. Lee, K. Z., Li, L. S., Kuo, C. P., Chen, C. I., Tien, M. Y., Huang, Y. J., Chuang, C. P., Wong, S. C., Cheng, S. S. (2010). Thermophilic bio-energy process study on hydrogen fermentation with vegetable kitchen waste. *Int. J. Hydrogen Energy*. *35*, 13458–13466.

85. VanFossen, A. L., Ozdemir, I., Zelin, S. L., Kelly, R. M. (2011). Glycoside hydrolase inventory drives plant polysaccharide deconstruction by the extremely thermophilic bacterium *Caldicellulosiruptor saccharolyticus*. *Biotechnol. Bioeng*. *108*, 1559–1569.

86. van de Werken, H. J., Verhaart, M. R., VanFossen, A. L., Willquist, K., Lewis, D. L., Nichols, J. D., Goorissen, H. P., Mongodin, E. F., Nelson, K. E., van Niel, E. W. J., Stams, A. J., Ward, D. E., de Vos, W. M., van der Oost, J., Kelly, R. M., Kengen, S. W. (2008). Hydrogenomics of the extremely thermophilic bacterium *Caldicellulosiruptor saccharolyticus*. *Appl. Environ. Microbiol*. *74*, 6720–6729.

87. Onyenwoke, R. U., Lee, Y. J., Dabrowski, S., Ahring, B. K., Wiegel, J. (2006). Reclassification of *Thermoanaerobiumacetigenum* as *Caldicellulosiruptor acetigenus* comb. nov. and emendation of the genus description. *Int. J. Syst. Evol. Microbiol*. *56*, 1391–1395.

88. Zeidan, A. A., van Niel, E. W. J. (2009). Developing a thermophilic hydrogen-producing coculture for efficient utilization of mixed sugars. *Int. J. Hydrogen Energy*. *34*, 4524–4528.

89. Hay, J. X. W., Wu, T. Y., Juan, J. C., Jahim, J. M. (2013). Biohydrogen production through photo fermentation or dark fermentation using waste as a substrate: overview, economics, and future prospects of hydrogen usage. *Biofuel. Bioprod. Bior. 7*, 334–352.

90. Lo, Y. C., Lu, W. C., Chen, C. Y., Chang, J. S. (2010). Dark fermentative hydrogen production from enzymatic hydrolysate of xylan and pretreated rice straw by *Clostridium butyricum* CGS5. *Bioresour. Technol. 101*, 5885–5891.

91. Cui, M., Yuan, Z., Zhi, X., Wei, L., Shen, J. (2010). Biohydrogen production from poplar leaves pretreated by different methods using anaerobic mixed bacteria. *Int. J. Hydrogen Energy. 35*, 4041–4047.

92. Su, X., Mackie, R. I., Cann, I. K. (2012). Biochemical and mutational analyzes of a multidomain cellulase/mannanase from *Caldicellulosiruptor bescii*. *Appl. Environ. Microbiol. 78*, 2230–2240.

93. Brentner, L. B., Peccia, J., Zimmerman, J. B. (2010). Challenges in developing biohydrogen as a sustainable energy source: implications for a research agenda. *Environ. Sci. Technol. 44*, 2243–2254.

94. Kim, M. S., Lee, D. Y. (2010). Fermentative hydrogen production from tofu processing waste and anaerobic digester sludge using microbial consortium. *Bioresour. Technol. 101*, S48-S52.

95. Babu, V. L., Mohan, S. V., Sarma, P. N. (2009). Influence of reactor configuration on fermentative hydrogen production during wastewater treatment. *Int. J. Hydrogen Energy. 34*, 3305–3312.

96. Fang, H. H., Zhang, T. (2002). Characterization of a hydrogen-producing granular sludge. *Biotechnology and Bioengineering. 78*, 44–52.

97. Chen, C. C., Lin, C. Y., Chang, J. S. (2001). Kinetics of hydrogen production with continuous anaerobic cultures using sucrose as the limiting substrate. *Appl. Microbiol. Biotechnol. 57*, 56–64.

98. Lee, K. S., Lo, Y. C., Lin, P. J., Chang, J. S. (2006). Improving biohydrogen production in a carrier-induced granular sludge bed by altering physical configuration and agitation pattern of the bioreactor. *Int. J. Hydrogen Energy. 31*, 1648–1657.

99. Lay, C. H., Wu, J. H., Hsiao, C. L., Chang, J. J., Chen, C. C., Lin, C. Y. (2010). Biohydrogen production from soluble condensed molasses fermentation using anaerobic fermentation. *Int. J. Hydrogen Energy. 24*, 13445–13451.

100. Hallenbeck, P. C. (2005). Fundamentals of the fermentative production of hydrogen. *Water Sci. Technol. 52*, 21–29.

101. Kim, I. S., Hwang, M. H., Jang, N. J., Hyun, S. H., Lee, S. T. (2004). Effect of low pH on the activity of hydrogen using methanogen in bio-hydrogen process. *Int. J. Hydrogen Energy. 29*, 1133–1140.

102. Oh, S. E., Lyer, P., Bruns, M. A., Logan, B. E. (2004). Biological hydrogen production using a membrane bioreactor. *Biotechnol. Bioeng. 87*, 119–127.

103. Lin, C. Y., Chang, R. C. (2004). Fermentative hydrogen production at ambient temperature. *Int. J. Hydrogen Energy. 29*, 715–720.

104. Levin, D. B., Chahine, R. (2010). Challenges for renewable hydrogen production from biomass. *Int. J. Hydrogen Energy. 35*, 4962–4969.

105. Levin, D. B., Carere, C. R., Cicek, N., Sparling, R. (2009). Challenges for biohydrogen production via direct lignocellulose fermentation. *Int. J. Hydrogen Energy. 34*, 7390–7403.

106. Kumar, S., Bhalla, A., Shende, R. V., Sani, R. K. (2012). Decentralized thermophilic biohydrogen: a more efficient and cost-effective process. *BioResources. 7*, 1–2.

107. Barreto, L., Makihira, A., Riah, K. (2003). The hydrogen economy in the twenty-first century: a sustainable development scenario. *Int. J. Hydrogen Energy. 28*, 267–284.

108. Chornet, E., Czernik, S., Wang, D., Gregoire, C., Mann, M. (1994). Biomass to hydrogen via pyrolysis and reforming. In proceedings of the 1994 DOE/NREL Hydrogen Program Review, 407–432. Livermore, California, NREL/CP-470-6431; Conf-9404194.

109. Liu, H., Grot, S., Logan, B. E. (2005). Electrochemically assisted microbial production of hydrogen from acetate. *Environ. Sci. Technol. 39*, 4317–4320.

110. Kotay, S. M., Das, D. (2008). Biohydrogen as a renewable energy resource-prospects and potentials. *Int. J. Hydrogen Energy. 33*, 258–263.

111. Panagiotopoulos, I. A., Bakker, R. R., Budde, M. A. W., Vrije, D. T., Claassen, P. A. M., (2009). Koukios, for example, Fermentative hydrogen production from pretreated biomass: a comparative study. *Bioresour. Technol. 100*, 6331–6338.

112. Sierra, R., Smith, A., Granda, C., Holtzapple, M. T. (2009). Producing fuels and chemicals from lignocellulosic biomass. *Society of Biological Engineering Special Section*, 10–18.

113. Kumari, M., Kumar, S. (2009). Biorefineries: India's future option for energy and chemical feedstock. *Asian J. Energy Environ. 10(3)*, 160–164.

114. Koutrouli, E. C., Kalfas, H., Gavala, H. N., Skiadas, I. V., Stamatelatou, K., Lyberatos, G. (2009). Hydrogen and methane production through two-stage mesophilic anaerobic digestion of olive pulp. *Bioresour Technol. 100*, 3718–3723.

115. Ntaikou, I., Kourmentza, C., Koutrouli, E. C., Stamatelatou, K., Zampraka, A., Kornaros, M., Lyberatos, G. (2009). Exploitation of olive oil mill wastewater for combined biohydrogen and biopolymers production. *Bioresour Technol. 100*, 3724–3730.

116. Ren, Z., Ward, T. E., Logan, B. E., Regan, J. M. (2007). Characterization of the cellulolytic and hydrogen-producing activities of six mesophilic Clostridium species. *J. Appl. Microbiol. 103*, 2258–2266.

117. Mohan, S. V., Mohanakrishna, G., Raghavulu, S. V., Sarma, P. N. (2007). Enhancing biohydrogen production from chemical wastewater treatment in anaerobic sequencing batch biofilm reactor (AnSBBR) by bioaugmenting with selectively enriched kanamycin resistant anaerobic mixed consortia. *Int. J. Hydrogen Energy. 32*, 3284–3292.

118. Zahedi, S., Sales, D., Romero, L. I., Solera, R. (2013). Hydrogen production from the organic fraction of municipal solid waste in anaerobic thermophilic acidogenesis: influence of organic loading rate and microbial content of the solid waste. Bioresource Technology, *129*, 85–91.

119. Uduman, N., Qi, Y., Danquah, M. K., Forde, G. M., Hoadley, A. (2010). Dewatering of microalgal cultures: A major bottleneck to algae-based fuels. *J. Renew. Sustain. Ener. 2*, 012701–15.

120. Kapdan, I. K., Kargi, F. (2006). Bio-hydrogen production from waste materials. *Enzyme Microb. Technol. 38*, 569–582.

121. Jayasinghearachchi, H. S., Sarma, P. M., Lal, B. (2012). Biological hydrogen production by extremely thermophilic novel bacterium *Thermoanaerobacter mathranii* A3N isolated from oil producing well. *Int. J. Hydrogen Energy. 37*, 5569–5578.

122. Ngo, T. A., Nguyen, T. H., Bui, H. T. V. (2012). Thermophilic fermentative hydrogen production by *Thermotoganeapolitana* DSM 4359. *Renew. Energ. 37*, 174–179.

123. Zhang, K., Ren, N. Q., Cao, G. L., Wang, A. J. (2011). Biohydrogen production behavior of moderately thermophile *Thermoanaerobacterium thermosaccharolyticum* W16 under different gas-phase conditions. *Int. J. Hydrogen Energy. 36*, 14041–14048.

124. Shaw, A. J., Hogsett, D. A., Lynd, L. R. (2009). Identification of the [FeFe]-hydrogenase responsible for hydrogen generation in *Thermoanaerobacterium saccharolyticum* and demonstration of increased ethanol yield via hydrogenase knockout. *J. Bacteriol. 191*, 6457–6464.

125. Abreu, A. A., Karakashev, D., Angelidaki, I., Sousa, D. Z., Alves, M. M. (2012). Biohydrogen production from arabinose and glucose using extreme thermophilic anaerobic mixed cultures. *Biotechnol. Biofuels. 5*, 6.

126. Lu, W., Fan, G., Zhao, C., Wang, H., Chi, Z. (2012). Enhancement of fermentative hydrogen production in an extreme-thermophilic (70 °C) mixed-culture environment by repeated batch cultivation. *Curr. Microbiol. 64*, 427–432.

127. Hniman, A., O-Thong, S., Prasertsan, P. (2011). Developing a thermophilic hydrogen-producing microbial consortia from geothermal spring for efficient utilization of xylose and glucose mixed substrates and oil palm trunk hydrolysate. *Int. J. Hydrogen Energy. 36*, 8785–8793.

128. Karadag, D., Puhakka, J. A. (2010). Enhancement of anaerobic hydrogen production by iron and nickel. *Int. J. Hydrogen Energy. 35*, 8554–8560.

129. Zhao, C., Karakashev, D., Lu, W., Wang, H., Angelidaki, I. (2010). Xylose fermentation to biofuels (hydrogen and ethanol) by extreme thermophilic (70 °C) mixed culture. *Int. J. Hydrogen Energy. 35*, 3415–3422.

130. Kongjan, P., Min, B., Angelidaki, I. (2009). Biohydrogen production from xylose at extreme thermophilic temperatures (70 °C) by mixed culture fermentation. *Water Res. 43*, 1414–1124.

131. Hasyim, R., Imai, T., Reungsang, A., O-Thong, S. (2011). Extreme-thermophilic biohydrogen production by an anaerobic heat treated digested sewage sludge culture. *Int. J. Hydrogen Energy. 36*, 8727–8734.

132. Geng, A., He, Y., Qian, C., Yan, X., Zhou, Z. (2010). Effect of key factors on hydrogen production from cellulose in a coculture of *Clostridium thermocellum* and *Clostridium thermopalmarium*. *Bioresour. Technol. 101*, 4029–4033.

133. Magnusson, L., Islam, R., Sparling, R., Levin, D., Cicek, N. (2008). Direct hydrogen production from cellulosic waste materials with a single-step dark fermentation process. *Int. J. Hydrogen Energy. 33*, 5398–5403.

134. Ho, K. L., Lee, D. J., Su, A., Chang, J. S. (2012). Biohydrogen from lignocellulosic feedstock via one-step process. *Int. J. Hydrogen Energy. 37*, 15569–15574.

135. Prakasham, R. S., Sathish, T., Brahmaiah, P., Rao, C. S., Rao, R. S., Hobbs, P. J. (2009). Biohydrogen production from renewable agri-waste blend: optimization using mixer design. *Int. J. Hydrogen Energy. 34*, 6143–6148.

CHAPTER 2

BIO-HYDROGEN PRODUCTION: CURRENT TRENDS AND FUTURE PROSPECTS

VINITA MISHRA,[1,]* and ISHA SRIVASTAVA[2]

[1]*Cell Biology Lab, School of Biological Sciences and Biotechnology, Indian Institute of Advanced Research, Koba Institutional Area, Gandhinagar, 382 007, Gujarat, India*

[2]*Department of Biotechnology, Delhi Technological University (Formerly Delhi College of Engineering), Shahbad Daulatpur, Main Bawana Road, Delhi-110 042, India*

*Corresponding author's E-mail: vinita.mbif@gmail.com

CONTENTS

ABSTRACT

Bio-hydrogen is gaining importance as a clean source of energy which can be used as fuel in future but the real problem is bio-hydrogen production at a large scale, so that it can be commercialized. Due to this limitation bio-hydrogen production has become an important area of research and gaining attention of researchers. These issues motivated us to compile this review. This review covers the processes of bio-hydrogen production, microorganisms involved, factors that limit the bio-hydrogen production, etc. In this chapter, we have proposed a hypothetical three-stage hybrid system by keeping in view the problems associated with bio-hydrogen production. This hypothetical model can be applied in designing a bioreactor to increase the yield and efficiency of bio-hydrogen production. The findings of this review will be helpful in overcoming the problems currently associated with the bio-hydrogen production and will help researchers in designing and developing more efficient and cost effective bio-hydrogen production techniques.

2.1   INTRODUCTION

In the present global scenario, major sources of energy in the world are fossil fuels (about 80% of the present demand) and the resources of fossil energy are decreasing day by day [1]. Nowadays Hydrogen (H_2) gas is considered as a source of future energy because of the fact that it is renewable and does not contribute in creating "greenhouse effect". H_2 gas produces a large amount of energy per unit weight by combustion (143 GJ/ tons) and can be easily converted into electricity with the help of fuel cells [1]. However, hydrogen is not a primary energy source, but rather it serves as a medium through which primary energy sources such as nuclear and/or solar energy can be stored, used and transmitted for the fulfillment of our energy needs [2]. Hydrogen is ecofriendly and does not cause any harm to human beings, so it is considered as a clean source of energy [3, 4]. In recent years, due to the requirement of hydrogen gas as a clean source of energy, development of cost-effective and efficient bio-hydrogen production technology, which can be converted into commercially viable hydrogen production technology, is gaining attention of researchers [5]. Hydrogen gas produced by biological processes due to the action of microorganisms is called bio-hydrogen, is a type of bio-fuel, like bio-ethanol, bio-diesel or bio-gas or bio-oil. Bio-fuels can be categorized into three classes on the basis of materials and technology they use; first generation (made from food crops), second generation (made from nonfood crops or wastes), third generation or advanced (made using microbe). Third generation bio-fuels are known to have several advantages over first and second generation bio-fuels for instance third generation bio-fuels are more economical in comparison to first generation bio-fuels because first generation bio-fuels have caused increase in food prices, third generation bio-fuels are more efficient in capturing sun light energy which is almost ten times higher than the second generation bio-fuels, which means that smaller areas of land will be required to produce enough fuel. Bio-hydrogen is an example of an advanced or third generation bio-fuel [3, 5]. In advanced bio-fuel technologies, microorganisms are allowed to grow in special type of bioreactors and provided with the energy and nutrients that they need for their growth including sunlight, waste organic material, CO_2 from the air or from conventional gas plants. As the microbes grow they produce bio-hydrogen [4, 5]. Demand of hydrogen gas is not limited to utilization as a source of energy. Hydrogen gas is widely used for hydrogenation of fats and oils in food industry, production of electronic devices, processing steel, desulfurization and reformulation of gasoline in refineries and is also used as feedstock for the production of chemicals [5].

2.2 BIO-HYDROGEN PRODUCTION

2.2.1 BIOPROCESSES OF BIO-HYDROGEN PRODUCTION

2.2.1.1 Biophotolysis of Water (Water Splitting) by Green Algae and Cyanobacteria

During the biophotolysis (Fig. 2.1), solar energy is converted into chemical energy (H_2 energy) by photosynthetic algae and cyanobacteria with the

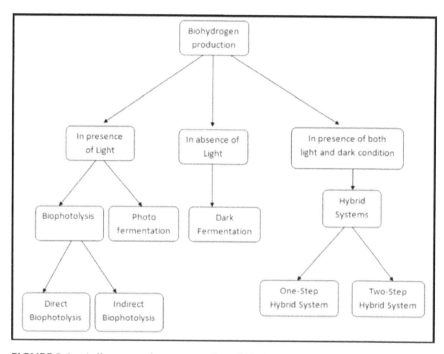

FIGURE 2.1   A diagrammatic representation of bio-hydrogen production.

help of hydrogenase and nitrogenase enzymes [2, 4] (Table 2.1). Hydrogenase and nitrogenase enzymes are highly sensitive towards presence of oxygen; therefore constant removal of oxygen must be performed for higher yield of hydrogen production via Biophotolysis [6].

Biophotolysis is possible in two ways:
1. Direct biophotolysis; and
2. Indirect biophotolysis

2.2.1.1.1 *Direct Biophotolysis*

Under anaerobic condition, green algae either use H_2 as electron donor in the process of CO_2 fixation or produces hydrogen (Fig. 2.1) [7]. In the process of direct biophotolysis photosystem II (PSII) absorbs the light energy, which are having wavelength lower than 680 nm as a result stream of

TABLE 2.1 Classification of Bio-Hydrogen Producing Microorganism

Process of Hydrogen Production	Class of Microorganism	Sub-Class of Microorganism	Genus of Microorganism	Hydrogen Source
Photosynthesis	Algae	Green algae	*Chlamydomonas sp.*	Water
			Scenedesmus sp.	
		Marine Green algae	*Chlorococcum sp.*	
			Playtmonas sp.	
			Chlorella sp.	
	Cyanobacteria	Marine cyanobacteria	*Anabaena sp.*	
			Oscillatoria sp.	
			Calothrix sp.	
		Marine unicellular cyanobacteria	*Cyanothece sp.*	
		Heterocystous cyanobacteria	*Anabaena sp.*	
			Nostoc sp.	
		Heterocyst filamentous cyanobacteria	*Anabaena sp.*	
		Non-nitrogen-fixing unicellular cyanobacteria	*Synechococcus sp.*	
			Microcystis sp.	
			Gloebacter sp.	
			Synechocystis sp.	
			Aphanocapsa sp.	
			Chroococcidiopsis sp.	
		Unicellular nondiazotrophic cyanobacteria	*Gloeocapsa sp.*	
		Unicellular/colony embedded in matrix	*Mycrocystis sp.*	
		Mat-building cyanobacteria	*Microcoleus sp.*	

TABLE 2.1   (*Continued*)

Process of Hydrogen Production	Class of Microorganism	Sub-Class of Microorganism	Genus of Microorganism	Hydrogen Source
Dark-Fermentation	Non-Photosynthetic Anaerobic Bacteria	Obligate	*Clostridia sp.* *Actinomyces sp.* *Porphyromonos sp.*	Carbohydrate Containing Organic Wastes
		Facultative	*Enterobacter sp.* *E. coli* *Hafnia sp.*	
	Non-Photosynthetic Aerobic Bacteria	–	*Aeromonos sp.* *Pseudomonos sp.* *Vibrio sp.*	
Photo-Fermentation	Photosynthetic anaerobic bacteria	Purple sulfur bacteria (Obligate anaerobe)	*Chromatium sp.* *Thiocapsa sp.*	Sulfates
		Purple-nonsulfur (PNS) bacteria (Facultative anaerobe)	*Rhodobacter sp.* *Rhodovulum sp.* *Rhodopseudomonas sp.* *Rhodospirillum sp.*	Organic Acids

electrons, and protons are generated from water molecule. Whereas photosystem I (PSI) is induced by the light with wavelength lower than 700 nm. Cytochrome *bf*, which behaves as chain of reductors, allows the transportation of electrons from PSII to PSI. Electrons from PSI are transferred to hydrogenase enzyme via ferrodoxine, which converts the proton (H^+) into molecular hydrogen (H_2) [7], (*see*, Eq. (1)).

$$2H_2O + \text{Light Energy} \rightarrow 2H_2 + O_2 \qquad (1)$$

2.2.1.1.2   Indirect Biophotolysis

Cyanobacteria produce H_2 from water by indirect photolysis. Indirect biophotolysis can be completed in two steps (*see*, Eqs. (2)–(3)). In the first step PSII and PSI are induced by light as it occurs in direct biophotolysis, but in this step CO_2 adsorption occurs due to presence of RuBisCO enzyme, which is fixed into glucose and oxygen, is produced as a result of photosynthesis (Fig. 2.1). In the second step decomposition of organic compound (glucose) occurs due to the presence of hydrogenase and nitrogenase enzymes in cyanobacteria [9]. Cyanobacterial species contains many enzymes, which are directly involved in H_2 metabolism and production. Out of these enzymes, nitrogenases enzyme catalyze the process of H_2 production, which is produced as byproduct during the reduction of nitrogen into ammonia (NH_3) [7, 9].

$$12H_2O + 6CO_2 + \text{Light Energy} \rightarrow C_6H_{12}O_6 + 6O_2 \qquad (2)$$

$$C_6H_{12}O_6 + 12H_2O + \text{Light Energy} \rightarrow 12H_2 + 6CO_2 \qquad (3)$$

The algal and cyanobacterial hydrogen production could be considered as a more economical and sustainable method of bio-hydrogen production as it uses water, which is a renewable resource and consume CO_2, which is an air pollutant [5].

2.2.1.2 Dark-Fermentation by Non-Photosynthetic Bacteria

Dark fermentation occurs in the absence of light with the help of non-photosynthetic anaerobic bacteria (Fig. 2.1, Table 2.1). During the dark fermentation of carbohydrate and organic compounds rich waste, which is an acidogenic phase of anaerobic digestion, hydrogen is produced as a byproduct [5] (*see*, Eq. (4)).

$$\text{Sugar} + \text{Dark (absence of light)} \rightarrow H_2 + CO_2 + \text{Organic Acids} \qquad (4)$$

2.2.1.3 Photo-Fermentation by Photosynthetic Bacteria

In the presence of light and anaerobic conditions, some photo-heterotrophic bacteria such as purple nonsulfur (PNS) bacteria (Fig. 2.1, Table 2.1) can convert organic acids (acetic, lactic and butyric) into hydrogen (H_2) and carbon dioxide (CO_2) [5]. Therefore, the organic acids produced during the acidogenic phase of anaerobic digestion of organic wastes can be converted to H_2 and CO_2 with the help of these photosynthetic anaerobic bacteria [1, 5, 7] (*see*, Eq. (5)).

$$\text{Organic acids} + \text{Light Energy} \rightarrow H_2 + CO_2 + \text{Carbon Compounds} \quad (5)$$

2.2.1.4   Hybrid Systems Using Photosynthetic and Fermentative Bacteria

Hybrid systems can be created by combining both photo and dark fermentation processes (Fig. 2.1) [1]. Hybrid system uses those substrates, which are useless and difficult to operate for photo fermentation. These substrates such as saccharides can be breakdown into simpler compounds, for instance acetic or butyric acids, during the process of dark fermentation in hybrid system, which can be further, operated by PNS bacteria in photo fermentation. Two different hybrid systems have been designed so far:

1. One-step hybrid system; and
2. Two-step hybrid system

2.2.1.4.1   *One-Step Hybrid System*

In one-step hybrid systems both dark and photo fermentative bacteria grow in a single culture medium, that is, in a single container, as a result of which the hydrogen production from both the processes almost occur simultaneously. One step hybrid system shows much higher yield and higher rate of hydrogen production in comparison to those obtained in photo or dark fermentation alone by using single culture [10, 11].

2.2.1.4.2   *Two-Step Hybrid System*

Two-step hybrid system can be applied for further increase in the yield and rate of hydrogen production by hybrid systems [11]. Two-step hybrid systems allow the use of waste materials of photo-fermentation process, which are containing elements having inhibitory effect on photo-fermentation (e.g., NH_4 ions) [12] and does not affect the microorganism involved in the dark fermentation. In two-step hybrid systems both photo and dark fermentative bacteria grow in separate containers because there are greater number of parameters, which can influence hydrogen production in photo-fermentation than in dark fermentation [13]. So, both the processes must be performed in separate bioreactors.

2.2.2 *BIOREACTOR FOR HYDROGEN PRODUCTION*

Based on the nature of H_2 evolving reactions, Microbial bioreactors can be divided into three separate groups [1, 5, 9]:
1. Photobioreactors based on H_2 photo-production.
    a. Photobioreactors incorporating green algae (photosynthesis)
    b. Photobioreactors incorporating cyanobacteria (photosynthesis)
    c. Photobioreactors incorporating photosynthetic bacteria (photo-fermentation)
2. Bioreactors based on dark anaerobic H_2 production by bacteria.
    a. Bioreactors based on fermentation (dark-fermentation)
    b. Bioreactors based on "water-gas shift-reaction" (dark-fermentation)
3. Hybrid bioreactors based on photo and dark fermentation.

2.2.3 *SPEED BREAKERS IN BIO-HYDROGEN PRODUCTION*

There are several extrinsic and intrinsic factors, which behave as speed breakers on the road of bio-hydrogen production. These factors become more critical when bio-hydrogen production is aimed at industrial scale. Some of these factors are discussed in the following sub-sections.

2.2.3.1 Extrinsic Factors

These are the external factors, which are either present in the environment or produced as a byproduct during the process of bio-hydrogen production.

2.2.3.1.1 Effect of Light

Light requirement for hydrogen production varies among the different groups of microorganisms. Several hydrogen producing microorganisms such as photosynthetic green algae, cyanobacteria, and purple sulfur bacteria produce hydrogen in the presence of light whereas nonphotosynthetic anaerobic bacteria do not require light for hydrogen production.

2.2.3.1.2 Effect of Temperature

For most of the cyanobacterial species, optimum temperature for hydrogen production is in the range of 30–40 °C and varies among the cyanobacterial species. For instance, *Nostoc* showed higher hydrogen production rates when it was cultured at 22 °C than at 32 °C [14], whereas optimum hydrogen production rate by *Nostoc muscorum* SPU004 was reported at 40 °C [15]. On the other hand optimum hydrogen production by *Anabaena variabilis* SPU 003 was observed at 30 °C [16, 17]. Optimum temperature for hydrogen production by photo-fermentation with the help of photosynthetic bacteria is between 30 and 35 °C [18–25].

2.2.3.1.3 Effect of pH

Several experimental studies reported that pH can affect bio-hydrogen production in terms of yield, production rate, content of biogas and types of organic acids produced. Reported optimum pH for bio-hydrogen production is in the range of pH 5.0- 6.0 [26–30]. Whereas some investigators observed optimum pH range in between 6.8 and 8.0 [31–34]. Most of the studies on effect of pH reported that pH decreases as the process proceeds due to the production of organic acids and the final pH was found to be in between 4.0–4.8 regardless of initial pH [29, 32–37]. This gradual decrease of pH results in decreased hydrogen production as pH affects the activity of hydrogenase, which is an iron-containing enzyme [38]. It was also observed in some studies that initial pH can affect the lag phase of microbial growth during hydrogen production process, low initial pH increases the lag period whereas high initial pH decreases lag period as a result decreased yield of hydrogen production [29, 32, 33]. So, to achieve controlled pH at the optimum level is necessary for bio-hydrogen production.

2.2.3.1.4 *Effect of Salinity*

It has been proved by experimental studies that salinity affects hydrogen production by microorganisms [39]. Generally fresh water cyanobacteria have shown to have lower rate of hydrogen production when the salt concentration increases. This may happen because of diversion of energy and reductants either for extrusion of Na^+ ions outside the cells or prevention of Na^+ influx [40].

2.2.3.1.5 *Effect of Carbon*

Carbon sources are known to influence the activity of nitrogenase enzyme as a result of which it also affects the rate of hydrogen production considerably [15]. Capabilities of electron donation by cofactor compounds to nitrogenase enzyme vary due to the presence of different carbon sources and thus influence hydrogen production [15].

2.2.3.1.6 *Effect of Nitrogen*

It has been reported that a number of inorganic nitrogenous compounds affect the rate of hydrogen production in different ways. It has been proved from experimental studies that nitrite, nitrate and ammonia have inhibitory effect on nitrogenase in *Anabaena variabilis* SPU003 and *Anabaena cylindrical* [15, 41]. It was observed that generally all externally added nitrogen sources have negative effect on synthesis of nitrogenase enzyme [42]. Although it was also observed in *Anabaena cylindrical* that addition of ammonium ion (0.2 mM $NH4^+$) at a given time point eventually suppresses rate of hydrogen production, but when it is periodically added in smaller amounts (0.1 mM NH_4Cl), doesn't show inhibitory effect on production rate of hydrogen [44].

2.2.3.1.7 Effect of Molecular Nitrogen

In some studies, considerable decrease in hydrogen production has been reported due to the presence of molecular nitrogen and conclusion was made that molecular nitrogen behaves as competitive inhibitor for hydrogen production, so it becomes necessary to remove molecular nitrogen for increased hydrogen production [44].

2.2.3.1.8 Effect of Micronutrients

Several studies showed the effect of micronutrients on hydrogen production such as cobalt (Co), copper (Cu), molybdenum (Mo), zinc (Zn), and nickel (Ni) [45]. Many of these micronutrients are responsible for the activity of nitrogenase enzyme and shown to have enhancement effect on bio-hydrogen production. For instance, *Anabaena variabilis* SPU003 was found to be very sensitive to Co, Cu, Mn, Zn, Ni, Fe ions and didn't show hydrogen production at lower concentrations (below 10 mM) of these microelements [46]. The enhancement in hydrogen production was also observed in a culture of *Anabaena cylindrica,* which was grown in a medium containing 5.0 mg of Ferric ions/ liter, this enhancement was almost two times of a culture, which was grown in a medium containing 0.5 mg of Ferric ions/ liter [44].

2.2.3.1.9 Effect of Oxygen

Both hydrogenases and nitrogenases are highly sensitive to oxygen, so the process of bio-hydrogen production must be carried out in strict anaerobic conditions [47].

2.2.3.1.10 Effect of Sulfur

Sulfur deprived condition is known to have positive impact on bio-hydrogen production. It has been proved by several studies that sulfur starvation

increases the rate of hydrogen production by providing anaerobic condition and reducing the loss of activity of hydrogenases in the presence of
oxygen [48, 49–52].

2.2.3.2 Intrinsic Factors

Intrinsic factors are the factors, which are present inside the microorganisms itself, and these factors such as genetic components and sensitive proteins, etc. can affect hydrogen production. Uptake Hydrogenase
enzyme is one of the several intrinsic factors that limits the process of bio-
hydrogen production [4, 6, 53]. Uptake hydrogenase enzyme, present in
photo-fermentative bacteria, inhibits the hydrogen yield by using hydrogen gas produced and thus behaves as an antagonist for nitrogenase [54].

2.2.4 LIMITATIONS AND CHALLENGES IN BIO-HYDROGEN PRODUCTION

There are number of limitations and challenges in bio-hydrogen production such as low hydrogen yield, slow rate of hydrogen production, strong
inhibitory effect of oxygen on hydrogenase and nitrogenase enzymes and
no waste utilization are the major limitations of hydrogen production by
algae and cyanobacteria [4–6]. Whereas yield of hydrogen gas and rate
of production are the limitations associated with dark fermentation and
requirement of large land area is the major limitation of bio-hydrogen production via photo fermentation due to the high requirement of light [1,
5, 54]. Due to these limitations bio-hydrogen production has become the
most challenging area of biotechnology. The future of bio-hydrogen production depends not only on research advancement, that is, improvement
in efficiency through genetically engineering microorganisms and/or the
development of bioreactors, but also on economic considerations (the cost
of fossil fuels), acceptance of bio-hydrogen production at social level, and
the development of hydrogen energy systems. The process requires special microorganisms, strict control of light and other environmental conditions, which can affect the yield and rate of bio-hydrogen production.
Sequential or combined bio-processes (hybrid systems) of dark and photo-

fermentations seem to be the most attractive approach resulting in high hydrogen yields for hydrogen production from carbohydrate rich wastes. Recently few hydrogen producing aerobic cultures such as *Aeromonos sp.*, *Pseudomonos sp.* and *Vibrio sp.* were identified [5], but using these bacterial species in hybrid culture requires a very thorough study because many of these species are pathogenic and as we are going to improve the synthesis of hydrogen gas, which is known as a source of future energy we can't create an environment which can be a threat for the future generations so, we need to be very careful in designing any such hybrid system.

2.3   A HYPOTHETICAL THREE STAGE HYBRID SYSTEM

In this chapter, a hypothetical three-stage hybrid system has been proposed, which may overcome the limitations associated with bio-hydrogen production. This hybrid system is designed in such a way that side products produced at one stage of hybrid system will be used at later stages so that the waste materials can be used up to the maximum possible extent, which will result in higher H_2 yield and lesser side production. We had also proposed a mixture of microorganisms which can be used in this hybrid system because it has been clear from previous studies that mixture of microorganisms is more effective in H_2 production rather than using single microorganism [55]. In this mixture of microorganisms, we had proposed combination of those microorganism which are known for highest H_2 production among their class till present (Table 2.2). Diagrammatic representation of three-stage hybrid system is given in Fig. 2.2.

2.3.1   ADVANTAGES OF THREE-STAGE HYBRID SYSTEM

The proposed three-stage hybrid system is trying to overcome limitation of bio-hydrogen production in different methodologies. The advantages of hybrid systems may be as follows:

1. As we know there is no waste utilization by algae and cyanobacteria [5], we are using algae and cyanobacteria at the end of this hybrid system where only small amount of waste material will remain unused after the dark and the photo fermentation respec-

TABLE 2.2 Proposed Microorganisms for Three Stage Hybrid System

Process of H_2 Production	Class of Microorganism	Name of Organism	H_2 Yield	Reference
DARK FERMENTATION	Anaerobic (Obligate and facultative) Bacteria	*Enterobactor cloacae* IIT-BT 08	2.2 mol/mol glucose	[56]
		C. pasteurium (dominant)	4.8 mol/mol sucrose	[57]
		Enterobactor cloacae IIT-BT 08	6 mol/mol sucrose	[56]
		Enterobactor aerogenes	1.09 mol/mol glucose	[58]
PHOTO FERMENTATION	Purple Photosynthetic Bacteria	Rhodopseudomonas palustris	2.8 mol/mol acetate	[59]
		Rhodobactor capsulatus	1.1 mol/mol acetate	[60]
		Rhodobactor capsulatus	2.8 mol/ mol butyrate	[60]
PHOTOSYNTHESIS	Green algae	Chlamydomonas reinhardtii		[5]
	Cyanobacteria	Anabaena cylindrica		[5]

tively and large amount of CO_2 and Water will remain present as a byproduct of dark and photo fermentation respectively, which is required for hydrogen production via photosynthesis and biophotolysis with the help of algae and cyanobacteria [5].

2. A mixture of microorganisms (Table 2.2) has been proposed for use at each different step of hypothetical hybrid system and these microorganisms are known to be highest H_2 yielding microorganisms so the yield will get increased by using them.

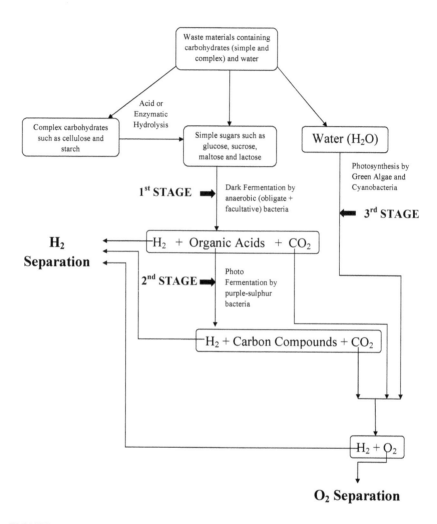

FIGURE 2.2   Proposed three-stage hybrid system.

2.4   CONCLUSIONS

Hydrogen gas cannot be used as an energy source naturally because it is not present in adequate amount in earth atmosphere so we need to produce it synthetically but the major problem associated with the hydrogen production is its commercialization. There are various methods available for the hydrogen production one of which is bio-hydrogen production. Bio-hydrogen production has been proved successful in improving the rate and yield of hydrogen production but still not up to the industrial level, so we need to come up with new methodologies which can improve the pace of bio-hydrogen production, one of the such new methodologies can be, designing a hybrid system by using existing bio-hydrogen producing processes and microorganisms, another efficient approach can be designing new microorganisms with increased bio-hydrogen production capacity through genetic engineering, in this approach *in silico* studies of genomic constitution of hydrogen producing microorganisms can speed up the process. This chapter suggests that high quality research and development activities are needed for overcoming the limitations associated with the bio-hydrogen production.

2.5   ABBREVIATIONS

H_2: Molecular hydrogen, H_2O: Water, CO_2: Carbon dioxide, NH_4: Ammonium, Ni: Nickel; Co: Cobalt, Cu: Copper, Zn: Zinc, Mn: Manganese, Na^+: Sodium ion, PNS: Purple Non Sulfur bacteria.

2.6   AUTHORS' CONTRIBUTIONS

V. Mishra and I. Srivastava contributed in literature survey and manuscript preparation. V. Mishra edited the final manuscript.

KEYWORDS

- **Bio-hydrogen production**
- **Biomass**
- **Bioreactors**
- **Hybrid systems**
- **Hydrogen yield**

REFERENCES

1. Basak, N., Das, D. (2006). The prospect of purple nonsulfur (PNS) photosynthetic bacteria for hydrogen production: the present state-of-the-art. *Springer Science+Business Media B. V.*
2. Das, D., Veziroglu, T. N. (2001). Hydrogen production by biological processes: a survey of literature. *Int. J. Hydrogen Energy. 26*, 13–28.
3. Bockris, J. O. M. (1981). The economics of Hydrogen as a fuel. *Int. J. Hydrogen Energy. 6*, 223–241.
4. Lindblad, P. (1999).Cyanobacterial H_2 Metabolism: Knowledge and Potential/Strategies for a Photobiotechnological Production of H_2. *Biotecnologia Aplicada. 16*, 115–144.
5. Kapdan, I. K., Kargi, F. (2006). Bio-hydrogen production from waste materials, *Enzyme and Microbial Technology. 38*, 569–582.
6. Das, D., Veziroglu, T. N. (2008). Advances in biological hydrogen production processes. *Int. J. Hydrogen Energy. 33*, 6046–6057.
7. Turker, L., Gumus, S., Tapan, A. (2008). Bio-hydrogen Production: Molecular Aspects, *J. Scientific and Industrial Research. 67*, 994–1016.
8. Kars, G., Gunduz, U., Yucel, M., Rakhely, G., Kovacs, K. L., Eroglu, I. (2009). Evaluation of hydrogen production by Rhodobacter sphaeroides O. U.001 and its hupSL deficient mutant Rusing acetate and malate as carbon sources. *Int. J. Hydrogen Energy. 34*, 2184–2190.
9. Markov, S. A. (2012). Hydrogen production in bioreactors: current trends. *Energy Procedia. 29*, 394 – 400.
10. Liu, B. F., Ren, N., Tang, J., Ding, J., Liu, W. Z., Xu, J. F., Cao, G. L., Guo, W. Q., Xie, G. J. (2010). Bio-hydrogen production by mixed culture of photo-and dark-fermentation bacteria. *Int. J. Hydrogen Energy. 35*, 2858–2862.
11. Argun, H., Kargi, F., Kapdan, I. K. (2009). Hydrogen production by combined dark and light fermentation of ground wheat solution. *Int. J. Hydrogen Energy. 34*, 4305–4311.
12. Azbar, N., Tuba, F., Dokoz, C. (2010). The effect of dilution and L-malic acid addition on bio-hydrogen production with Rhodopseudomonas palustris from effluent of an acidogenic anaerobic reactor. *Int. J. Hydrogen Energy. 35*, 5028–5033.
13. Ozmihci, S., Kargi, F. (2010). Bio-hydrogen production by photo-fermentation of dark fermentation effluent with intermittent feeding and effluent removal. *Int. J. Hydrogen Energy. 35*, 6674–6680.
14. Ernst, A., Kerfin, W., Spiller, H., Boger, P. (1979). External factors influencing light-induced H2 evolution by the blue-green algae, Nostoc muscorum. *Z. Naturforsch. 34*, 820–825.
15. Madamwar, D., Garg, N., Shah, V. (2000). Cyanobacterial hydrogen production. *World J. Microbiol. Biotechnol. 16*, 757–767.
16. Serebryakova, L. T., Sheremetieva, M. E., Lindblad, P. (2000). H2-uptake and evolution in the unicellular cyanobacterium Chroococcidiopsis thermalis CALU 758. *Plant Physiol. Biochem. 38*, 525–530.
17. Moezelaar, R., Stal, L. J. (1994). Fermentation in the unicellular cyanobacterium Microcystis PCC7806. *Arch. Microbiol. 162*, 63–69.

18. Federov, A. S., Tsygankov, A. A., Rao, K. K., Hall, D. O. (1998). Hydrogen photoproduction by Rhodobacter sphaeroides immobilized on polyurethane foam. *Biotechnol. Lett. 20*, 1007–1009.
19. Eroglu, I., Aslan, K., Gunduz, U., Yucel, M., Turker, L. (1999). Substrate consumption rate for hydrogen production by Rhodobacter sphaeroides in a column photobioreactor. *J. Biotechnol. 70*, 103–113.
20. Kim, M. S., Baek, J. S., Lee, J. K. (2006). Comparison of H2 accumulation by Rhodobacter sphaeroides KD131 and its uptake hydrogenase and PHB synthase deficient mutant. *Int. J. Hydrogen Energy. 31*, 121–127.
21. Zhu, H., Ueda, S., Asada, Y., Miyake, J. (2002). Hydrogen production as a novel process of wastewater treatment—studies on tofu wastewater with entrapped R. sphaeroides and mutagenesis. *Int. J. Hydrogen Energy. 27*, 1349–1357.
22. He, D., Bultel, Y., Magnin, J. P., Roux, C., Willison, J. C. (2005). Hydrogen photosynthesis by Rhodobacter capsulatus and its coupling to PEM fuel cell. *J. Power Sources. 141*, 19–23.
23. Oh, Y. K., Scol, E. H., Lee, E. Y., Park, S. (2002). Fermentative hydrogen production by new chemoheterotrophic bacterium Rhodopseudomonas palustris P4. *Int. J. Hydrogen Energy. 27*, 1373–1379.
24. Yokoi, H., Mori, S., Hirose, J., Hayashi, S., Takasaki, Y. (1998). H_2 production from starch by mixed culture of Clostridium butyricum and Rhodobacter sp M-19. *Biotechnol. Lett. 20*, 895–899.
25. Lee, C. M., Chen, P. C., Wang, C. C., Tung, Y. C. (2002). Photohydrogen production using purple nonsulfur bacteria with hydrogen fermentation reactor effluent. *Int. J. Hydrogen Energy. 27*, 1309–1313.
26. Fang, H. H. P., Liu, H. (2002). Effect of pH on hydrogen production from glucose by mixed culture. *Bioresour. Technol. 82*, 87–93.
27. Lay, J. J., Lee, Y. J., Noike, T. (1999). Feasibility of biological hydrogen production from organic fraction of municipal solid waste. *Water Res. 33*, 2579–2586.
28. Lay, J. J. (2000). Modeling and optimization of anaerobic digested sludge converting starch to hydrogen. *Biotechnol. Bioeng. 68*, 269–278.
29. Khanal, S. K., Chen, W. H., Li, L., Sung, S. (2004). Biological hydrogen production: effects of pH and intermediate products. *Int. J. Hydrogen Energy. 29*, 1123–1131.
30. Chen, C. C., Lin, C. Y., Chang, J. S. (2001). Kinetics of hydrogen production with continuous anaerobic cultures using sucrose as limiting substrate. *Appl. Microbiol. Biotechnol. 57*, 56–64.
31. Collet, C., Adler, N., Schwitzgu'ebel, J. P., P'eringer, P. (2004). Hydrogen production by Clostridium thermolacticum during continuous fermentation of lactose. *Int. J. Hydrogen Energy. 29*, 1479–1485.
32. Liu, G., Shen, J. (2004). Effects of culture medium and medium conditions on hydrogen production from starch using anaerobic bacteria. *J. Biosci. Bioeng. 98*, 251–256.
33. Zhang, T., Liu, H., Fang, H. H. P. (2003). Bio-hydrogen production from starch in wastewater under thermophilic conditions. *J. Environ. Manag. 69*, 149–156.
34. Lay, J. J. (2001). Bio-hydrogen generation by mesophilic anaerobic fermentation of microcrystalline cellulose. *Biotechnol. Bioeng. 74*, 281–287.
35. Yokoi, H., Saitsu, A. S., Uchida, H., Hirose, J., Hayashi, S., Takasaki, Y. (2001). Microbial hydrogen production from sweet potato starch residue. *J. Biosci. Bioeng. 91*, 58–63.

36. Liu, H., Zhang, T., Fang, H. P. P. (2003). Thermophilic H$_2$ production from cellulose containing waste water. *Biotechnol. Lett. 25,* 365–369.
37. Morimoto, M., Atsuko, M., Atif, A. A. Y., Ngan, M. A., Fakhru'l-Razi, A., Iyuke, S. E. et al. (2004). Biological production of hydrogen from glucose by natural anaerobic microflora. *Int. J. Hydrogen Energy. 29,* 709–713.
38. Dabrock, B., Bahl, H, Gottschalk, G. (1992). Parameters affecting solvent production by Clostridium pasteurium. *Appl. Environ. Microbiol. 58,* 1233–1239.
39. Shah, V., Garg, N., Madamwar, D. (2003). Ultrastructure of the cyanobacterium Nostoc muscorum and exploitation of the culture for hydrogen production. *Folia Microbiol. (Praha). 48,* 65–70.
40. Rai, A. K., Abraham, G. (1995). Relationship of combined nitrogen sources to salt tolerance in freshwater cyanobacterium Anabena doliolum. *J. Appl. Bacteriol. 78,* 501–506.
41. Lambert, G. R., Daday, A., Smith, G. D. (1979). Hydrogen evolution from immobilized cultures of cyanobacterium Anabena cylindrica. *FEBS Lett. 101,* 125–128.
42. Rawson, D. M. (1985). The effects of exogenous aminoacids on growth and nitrogenase activity in the cyanobacterium Anabena cylindrica PCC 7122. *J. General Microbiol. 134,* 2549–2544.
43. Jeffries, T. W., Timourien, H., Ward, R. L. (1978). Hydrogen production by Anabaena cylindrica: Effect of varying ammonium and ferric ions, pH and light. *Appl. Env. Microbiol. 35,* 704–710.
44. Lambert, G. R., Smith, G. D. (1977). Hydrogen formation by marine Bluegreen algae. *FEBS Lett. 83,* 159–162.
45. Ramchandran, S., Mitsui, A. (1984). Recycling of hydrogen photoproduction system using an immobilized marine blue green algae Oscillatoria sp. Miami BG7, solar energy and seawater [abstract]. *VII Int. Biotechnology Symposium.* 183–184.
46. Moezelaar, R., Bijvank, S. M., Stal, L. J. (1996). Fermentation and sulfur reduction in the mat-building cyanobacterium Microcoleus chtonoplastes. *Appl. Environ. Microbiol. 62,* 1752–1758.
47. Fay, P. (1992). Oxygen relations of nitrogen fixation in cyanobacteria. *Microbiol Rev. 56,* 340–373.
48. Antal, T. K., Lindblad, P. (2005). Production of H$_2$ by sulfur-deprived cells of the unicellular cyanobacteria Gloeocapsa alpicola and Synechocystis sp. PCC 6803 during dark incubation with methane or at various extracellular pH. *J. of Appl. Microbiol. 98,* 114–120.
49. Guan, Y., Deng, M., Yu, X., Zang, W. (2004). Two stage photo-production of hydrogen by marine green algae Platymonas subcordiformis. *Biochem. Eng. J. 19,* 69–73.
50. Winkler, M., Hemschemeier, A., Gotor, C., Melis, A., Happer, T. (2002). [Fe]-hydrogenases in green algae: photo-fermentation and hydrogen evolution under sulfur deprivation. *Int. J. Hydrogen Energy. 27,* 1431–1439.
51. Melis, A., Zhang, L., Forestier, M., Ghirardi, M. L., Seibert, M. (2000). Sustained photohydrogen production upon reversible inactivation of oxygen evolution in the green algae Chlamydomonas reindhardtii. *Plant Physiol. 122,* 127–135.
52. Laurinavichene, T. V., Tolstygina, I. V., Galiulina, R. R., Ghirardi, M. L., Seibert, M., Tsygankov, A. A. (2002). Dilution methods to deprive Chlamydomonas reinhardtii cultures of sulfur for subsequent hydrogen photoproduction. *Int. J. Hydrogen Energy. 27,* 1245–1249.

53. Koku, H., Eroglu, I., Gunduz, U., Yucel, M., Turker, L. (2002). Aspects of metabolism of hydrogen production by Rhodobacter sphaeroides. *Int. J. Hydrogen Energy. 27*, 1315–1329.

54. Seifert, K., Thiel, M., Wicher, E., Włodarczak, M., Łaniecki, M. (2013). Microbiological Methods of Hydrogen Generation. www.intechopen.com. 223–250.

55. Chun-xiang, Q., Lu-yuan, C., Hui, R., Xiao-ming, Y. (2011). Hydrogen production by mixed culture of several facultative bacteria and anaerobic bacteria. *Progress in Natural Science: Materials Int. 21,* 506–511.

56. Kumar, N., Das, D. (2000). Enhancement of hydrogen production by Enterobacter cloacae IIT-BT 08. *Process. Biochem. 35,* 589–593.

57. Lin, C. Y., Chang, R. C. (2004). Fermentative hydrogen production at ambient temperature. *Int. J. Hydrogen Energy, 29,* 715–720.

58. Fabiano, B., Perego, P. (2002). Thermodynamic study and optimization of hydrogen production by Enterobacter aerogenes. *Int. J. Hydrogen Energy, 27,* 149–156.

59. Oh, Y. K., Scol, E. H., Kim, M. S., Park, S. (2004). Photoproduction of hydrogen from acetate by a chemoheterotrophic bacterium Rhodopseudomonas palustris P4, *Int. J. Hydrogen Energy, 29,* 1115–1121.

60. Fang, H. H. P., Liu, H., Zhang, T. (2005). Phototrophic hydrogen production from acetate and butyrate in wastewater. *Int. J. Hydrogen Energy, 30,* 785–793.

PART 2

MICROBIAL FUEL CELLS

CHAPTER 3

MICROBIAL FUEL CELLS: A PROMISING ALTERNATIVE ENERGY SOURCE

M. J. ANGELAA LINCY,[1] B. ASHOK KUMAR,[2] V. S. VASANTHA,[3] and P. VARALAKSHMI[1]

[1]*Department of Molecular Microbiology, School of Biotechnology, Madurai Kamaraj University, Madurai, Tamil Nadu, India*

[2]*Department of Genetic Engineering, School of Biotechnology, Madurai Kamaraj University, Madurai, Tamil Nadu, India*

[3]*Department of Natural Products, School of Chemistry, Madurai Kamaraj University, Madurai, Tamil Nadu, India*

CONTENTS

ABSTRACT

The Indian sub-continent is luxuriously bestowed with all wealth in the form of population, land and ocean. Being placed as the second highly populated nation in the world, India is under the urge to nurture the needs of its population. Its various demands such as the demand for food, employment, economy, etc., can be completely resolved if the country is adequately supplied with energy. But adequate supplying of energy to such a vast population is quiet an impossible task, which demands seek for alternative energy sources. The widely known alternative energy sources are generally, renewable energy sources such as solar cells, fuel cells and wind power. At this moment, several fuel cell types based on hydrogen and methanol work appropriately, and applications already exist for, for example, portable computers. However, the question can be raised whether this energy generation is really sustainable. Furthermore, the customer may not like to carry hydrogen gas (even captured within a metal hydride matrix) or methanol. But, Microbial fuel cells can operate on a large variety of substrates that are readily available, even in any supermarket. Substrates such as plain sugar and starch are easy to store, contain more energy than any other feed type per unit of volume, and are easy to dose. Furthermore, they have a more 'green' image than, for example, methanol. Moreover, MFCs can be developed that are environmentally friendly in terms of material composition.

When microorganisms function as biocatalysts that motivate the degradation of organic materials to produce electrons, which travel via an electric circuit, then the fuel cell is termed as a microbial fuel cell. This chapter covers an introduction to MFCs, state of art of MFCs and wide applications of MFCs technology. If the development of MFCs leads to a product that has a reasonable (read: usable) power output per unit of MFC volume, it will be a viable product. A customer will accept a larger battery, and a larger feeding tank, provided the feeding is easy to perform and has a green and safe label.

3.1 INTRODUCTION

The Indian sub-continent is luxuriously bestowed with all wealth in the form of population, land and ocean. Being placed as the second highly populated nation in the world, India is under the urge to nurture the needs of its population. Its various demands such as the demand for food, employment, economy, etc., can be completely resolved if the country is adequately supplied with energy. But adequate supplying of energy to such a vast population is quiet an impossible task, which demands seek for alternative energy sources. The widely known alternative energy sources are generally, renewable energy sources such as solar cells, fuel cells and wind power.

Fuel cells are electrochemical devices that convert the chemical energy stored in a fuel directly into electrical power. Fuel cells have various advantages compared to conventional power sources, such as internal combustion engines or batteries. Benefits include: higher efficiency compared than diesel or gas engine, their operation is free from noise and CO_2 pollution and do not need conventional fuels such as oil or gas and can therefore reduce economic dependence on oil producing countries and the main fuel (H_2) can be produced from the natural source water, etc.

The conventional batteries have several drawbacks:
- they need to be charged for several hours in order to be used.
- they are environmentally unfriendly due to the heavy metal content.
- one needs electricity to power them up.

Therefore, different types of fuel cells are developed as follows:
- alkaline fuel cells;
- proton exchange membrane fuel cells;
- direct methanol fuel cells;
- phosphoric acid fuel cells;
- molten carbonate fuel cells;
- solid oxide fuel cells.

Alkaline fuel cell is one of the most developed fuel cell technologies. They use alkaline electrolytes such as potassium hydroxide in water and operate at 70°C. They employ cheaper, non noble metal catalysts such as nickel and silver [1, 2]. Proton exchange membrane fuel cells also known as Polymer electrolyte membrane fuel cells (PEM) are a type of fuel cell being developed for transport applications. They employ a proton con-ductive, water based, acidic polymer membrane as an electrolyte [3] and operate at relatively low temperatures. Direct methanol fuel cells are a sub category of Proton exchange membrane fuel cells, the fuel being metha-nol. It makes use of a polymer membrane as an electrolyte and operates between 60°C and 130°C [4].

Phosphoric acid fuel cells employ highly concentrated or pure liquid phosphoric acid as the electrolyte. The catalyst being finely dispersed platinum, they operate at 150°C to 220°C. Molten carbonate fuel cells represent a high temperature technology of fuel cells [5]. They employ molten carbonate salt as an electrolyte and do not require a metal cata-lyst. They operate at around 650°C. Solid oxide fuel cells are another type of high temperature fuel cells whose operating temperature ranges from 800°C–1000°C [5]. They use a solid ceramic such as stabilized zirconium oxide as the electrolyte and do not need a metal catalyst. They are widely used for stationary power generation.

First experimental evidence of bioelectricity was demonstrated by Luigi Galvani in the late eighteenth century by connecting the legs of a frog to a metallic conductor [6, 7]. The potential of bioelectricity was further experimented by Micheal C. Potter in 1911, who designed the first microbial fuel cell. In 1931, a potentiostat-poised half-cell was operated by Barnett Cohon and a current of 0.2 mA was obtained by applying +0.5 V [8].

In 1980s, electricity was successfully retrieved from MFCs by a Brit-ish researcher, H. Peter Bennetto by employing pure cultures of bacteria to catalyze the oxidation of organics and use artificial electron mediators to facilitate electron transfer in the anode [9–11]. As on date, fuel cells can vary from tiny cells, producing few watts electricity to large power plants producing megawatts [12] and are classified into different types based on the various electrolytes employed.

Conventional microbial fuel cell is a two-chamber system, consisting of anode and cathode chambers that are separated by a solid or a liquid electrolyte that transmits electrically charged particles across them [12]. At the anode end, the fuel gets oxidized and releases electrons. At the cathode, oxygen reduction occurs. The electrolyte conducts ions from one electrode to the other, inside a fuel cell. The catalyst speeds up a chemical reaction. The reformer extracts pure hydrogen from hydrocarbons.

Microbial fuel cells have become a promising alternative for electricity generation, biomass cultivation and wastewater treatment. If the microbial fuel cells make use of solar energy to generate electricity then they are called as Photo microbial fuel cells [13]. If they use photosynthetic algae for producing electricity, then they are called as Photosynthetic algal microbial fuel cells [14].

Microbial fuel cells are believed to serve as a viable technological alternative to conventional wastewater treatment. They can harvest electricity from the energy available in organic wastewater [15]. Microorganisms like *Chlorella vulgaris* [13, 14], *Leptothrix discophora* [6], *Klebsiella pneumonia* [16], *Thiobaccillus ferrooxidance* [17], *Pseudomonas fluroscens* [18], *Rhodospirillum rubrum* [19], *G. metallireducens* [20], *D. desulfuricans* [21] and other aerobic and anaerobic bacteria have been reported to be employed in microbial fuel cells so far. Few other microorganisms like cyanobacteria and some microalgae are also believed to play remarkable roles if employed in microbial fuel cells. Improvising the functioning of these cells to give exponential yields will be the expected future of the field.

Fuel cells are electrochemical devices that convert the chemical energy stored in a fuel directly into electrical power. They are considered to be highly reliable because of their high efficiency and low or zero emission [22]. When microorganisms function as biocatalysts that motivate the degradation of organic materials to produce electrons, which travel via an electric circuit, then the fuel cell is termed as a microbial fuel cell [6].

Bacteria grow by catalyzing chemical reactions and harnessing and storing energy in the form of adenosine triphosphate (ATP). In some bacteria, reduced substrates are oxidized and electrons are transferred to respiratory enzymes by NADH, the reduced form of nicotinamide adenine dinucleotide (NAD). These electrons flow down a respiratory chain—a series of enzymes that function to move protons across an internal membrane—creating a proton gradient. The protons flow back into the cell through the enzyme ATPase, creating 1 ATP molecule from 1 adenosine diphosphate for every 3–4 protons. The electrons are finally released to a soluble terminal electron acceptor, such as nitrate, sulfate, or oxygen. In the first type of MFC, the electrons are transferred from the bacteria to the anode without the intervention of an intermediate fermentation product [23].

First experimental evidence of bioelectricity was demonstrated by Luigi Galvani in the late eighteenth century by connecting the legs of a frog to a metallic conductor [7]. The potential of bioelectricity was further experimented by Micheal C. Potter in 1911, who designed the first microbial fuel cell. He demonstrated current flow between two electrodes emerged from a bacterial culture and in a sterile medium [24]. In 1931, a potentiostat-poised half cell was operated by Barnett Cohon and a current of 0.2 mA was obtained by applying +0.5 V [8].

In 1980s, electricity was successfully retrieved from MFCs by a British researcher, H. Peter Bennetto. He employed pure cultures of bacteria to catalyze the oxidation of organics and use artificial electron mediators to facilitate electron transfer in the anode [9–11, 91]. As on date, fuel cells can vary from tiny cells, producing few watts electricity to large power plants producing megawatts [12] and are classified into different types based on the various electrolytes employed.

3.2   COMPONENTS OF A FUEL CELL

Conventional microbial fuel cell is a two-chamber system, consisting of anode and cathode chambers that are separated by a solid or a liquid electrolyte that transmits electrically charged particles across them [12]. At the anode end, the fuel gets oxidized and releases electrons. At the cathode, oxygen reduction occurs. The electrolyte conducts ions from one electrode to the other, inside a fuel cell. The catalyst speeds up a chemical reaction. The reformer extracts pure hydrogen from hydrocarbons.

3.2.1 TYPES OF FUEL CELLS

3.2.1.1 Alkaline Fuel Cells

Alkaline fuel cell (AFC) also known as Bacon fuel cell after its British inventor, is one of the most developed fuel cell technologies. The AFC use alkaline electrolytes such as potassium hydroxide in water (Fig. 3.1). They are known to be the best performing fuel cells of the existing conventional hydrogen-oxygen fuel cells operable at temperatures below 200°C [25]. Typical operating temperatures are around 70°C. They employ cheaper, non noble metal catalysts such as nickel and silver [1, 2] and can offer high electrical efficiency.

AFC are advantageous over other fuel cells that make use of precious metal catalysts, like DMFCs, in the following ways:

1. the slow electrode kinetics problem encountered by the other cells is not found I alkaline fuel cells because it is a well established fact that the electrode kinetics of oxygen reduction is enhanced in an alkaline medium;
2. they use non precious metals as catalysts such as silver catalysts [26], nickel catalysts [27] and perovskite type oxides [28], which are less prone to methanol crossover; and

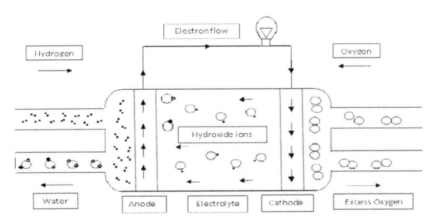

FIGURE 3.1 Schematic diagram of an Alkaline fuel cell.

3. CO poisoning of Platinum catalyst at low temperatures is another problem encountered by other fuel cells but not alkaline fuel cells.

The only problem with these fuel cells is the progressive carbonation of the alkaline solution, due to carbon dioxide release from the product of the fuel. This reduces the pH of the alkaline solution, thus decreasing the reactivity for the electro oxidation of methanol ultimately. Research is to be established in this area of alkaline fuel cells that could rectify this carbonation problem.

3.2.1.2 Proton Exchange Membrane Fuel Cells

Proton exchange membrane fuel cells also known as Polymer electrolyte membrane fuel cells, are a type of fuel cell being developed for transport applications as well as for stationery fuel cell applications. Figure 3.2 shows the schematic diagram of a Proton exchange membrane fuel cell. PEM cell uses a proton conductive, water based, acidic polymer membrane as an electrolyte [3]. The polymer membrane used is impermeable to gases but it conducts protons [29]. Among the membranes

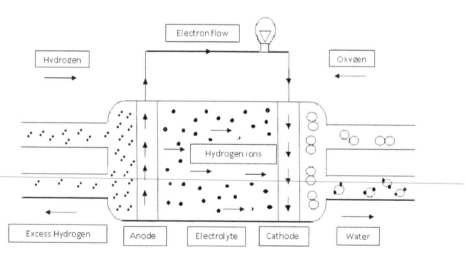

FIGURE 3.2 Schematic diagram of Proton exchange membrane fuel cell.

of hydrogen fuelled Proton exchange membrane fuel cells, perfluoro-sulfonic acid based membranes, mostly DuPont's Nafion [30], show complete domination upon other commercialized membranes [31].

Proton exchange membrane fuel cells operate at relatively low temperatures (below 100°C). The challenges faced while operating these cells under low temperature include fuel impurity, CO tolerance and heat rejection [30]. Their shortcomings could be surmounted at a functioning at high temperatures. High temperature may be till 200°C. But the primary problem in operating at high temperatures is that, the Relative humidity is directly proportional to the Nafion Membrane conductivity of the cells. Therefore, whenever there is a fall in the relative humidity of the cell, the conductivity of the Nafion membrane also drops down. The cells cannot yield high membrane conductivity at a range of (> 0.1 S cm^{-1}) unless an external humidification sub system is functioned with the fuel cell system [32, 33]. The electrical output is generally variable and they are ideal for vehicles. It has been reported that they show 50% higher fuel conversion efficiency than a car engine. Under favorable conditions, their efficiency can be almost 60% where a car engine has achieved only 30%.

However, there are certain factors that affect the performance of PEM-FCs. They are as follows:

1. Partial pressures of reactants and pressure variations across the membrane, while the cell is functioning.
2. Hydrogen supply at the anode.
3. Oxygen starvation at the cathode.
4. Dehydrated membrane conductivity.
5. On the other hand, presence of too much water in the back plate channels also depletes the cell efficiency [34].
6. Hydrophobicity of the membrane surface [35].
7. Dehydration at the anode [36].
8. Lack of assistance to diffuse back the water produced at the cathode.

Looking upon the above-mentioned factors, the present shortcomings of these fuel cells could be surmounted and these could be argued to be the most simple and highly efficient ones [37].

3.2.1.3 Direct Methanol Fuel Cells

Direct methanol fuel cells are a sub category of Proton exchange membrane fuel cells, the fuel being methanol (Fig. 3.3). The structure of a Direct methanol fuel cell is given below. It is considered to be advantageous over the other types of cells because of the use of liquid fuel (methanol), easy refueling, low cost and ease in transport of methanol [4]. Moreover, methanol is energy dense, yet reasonably stable at all environmental conditions. However, methanol is highly volatile and toxic, which may pose severe troubles when used in portable DMFC electronic devices. Apart from this, CO poisoning of the platinum catalysts and methanol cross over are the other major issues with DMFC. Researches are being carried out upon electro – oxidation, with polyhydric alcohols in alkaline solutions using Platinum electrode and Ethylene glycol has been found to show highest reactivity among the alcohols examined in KOH and K_2CO_3 solutions. Ethylene glycol is also believed to overcome the carbonation problem, due to CO_2 release from the product of the fuel. It is much less toxic and volatile than methanol and is supposed to be safer.

DMFCs make use of a polymer membrane as an electrolyte like the PEMFC. The most effective catalysts are found to be carbon supported or unsupported platinum-ruthenium at anode and platinum at cathode [38]. They operate in temperatures between 60°C and 130°C and possess a theoretical energy density of 6094 W h/kg and a practical harvest of

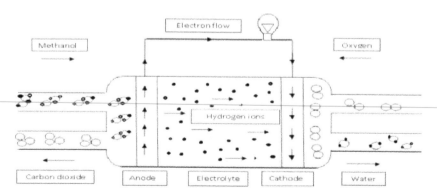

FIGURE 3.3   Schematic diagram of Direct methanol fuel cell.

1000–3000 W h/kg [39]. Their compact cell design makes them suitable for various stationary and nonstationary applications [4].

Regarding DMFCs for transportation or portable power applications, long-term stability of the anode and methanol cross over are the problems that affect the performance of the cells. Concentrating on fuel utilization, overall anode performance, conversion efficiency of the cell and platinum loading, may help improving electricity generation and stability. It has been reported that DMFC cathodes, operating on ambient air at 60°C, can deliver high performance of more than 0.85 V versus RHE at 0.100 A cm^{-2} [40]. For the cells operating at high temperature of 148°C, Nafion – Silica composite membranes doped with phosphotungstic acid has been found suitable. It shows a maximum power density of 400 mW cm^{-2} in the presence of oxygen feed and 250 mW cm^{-2} in the presence of air feed [41]. Further research is to be undertaken in this area of improving DMFCs for portable power applications.

3.2.1.4 Phosphoric Acid Fuel Cells

Phosphoric acid fuel cells developed in the mid 1960s were named after their working electrolyte, which is highly concentrated or pure liquid phosphoric acid, saturated in silicon carbide mix. They use finely dispersed platinum catalyst on carbon. Usually phosphoric acid is surrounded by a couple of porous matrix embedded in a ceramic matrix [42]. Phosphoric acid is a bad conductor and the cell needs 150°C to 220°C to work effectively [5]. Being the first commercialized fuel cells, they are extensively employed in power generators to generate 100 kW to 400 kW [43]. They have a wide range of applications in central heating plant, hospitals, dormitory, pools, office buildings, laundry and kitchen [44]. Their only disadvantage is they cannot be compatible with high temperature thermal steam applications.

There are a number of benefits in using phosphoric acid fuel cells in common use. They are as follows:

1. They are environmentally beneficial because, they emit only water, which is also very pure. They are circulated inside the fuel cell system as a coolant and liberation of a little amount of water into the environment will not be pollution causing.

2. They are highly fuel efficient with a higher heating value of 36% and a lower heating value of 40%.
3. They can be configured in such a way that they provide power back up.
4. They are low noise emitters that are too low when compared to traditional combustion technologies.
5. They are flexible enough to be sited in urban areas.

Studies have shown that addition of small amounts of fluorinated organic compounds or silicone compound, having a vapor pressure less than that of phosphoric acid, to the phosphoric acid electrolyte of a phosphoric acid fuel cell enhances oxygen solubility and diffusivity of the electrolyte. Moreover, there is a fact that decreasing polarization of the cathode increases efficiency of the cell. Using modified electrolyte increases cell efficiency by decreasing cathodic polarization [45].

3.2.1.5 Molten Carbonate Fuel Cells

Molten carbonate fuel cells represent a high temperature technology of fuel cells [5]. They use molten carbonate salts of alkali metals as electrolyte. Their anode or cathode is made of nickel oxide or cobalt oxide or rare earth element oxide [46]. Their operating temperature is as high as 650°C. That is why, rather than using a precious metal like platinum, a relatively cheap metal such as nickel is preferred as the catalyst. They can run on hydrocarbon fuels such as methane, natural gas or coal-reformed gases. Since their efficiency has been reported to be 45% which can also be raised up to 60–70% they could be considered as potential alternatives for petroleum [48].

In spite of all their advantages, a molten carbonate fuel cell using an alkali metal carbonate electrolyte encounters the following problems.

1. The presence of oxygen forms cracks on the electrode, particularly when sintered porous nickel plate is used as a cathode.
2. The particle size of the sintered porous nickel plate has decelerating effects on the output when the particle size is large.
3. Usage of a sintered porous nickel plate in the anode decreases the surface area of electrode by the sintering phenomenon of nickel and thus decreasing the cell output.

These shortcomings could be surmounted by finding an alternative electrolyte that is devoid of the above said disadvantages and is highly active, even at high operating temperatures.

3.2.1.6 Solid Oxide Fuel Cells

Solid oxide fuel cells are another type of high temperature fuel cells whose operating temperature ranges from 700°C–1000°C (Solid oxide fuel cell, spotlight). They use a solid ceramic such as stabilized zirconium oxide, oxygen ion, a proton or a mixed oxygen ion–proton conductor, which is gas impermeable and an electronic insulator [49] and do not need a metal catalyst. Their efficiency is reported to be 60%, theoretically up to 80% [5] and is widely used for stationary power generation.

Solid oxide fuel cells are highly advantageous over other fuel cells in the following aspects:

1. they are highly performative even at a temperature range of 1000°C and the electrode activity is extremely high.
2. they do not require the use of expensive catalysts such as platinum.
3. scale of equipments does not generally influence the fuel cell efficiency.
4. they have high-energy conversion capacity due to low polarization and high output voltage.

The very few disadvantages of SOFC is the ohmic resistance exhibited by the fuel electrode and air electrode, which influences power generation [50] and over voltage with Hydrogen and Carbon monoxide [51]. Studies are being carried out to rectify these discrepancies and to bring out the efficacy of the fuel cell at it's fullest.

3.2.1.7 Microbial Fuel Cells

Microbial fuel cells are devices that convert the chemical energy present in organic compounds to electrical energy through microbial catalysts [52]. They employ bacteria on the anode to carry out oxidation of organic matters and bacteria or microalgae on the cathode to undergo reduction.

3.2.2   TYPES OF MFC BASED ON THE REACTOR MODELS

MFC can be classified in to two types based on the reactor models:

In the first type of MFC, the electrons are transferred from the bacteria to the anode without the intervention of an intermediate fermentation product [23]. Since microorganisms act as a catalyst in the transfer of electrons from the substrate to the anode, the selection of a high performing microbial consortium (either pure or mixed culture) is of crucial importance in these microbial fuel cells. The electron transfer from the bacterium to the anode can proceed in a direct way from the bacterial membrane to the anode surface or indirectly by means of a mediator (Fig. 3.4).

Both transfers through bacterial contact with the electrode and through soluble shuttles can be regarded as mediated. In the first case, a bacterial redox enzyme, immobilized in the cell wall, provides the electron transfer. Examples of such bacteria are *Geobacter sulfurreducens* [53] and *Rhodoferax ferrireducens* [54]. When a soluble mediator is used, the electrons are shuttled by mediator molecules between the redox enzyme(s) of the bacteria and the electrode surface, thereby facilitating electron transport

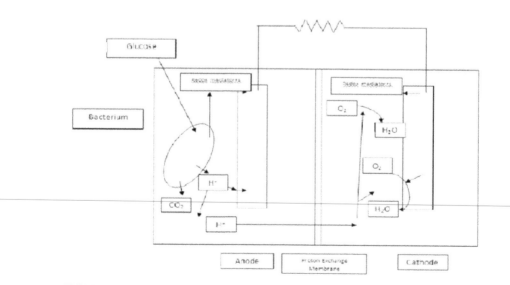

FIGURE 3.4   Schematic diagram of a microbial fuel cell.

[55]. Mediators are typically redox molecules (e.g., ubiquinones, dyes and metal complexes) that can form reversible redox couples, are stable in both oxidized and reduced form, are not biologically degraded and are not toxic towards the microbial consortium [56].

3.2.3 COMPONENTS OF A MICROBIAL FUEL CELL

Conventional microbial fuel cells contain an anode, a cathode, proton exchange membrane and a resistor through which the electrons pass to the anode. The anode is generally immobilized with a bacterial consortium [13] and the cathode compartment contains the microbial source. At times, the anode compartment is loaded with the fuel source or the organic material to be oxidized [14]. After oxidation, the protons pass through the proton exchange membrane to the cathode, where they are reduced into water.

Second type of microbial cells, which operates with a trielectrode system, is also employed in wastewater treatment and other applications. These microbial fuel cells contain three electrodes: the working electrode, the counter electrode and the reference electrode. Working electrode is the cathode that harbors the microalgae or bacteria. It may be glassy carbon electrode or platinum electrode upon which the microalgae or bacteria are immobilized. Counter electrode functions as the conductor of electricity. Reference electrode is the standard or the electrode of reference. Silver or silver in potassium chloride are the widely employed reference electrodes. The current generated is read with the help of a cyclic voltmeter.

Ideal MBFC consists of two chambers made up of glass, polycarbonate or Plexiglas with two electrodes. Very large surface areas are needed for supporting the biofilm, and the structure must be able to bear the weight of the water and biofilm. Electrode materials range from carbon cloth and carbon paper, to graphite rods, plates, granules, and RVC. Cathodes are made from the same materials, but they also contain precious metals, such as platinum, when oxygen is used as the electron acceptor. Platinum can be replaced by cobalt– and iron–organic-mixture catalysts, although the longevity of such materials is not well studied [57, 58].

3.2.4   APPLICATIONS OF MICROBIAL FUEL CELLS

Microbial fuel cells have become a promising alternative for electricity generation, biomass cultivation and wastewater treatment. If the microbial fuel cells make use of solar energy to generate electricity then they are called as Photo microbial fuel cells [13]. If they use photosynthetic algae for producing electricity, then they are called as Photosynthetic algal microbial fuel cells [14]. Therefore, Photo microbial fuel cells and Photosynthetic microbial fuel cells could be considered as the two different types of Microbial fuel cells.

3.2.5   MICROBIAL FUEL CELLS IN ELECTRICITY GENERATION

Though the launch of microbial fuel cells dates back to the late eighteenth century when Luigi Galvani first reported bioelectricity by connecting frog's legs to a metallic conductor [7], it is in the recent times that, they have been developed to harvest electricity, thus providing opportunities for practical applications [59–63].

3.2.6   MECHANISM OF ELECTRICITY PRODUCTION

The oxidation and reduction reaction that takes part in the anode and cathode chamber of a microbial fuel cell contributes to the flow of electrons, that pass through a resistance to produce current [64]. A three-chambered microbial fuel cell is built with the anode immobilized with a bacterial consortium and the cathode containing any microorganism such as the bacteria or microalgae [65–67]. Therefore, the metabolic pathways undertaken by the bacterium are solely responsible for the microbial electron and proton flow [68–70].

To understand how an MFC produces electricity, we must understand how bacteria capture and process energy. Bacteria grow by catalyzing chemical reactions and harnessing and storing energy in the form of adenosine triphosphate (ATP). In some bacteria, reduced substrates are oxidized and electrons are transferred to respiratory enzymes by NADH, the reduced form of nicotinamide adenine dinucleotide (NAD).

These electrons flow down a respiratory chain—a series of enzymes that function to move protons across an internal membrane—creating a proton gradient. The protons flow back into the cell through the enzyme ATPase, creating 1 ATP molecule from 1 adenosine diphosphate for every 3–4 protons. The electrons are finally released to a soluble terminal electron acceptor, such as nitrate, sulfate, or oxygen.

However the anode potential determines the metabolism of the bacteria [71]. When the current of the MFC increases, it eventually decreases the anode potential, thus forcing the bacteria to deliver electrons through more reduced complexes. At high anodic potentials the bacteria can use the respiratory chain in an oxidative metabolism. Normally, the current from a MFC is inhibited by the various inhibitors of the respiratory chain. At the same time, processes using oxidative phosporylation yield 65% high efficiency MF cells [61].

At low anode potentials and in the presence of an alternate electron acceptor such as sulfur, the electrons tend to be deposited onto these components and methanol production is high, indicating that, a bacterium does not use the anode [68, 72]. If no electrons acceptors like sulfur or nitrate are present, then fermentation will be the main reaction yielding a maximum of one-third of the hexose substrate electrons [73]. Thus, optimization of the anode potential can hike up the efficiency of a microbial fuel cell.

The performance of a MFC is determined by many factors like reactor configuration, pH, seed culture or substrate, electrolyte resistance and so on [52]. In spite of that they convert a wide array of electron donors with effective energy generation at low and moderate temperatures, even when the electron donor is provided at low concentrations. This makes them one of the incomparable alternative fuel sources.

3.2.7 MICROBIAL FUEL CELLS IN HYDROGEN PRODUCTION

MFCs can also be modified to produce hydrogen gas (H_2) by removing oxygen at the cathode and adding in a small voltage via the bioelectrochemically assisted microbial reactor (BEAMR) process or the biocatalyzed electrolysis process [43–46]. Bacteria produce an anode working potential of ~ –0.3 V. The protons and electrons that are produced at the anode can combine at the cathode to produce H_2 with only an additional total cell potential of 0.11 V [44]. In practice, however, 0.25 V or more must be put into the circuit to make H_2, because of overpotential at the cathode [74–77].

3.2.8   MICROBIAL FUEL CELLS IN WASTEWATER TREATMENT

Microbial fuel cells are believed to serve as a viable technological alternative to conventional wastewater treatment. Conventional wastewater treatments for removal of organic pollutants are energy and cost intensive because of the requirement of aeration and disposal of excess sludge generated in the process [78]. MFCs can harvest electricity from the energy available in organic wastewater [15]. They are built with cost effective systems such as activated carbon cathodes and graphite fiber brush anodes [79–81]. More specifically, PAMFCs, which are constructed with an immobilized microalga in the cathode, can generously contribute bioelectricity production, biomass production and wastewater treatment [14].

Though conventional MFCs are dual chambered, the problem with the dual chambered MFCs is the cost involved. Generally the oxygen reduction reaction is poor at the non-catalyzed cathodes. Therefore in order to improve their efficiency, costly catalysts are required which make them practically inapplicable directly on field [15]. Therefore MFCs are constructed with Separator electrode assembly (SEA) or widely Spaced electrodes (SPA). SEA has been found to reduce internal resistance when compared with SPA [81, 82]. But when it comes for Refinery wastewater (RW) treatment, SPA is found to be effective than SEA. Because, the separator electrode present in SPA reduces the overall oxygen transfer into the anode chamber, which affects overall COD removal. Whereas the SPA MFCs have higher organic removal with larger difference [82]. A report states that in the U.S, the cost of treating 33 billion gallons of wastewater is around $25 billion a year. MFCs could curb this cost by producing electricity with the organic matters dissolved in the wastewater plants [83]. Thus MFCs are not only cost effective but also productive.

3.2.9 ENVIRONMENTAL SENSORS

Data on the natural environment can be helpful in understanding and modeling ecosystem responses, but sensors distributed in the natural environment require power for operation. MFCs can possibly be used to power such devices, particularly in river and deep-water environments where it is difficult to routinely access the system to replace batteries. Sediment fuel cells are being developed to monitor environmental systems such as creeks, rivers, and the ocean [23, 84–86]. A linear correlation between the columbic of MFC and strength of organic matter in wastewater makes MFC a possible BOD sensor [87].

3.2.10 BIOREMEDIATION

The MFC is not used to produce electricity; instead, power can be put into the system to drive desired reactions to remove or degrade chemicals, such as converting soluble U (VI) to insoluble U (IV) [86]. Bacteria are not only able to donate electrons to an electrode but can also accept electrons from the cathode. By poising the electrodes at –500 mV, Gregory et al. were able to precipitate uranium directly onto a cathode because of bacterial reduction. Nitrate can also be converted to nitrite when electrodes are used as electron donors [20].

3.2.11 ORGANISMS EMPLOYED IN FUEL CELLS

MFCs have gained the attention of many researchers throughout the World in the present scenario. Different types of researches are being carried out to bring out the potency of these cells. So far, Microorganisms like *C. vulgaris* [13, 14], *L. discophora* [6], *K. pneumonia* [16], *T. ferrooxidance* [17], *P. fluroscens* [18], *R. rubrum* [19], *G. metallireducens* [20], *D. desulfuricans* [21] and other aerobic and anaerobic bacteria have been reported to be employed in microbial fuel cells. Few other microorganisms like cyanobacteria and some microalgae are also believed to play remarkable roles if employed in microbial fuel cells. Improvising the functioning of these cells to give exponential yields will be the expected future of the field. Apart from their application in bioelectricity production, wastewater treatment and biomass production, MFCs also contribute to biohydrogen production via biohydrolysis [88], bioremediation [89], biosensors [87], in-situ power source for remote areas [90].

3.3 CONCLUSION

Though MFCs have shown remarkable increase in power outputs over the past few years, they could not be considered as energy supplying ways. Optimization and identification of sufficient sturdy materials and alternatives membranes for conventional systems are still required in order to overcome the shortcomings still present. More study in order to expand the applications of MFCs beyond power generation is our utmost need.

KEYWORDS

- **Microbial fuel cells**
- **Polymer electrolyte membrane fuel cells**
- **Substrates**
- **Wastewater treatment**

REFERENCES

1. Schulze, M., Gulzow, E. (2004). Degradation of nickel anodes in alkaline fuel cells. *J. Pow. Sources. 127*, 252–263.
2. Wagner, N., Schulze, M., Gulzow, E. (2004). Long term investigation of silver cathodes for alkaline fuel cells. *J Pow. Sourc. 127*, 264–272.
3. Barbir, F. (1997). Fuel Cell Tutorial, Presented at Future Car Challenge Workshop, Dearborn, MI, October, 25–26.
4. Hacquard, A. Thesis submitted to the faculty of Worcester Polytechnic Inst, 2005.
5. Georgi, L., Leccese, F. (2013). *The open fuel cells jour. 6*, 1–20.
6. He, Z., Angenent, L. T. (2006). Application of bacterial biocathodes in microbial fuel cells. *Electroanalysis. 18*, 19 – 20, 2009 – 2015.
7. Piccolino, M. (1998). Animal electricity and the birth of electrophysiology: the legacy of Luigi Galvani. *Brain. Res. Bull. 46*, 381.
8. Cohen, B. The bacterial culture as an electrical half cell. *J.Bacteriol.*1981, *21*, 18.

9. Bennetto, H., Stirling, J. L., Tanaka, K., Vaga, C. A. (1983).Anodic reactions in microbial fuel cells. *Biotechnol. Bioeng. 25*, 559- 568.

10. Bennetto, H. P., Delaney, G. M., Mason, J. P., Roller, S. D., Stirling, J. L., Thurston, C. F. (1985). Electron-transfer coupling in microbial fuel cell. *Biotechnol. Lett. 7*, 699.

11. Roller, S. D., Bennetto, H. P., Delaney, G. M., Mason, J. R., Stirling, J. L., Thurston, C. F. (1984). Electron-transfer coupling in microbial fuel cells to performance of fuel cells containing selected microorganism—mediator—substrate combinations. *J.Chem. Tech. Biotechnol. 34* B, 3.

12. www.fuelcelltoday.com

13. Gouveia, L., Neves, C., Sebastião, D., Nobre, B. P., Matos, C. T. (2014). Effect of light on the production of bioelectricity and added value microalgae biomass in a photosynthetic alga microbial fuel cell. *Biores. Tech. 154*, 171–177.

14. He, H., Zhou, M., Yang, J., Hu Y., Zhao, Y. (2014). Simultaneous wastewater treatment, electricity generation and biomass production by an immobilized photosynthetic algal microbial fuel cell. *Biopro. Biosyst. Eng. 37(5)*, 873- 880.

15. Jadhav, D. A., Ghadge, A. N., Mondal, D., Ghangrekar, M. M. (2014). Comparison of oxygen and hypochlorite as cathodic electron acceptor in microbial fuel cells. *Biosourc. Technol. 154*, 330–335.

16. Rhoads, A., Beyenal, H., Lewandowski, Z. (2005). A microbial fuel cell using anaerobic respiration as an anodic reaction and biomineralized manganese as a cathodic reactant. *Environ. Sci.Technol. 39*, 4666–4671.

17. Lovley, D. R. (1991). Dissimilatory Fe (III) and Mn (IV) reduction. *Microbiol. Rev. 55*, 259–287.

18. Nemati, M., Harrison, S. T. L., Hansford, G. S., Webb, C. (1998). Biological oxidation of ferrous sulfate by Thiobacillus ferrooxidans: a review on the kinetic aspects. *Biochem. Eng. J. 1*, 171.

19. Berk, R. S., Canfield, J. H. (1964). Bioelectrochemical energy conversion. *Appl. Microbiol. 12*, 10.

20. Gregory, K. B., Bond, D. R., Lovley, D. R. (2004). Graphite electrodes as donors for anaerobic respiration. *Environ. Microbiol. 6*, 596–604.

21. Goldner, B. H., Otto, L. A., Canfield, J. H. (1963). Applications of bacteriological processes to the generation of electrical power. *Dev. Ind. Microbiol. 4*, 70.

22. Liu, H., Song, C., Zhang, L., Zhang, J., Wang, H., Wilkinson, D. (2006). A review of anode catalysis in the direct methanol fuel cell. *J. Pow. Sourc. 155*, 95–110.

23. Tender, L. M., Reimers, C. E., Stecher, H. A., Holmes, D. E., Bond, D. R., Lowy, D. A., Pilobello, K., Fertig, S. J., Lovley, D. R. (2002). Harnessing microbially generated power on the seafloor. *Nat. Biotechnol. 20*, 8, 821–825.

24. Lewis, K. (1966). Symposium on bioelectrochemistry of microorganisms. IV. Biochemical fuel cells. *Bacteriol.Rev. 30*, 101–113.

25. Kreuer, K. D. Fuel cells: Selected entries from the encyclopedia of sustainability science and technology, 2013.

26. Miao, F., Tao, B., Haochu, J. (2013). Nonenzymatic alkaline direct glucose fuel cell with a solon Microchanel plate supported electrocatalytic electrode, *J. Fuel cell Sci. Technol. 10*, 4, 041003.

27. EG & G Services Inc. Fuel Cell Handbook, 5th ed., UD Department of Energy office of fossil energy national energy technology laboratory Inc, : Morgantown, West Virginia, USA, 2000.
28. Reimers, C. E., Girguis, P., Stecher III, H. A., Tender, L. M., Ryckelynck, N., Whaling, P. (2006). Microbial fuel cell energy from an ocean cold seep. *Geobiol. 4,* 123–136.
29. Barbir, F. PEM fuel cells: Theory and practice, Elsevier academic press, New York, 2005.
30. Yuyan, S., Yin, G., Wang, Z., Gao, Y. (2007). Proton exchange membrane fuel cell from low temperature to high temperature: Material challenges. *J. Pow. Sourc. 167,* 235- 242.
31. Neburchilov, V., Martin, J., Wang, H., Zhang, J. (2007). A review of polymer electrolyte membranes for direct methanol fuel cells. *J. Pow. Sourc. 169,* 221–238.
32. Mathias, M. F., Makharia, R., Gasteiger, H.A, ; Conley, J. J., Fuller, T. J., Gittleman, S. S., Kocha, S. S., Miller, D. P., Mittelsteadt, C. K., Xie, T., Yan, S. G., Yu, P. T. (2005). Two fuel cell cars in every garage? *Inter. 14,* 24.
33. Gasteiger, H. A., Mathias, M. F., Murthy, M., Fuller, T. F., Van, Z. (2002). Proceedings of the Third International Symposium on Proton Conducting Membrane Fuel Cells, 1.
34. Voss, H. H., Wilkinson, D. P., Pickup, P. G., Johnson, M. C., Basura, V. (1995). Low cost electrodes for proton exchange membrane fuel cells. *Electrochem. Act. 40,* 321.
35. Zawondzinski, T. A., Springer, T. A., Uribe, F., Gottesfeld, S. (1993). Characterization of polymer electrolytes for fuel cell applications. *Sol. Stat. Ton. 60,* 199–211.
36. Ge, S., Li, X., Hsing, I. (2005). Internally humidified polymer electrolyte fuel cells using water absorbing sponge. *Electrochem. Act. 50,* 1909–1916.
37. Vishnyakov, V. M. (2006). Proton exchange membrane fuel cells. *Vacc. 80,* 1053–1065.
38. Arico, A. S., Baglio, V., Antonucci, V. Electrocatalysis of Direct methanol fuel cells. Edited by Hausan Liu and Jiugisn Zhang. 2009.
39. McGrath, M. K., Prakash, S., Olah, A. G. (2004). Direct methanol fuel cells. *J. Ind. Eng. Chem. 10,* 1063–1080.
40. Thomas, C. S., Ren, Z., Gottesfeld, S., Zelenay, P. (2002). Direct methanol fuel cells: Progress in cell performance and cathode research. *Electrochem. Act. 47,* 3741–3748.
41. Staiti, P., Arico, A. S., Baglio, V., Lufrano, F., Passalacqua, E., Antonucci, V. (2001). Hybrid Nafion – silica membranes doped with heteropolyacids for application in direct methanol fuel cells. *Sol. Stat. Ion. 145 Z,* 101–107.
42. Larmine, J., Dicks, A. Fuel cell systems explained, 2nd edition; 2003.
43. Sotouchi, H., Hagiwara, A. Energy carriers and conversion systems – Vol II: Phophoric acid fuel cells – 1998.
44. Holcomb, F. H., Binder, M. J., Taylor, W. R., Torrey, J. M., Westerman, J. F. Phophoric acid fuel cells. Construction Engineers Research Laboratory Publications. 2000.
45. Bjerrum, N. J., Idhusvej, R. Z., Gang, X., Gade, S. H., Hans, A., Ijuler, I., Dronningvie., Olsen, C., Plantageve., Berg, W. R., Frenderupvej. (1992). Text book of Phosphoric acid fuel cells. 903.
46. Takeuchi, M., Katsuta., Okada, H., Tobita, H., Kitaibaraki, Okabe, S., Matsuda, S., Ibaraki., Tonami, M., Tamura, K., Nakajima, F. (1982). Molten carbonate fuel cell. 362.

47. Takizawa, N. H., Text book of fuel cells. 1998. 431.

48. Subramanian, N., Haran, B. S., Ganesan, P., White, R. E., Popov, B. N. (2003). Analysis of Molten Carbonate Fuel Cell Performance Using a Three-Phase Homogeneous Model. *J. Elec.Chem. Soc. 150*, 1, A 46–A56.

49. Zuo, C., Liu, M., Liu, M. (2012). Solid oxide fuel cells. *Sprin.* 978-1-4614-1957-0, 2.

50. Ishihara, T., Misawa, H. (1991). Solid oxide fuel cells. 671.

51. Anzai, I, ; Matsuoka, S., Uehara, J. (1995).Solid oxide fuel cell. 489.

52. Kim, S., Chae, K. J., Choi, M. J., Verstraete, W. (2008). Microbial fuel cells: recent advances, bacterial communities and application beyond electricity generation. *Environ. Eng. Res. 13*, 2, 51–65.

53. Bond, D. R., Lovley, D. R. (2003). Electricity production by *Geobacter sulfurreducens* attached to electrodes. *Appl. Environ. Microbiol. 69*, 3, 1548–1555.

54. Chaudhuri, S. K., Lovley, D. R. (2003). Electricity generation by direct oxidation of glucose in mediatorless microbial fuel cells. *Nat. Biotechnol. 21*, 1229–1232.

55. Roller, S. D., Bennetto, H. P., Delaney, G. M., Mason, J. R., Stirling, J. L., Thurston, C. F. (1984). Electron-transfer coupling in microbial fuel-cells: Comparison of redox-mediator reduction rates and Respiratory rates of bacteria. *J. Chem. Technol. Biotechnol. B- Biotechnology. 34*, 1, 3–12.

56. Park, D. H., Zeikus, J. G. (2003). Improved fuel cell and electrode designs for producing electricity from microbial degradation. *Biotechnol. Bioeng. 81*, 3, 348–355.

57. Cheng, S., Liu, H., Logan, B. E. (2006). Power Densities Using Different Cathode Catalysts (Pt and CoTMPP) and Polymer Binders (Nafion and PTFE) in Single Chamber Microbial Fuel Cells. *Environ. Sci. Technol. 40*, 364–369.

58. Zhao, F., Hamisch, F., Schroer, U., Scholz, F., Bogdanoff, P., Hermann, I. (2006). Challenges and constraints of using oxygen cathodes in microbial fuel cells. *Environ. Sci. Technol. 40*, 5193.

59. Liu, H., Logan, B. E. (2004).Electricity generation using an air-cathode single chamber microbial fuel cell in the presence and absence of a proton exchange membrane. *Environ. Sci. technol. 38*, 4040–4046.

60. Liu, H. (2004). Production of electricity during wastewater treatment using a single chamber microbial fuel cell. *Environ. Sci. technol. 38*, 2281–2285.

61. Rabaey, K. (2003). A microbial fuel cell capable of converting glucose to electricity at high rate and efficiency. *Biotechnol. Lett, 25*, 1531–1535.

62. Rabaey, K. (2004). Biofuel cells select for microbial consortia that self mediate electron transfer. *Appl. Environ. Microbial. 70*, 5373–5382.

63. Schroder, A. (2003). Generation of microbial fuel cells with current outputs boosted by more than one order of magnitude. *Angew. Chem. Int. Ed. Engl. 42*, 2880–2883.

64. Strik, D. P. B. T. B., Hamelers, H. R. M., Snel, J. F. H., Buisman, C. J. N. (2008). Green electricity production with living plants and bacteria in a fuel cell. *Int. J. Ener. Res. 32*, 870–876.

65. Liu, H., Grot, S., Logan, B. E. (2005). Electrochemically Assisted Microbial Production of Hydrogen from Acetate. *Environ. Sci. Technol. 39*, 4317–4320.

66. Holmes, D. E., Bond, D. R., O'Neill, R. A., Reimers, C. E., Tenderad, L. R., Lovley, D. R. (2004). Microbial communities associated with electrodes harvesting electricity from a variety of aquatic sediments. *Microb.Ecol. 48*, 178–190.

67. Hasvold, O., Henribsen, H., Melvaer, E., Citi, G., Johansen, B. O., Kjonigsen, T., Galetti, R. (1997). Behavior of metal ions in microbial fuel cells. *J. Power. Sourc. 65*, 253.

68. Kim, B. H. (2004). Enrichment of microbial community generating electricity using a fuel cell type electrochemical cell. *Appl. Microbiol. Biotechnol. 63*, 672–681.

69. Lee, J. Y. (2003). Use of acetate for enrichment of electrochemically active microorganisms and their 16 s r DNA analyzes. *FEMS Microbiol. Lett. 223*, 185–191.

70. Rabaey, K., Lissens, G., Verstraete, W. Microbial fuel cells: Performances and perspectives. Text book of Microbial fuel cells.

71. Rabaey, K., Verstraete, W. (2005). Microbial fuel cells: novel biotechnology for energy generation. *Trend. In Biotechnol. 23*, 6.

72. Kim, J. R. Evaluation procedures to acclimate a microbial fuel cell for electricity production. *Appl. Microbiol. Biotechnol.* 2014.

73. Logan, B. E. (2004). Extraction of hydrogen electricity from renewable resources. *Environ. Sci. Technol.* 160A–167A.

74. Liu, H., Chang, S., Logan, B. E. Production of electricity from acetate or butyrate using a single chamber microbial fuel cell. *Environ. Sci.Technol.* (2005). *39*, 658–662.

75. Logan, B. E., Grot, S. A Bioelectrochemically Assisted Microbial Reactor (BEAMR) that Generates Hydrogen Gas.U. S. Patent Application 60/588,022, 2005.

76. Rozendal, R. A., Buisman, C. J. N. Bio-Electrochemical Process for Producing Hydrogen. International Patent WO-2005–005981, 2005.

77. Rozendal, R. A. Principle and Perspectives of Hydrogen Production through Biocatalyzed Electrolysis. *Int. J. Hydrogen Energy* (2006). in press, doi 10.1016/hydene.2005.12.006.

78. Du, Z., Li, H., Gu, T. (2007). A state-of-the-art review on microbial fuel cells; a promizing technology for wastewater treatment and bioenergy. *Biotechnol. Adv. 25*, 5, 464–482.

79. Dong, H., Yu, H., Wang, X., Zhou, Q., Feng, J. (2012). Novel structure of scalable air cathode without Nafion and platinum by rolling activated carbon and PTFE as catalyst layer in MFCs. *Wat. Res. 46*, 17, 57777–57787.

80. Logan, B. E., Cheng, S., Watson, V., Estadt, G. (2007). Graphite fiber bush anodes for increased power production in air-cathode microbial fuel cells. *Environ. Sci. Technol. 41*, 9, 3341–3346.

81. Zhang, F., Cheng, S., Pant, D., Bogaert, G. V., Logan, C. E. (2009). Power generation using an activated carbon and metal mesh cathode in a MFC. *Electrochem. Commun. 11*, 2177–2179.

82. Zhang, F., Ahn, Y., Logan, B. E. (2014). Treating refinery wastewaters in microbial fuel cells using separator electrode assembly or spaced electrode configurations. *Biores. Technol. 152*, 46–52.

83. www.engr.psu.edu/ce/ENVE/logan.htm

84. Reimers, C. E. (2001). Harvesting Energy from the Marine Sediment–Water Interface. *Environ. Sci. Technol. 35*, 192–195.

85. Shantaram, A. (2005). Wireless Sensors Powered by Microbial Fuel Cells. *Environ. Sci. Technol. 39*, 5037–5042

86. Gregory, K. B., Lovley, D. R. (2005). Remediation and Recovery of Uranium from Contaminated Subsurface Environments with Electrodes. *Environ. Sci. Technol. 39,* 8943–8947.

87. Chang, I. S., Jang, J. K., Gil, G. C., Kim, M., Kim, H. J., Cho, B. W., Kim, B. H. (2004). Continuous determination of biochemical oxygen demand using microbial fuel cell type biosensor. *Biosens. Bioelectron. 19,* 6, 607–613.

88. Rozendal, R. A., Hamelers, H. V. M., Molenkmp, R. J., Buisman, J. N. (2007). Performance of single chamber biocatalyzed electrolysis with different types of ion exchange membranes, *Water Res. 41,* 9, 1984–1994.

89. Lovley, D. R. (2006). Microbial fuel cells: novel microbial physiologies and engineering approaches. *Curr. Opin. Biotechnol. 17,* 3, 327–332.

90. Logan, B. E., Regan, J. M. (2006). Microbial challenges and applications. *Environ. Sci. Technol. 40,* 17, 5172–5180.

CHAPTER 4

CHALLENGES TO AND OPPORTUNITIES IN MICROBIAL FUEL CELLS

J. JAYAPRIYA,[1] and V. RAMAMURTHY[2]

[1]*Department of Chemical Engineering, A.C. Tech, Anna University, Chennai – 600025, India*

[2]*Department of Biotechnology, PSG College of Technology, Coimbatore – 641004, India*

CONTENTS

ABSTRACT

A chemical fuel cell is an electrochemical device that produces electrical energy from any fuel such as hydrogen, methane, propane, methanol, etc., using oxygen or an oxidizing agent. Microbial fuel cell (MFC) is one type of fuel cell, where the conversion of the chemical energy from the fuel to electrical energy is achieved by the catalytic action of microorganisms. A typical MFC is shown schematically in Fig. 4.1. It consists of anodic and cathodic compartments separated by a proton exchange membrane (PEM). Bacteria in the anodic chamber oxidize organic matter and transfer electrons to the cathode through an external circuit producing current. Protons diffuse through the solution across the PEM to the cathode where they combine with oxygen and electrons to form water. Unlike chemical fuel cells, MFCs operate under ambient operational temperature and pressure, also at neutral pH.

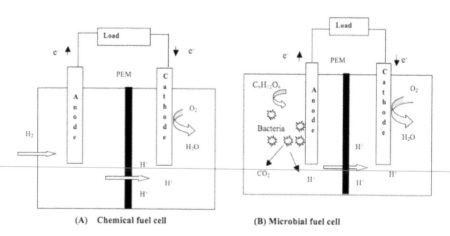

(A)   Chemical fuel cell   (B) Microbial fuel cell

FIGURE 4.1   Schematic representations of fuel cells. (A) Chemical fuel cell (B) Microbial fuel cell.

4.1 TYPES OF MICROBIAL FUEL CELLS

MFCs can be characterized as two classes depending on how electrons are transferred from the microorganisms to the electrode (anode): mediator-employing MFCs, where electron shutters or mediators are made available in the system, and mediator-less MFCs, where no extraneous mediators are deemed necessary. Typically exogenous mediators used for enhancing the electron-transfer rate in MFC are synthetic dyes such as neutral red [1], 2-hydroxy-1,4-naphthoquinone [2], methylene blue [3], and metal organics such as Fe (III) EDTA [4].

On the other hand in mediator-less MFC, the microorganisms typically make use of electrochemically active membrane-associated cytochromes, conductive pili or secrete redox-mediating molecules that facilitate electron transfer. Many such microorganisms are capable participants in the process, some of which include metal reducing bacteria *Geobacter metallireducens* [5], *Rhodoferax ferrireducens* [6], *Shewanella putrefaciens* [7] and *Aeromonas hydrophila* [8]. Secreted soluble redox shuttles by a microbial community consisting primarily of three bacteria, *Alcaligenes fecalis, Enterococcus faecium* and *Pseudomonas aeruginosa* leads to electricity generation [9].

4.2 PERFORMANCE OF MICROBIAL FUEL CELLS

4.2.1 IDEAL PERFORMANCE

Electricity is generated in an MFC only if the overall reaction is thermodynamically favorable. The maximum amount of electrical energy (or work done) that can be delivered by an electrochemical cell depends on the change in Gibbs free energy. It can be calculated as

$$\Delta G = \Delta G^\circ + RT \ln (\gamma) \qquad (1)$$

where ΔG (J) is the Gibbs free energy for the specific conditions, ΔG° (J) is the Gibbs free energy under standard conditions usually defined as 298 K, 1 bar and 1 M, R ($8.314 \text{ J mol}^{-1} \text{K}^{-1}$) is the universal gas constant, T (K) is the absolute temperature and γ (unit less) is the reaction quotient calculated as the activities of the products divided by those of the reactants. For MFC calculations, it is more convenient to evaluate the reaction in terms of the overall cell electromotive force (emf) cell [10–12]. E_{emf} (V), defined as the potential difference between the cathode and anode. This is related to the work, (J), produced as

$$Work = E_{emf} Q = -\Delta G \tag{2}$$

where $Q = nF$ is the charge transferred in the reaction, expressed in Coulomb (C), which is determined by the number of electrons exchanged in the reaction, n is the number of electrons transferred per mol, and F is Faraday's constant (96,481 C mol^{-1}). Hence, the cell emf can be calculated as,

$$E_{emf} = -\Delta G / nF \tag{3}$$

Thus the overall reaction is

$$E_{emf} = E^{\circ}_{emf} - RT/nF \ln (\gamma) \tag{4}$$

where E°_{emf} (V) is the cell electromotive force at standard conditions, as defined earlier. The cell emf is a thermodynamic value that does not take into account internal losses. In practice, under open circuit conditions the measured voltage (open circuit voltage) should in principle be exactly the equilibrium voltage (E_{emf}).

4.2.1.1 Actual Performance

The actual cell potential is always lower than its ideal potential because of irreversible losses. In general, the difference between the actual (measured) cell potential and the ideal potential is referred to as overvoltage [13]

$$E_{cell} = E_{emf} - \eta_{act} - \eta_{iR} - \eta_{diff} \tag{5}$$

where η_{act} (V) is the activation overpotential due to slow electrode reactions; η_{iR} (V) is the overpotential due to ohmic resistances in the cell and

η_{diff} (V) is the overpotential due to mass diffusion limitations. In MFCs, the measured cell voltage is a linear function of the current and it can be simply described as

$$E_{cell} = OCV - R_{Int} \qquad (6)$$

where R_{Int} is the sum of all internal losses of the MFC.

4.2.2 FACTORS THAT DECREASE CELL VOLTAGE

4.2.2.1 Activation Losses

Activation polarization is related to the loss of overall voltage at the expense of forcing oxidation/reduction reactions that occur during the transfer of electrons from or to a mediator reacting at the electrode surface [14]. This loss only occurs at low current densities in low temperature MFC. The extent of the activation losses can be described by the Tafel equation [15]

$$\Delta V = A \log (i/ i_0) \qquad (7)$$

where ΔV is overvoltage; A (V) is the Tafel slope, i (Am^{-2}) is current density; i_0 (Am^{-2}) is exchange current density, the current density at which the overpotential is zero. Activation losses can be reduced by increasing the electrode surface area, improving electrode catalysis and enriching biofilm on the electrode (s).

4.2.2.2 Ohmic Losses

The ohmic losses (or ohmic polarization) in an MFC include both the resistance to the flow of electrons through the electrodes /interconnections and the resistance to the flow of ions through the PEM/anodic/cathodic electrolytes [14, 16]. Ohmic losses can be reduced by minimizing the electrode spacing, using a membrane with a low resistivity, increasing solution conductivity to the maximum tolerated by the bacteria.

4.2.2.3   Concentration Losses

Concentration losses occur mainly at high current densities due to depletion of reduced species near the electrode (anode). This increases the ratio between the oxidized and the reduced species at the electrode surface, which can produce an increase in the electrode potential. At the cathode side the reverse may occur, causing a drop in cathode potential due to the accumulation of product. Ultimately these effects inhibit further reaction and the cell voltage drops to zero. Mass transport limitations in the bulk fluid can limit the substrate flux to the biofilm, which is a separate type of concentration loss [13].

4.2.2.4   Fuel Crossover and Internal Currents

This type of loss is usually due to unused fuel passing through the electrolyte and stray currents due to electron conduction through the electrolyte.

4.2.2.5   Bacterial Metabolic Losses

To generate metabolic energy, bacteria transport electrons from a substrate at a low potential through the electron transport chain to the final electron acceptor (such as, oxygen or nitrate) at a higher potential. In an MFC, the anode is the final electron acceptor and its potential determines the energy gain for the bacteria [13]. The higher the difference between the redox potential of the substrate and the anode potential, the higher the possible metabolic energy gain for the bacteria, but the lower the maximum attainable MFC voltage. To maximize the MFC voltage, therefore, the potential of the anode should be kept as low (negative) as possible. However, if the anode potential becomes too low, electron transport will be inhibited and fermentation of the substrate (if possible) may provide greater energy for the microorganisms.

4.2.3   TECHNICAL CHALLENGES

MFCs are operated under a range of conditions that include differences in system configuration, substrates, microorganism, anode/cathode materials, electrode surface area, mediators, cation exchange membrane, pH and operating conditions. Recent progress in understanding of system architecture, membranes and anode/cathode materials for MFCs has been reviewed, highlighting the characteristics, modifications, and processing of the electrode materials for MFCs. How bacteria use the soluble synthetic redox mediators and to what extent they generated the power output in MFCs has also been reviewed. It is generally accepted that MFCs show great promise as a sustainable biotechnological solution to future energy needs. However, it is yet to be commercially exploited since there are many hurdles to be overcome.

4.2.3.1 MFC Reactor Configurations

The basic MFC design comprises of anodic and cathodic compartments separated by a selective partition, such as, PEM (Fig. 4.2a). H-shaped MFC is widely used and are generally used to examine power generation with different electrode materials, membranes, type of microbial consortia with different substrates [13]. High internal resistance that reduces the power produced in this design is mainly due to the large electrode spacing and small membrane area. In order to reduce the internal resistance, cubic MFC with a larger membrane area and less electrode spacing could be used (Fig. 4.2b). So far, these designs were operated in batch mode, while

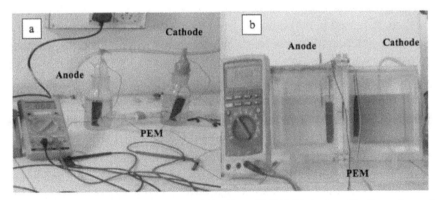

FIGURE 4.2 H shaped MFC (A) and Cube Shaped MFC (B) designed by Jayapriya and Ramamurthy [24].

the treatment of wastewater required a flow-through mode. Liu et al. [17] designed a single chamber MFC as shown in Fig.4.3a containing eight graphite electrodes (anodes) and a single air cathode which generated a maximum power density of 0.026 W.m^{-2} while removing up to 80% of the COD of the wastewater. No aeration was required for cathode chamber in this system that made it cost effective when compared to H-shaped MFC. Another single chambered MFC prototype comprising a flat plate was designed by Min et al. [18] consisting of a single channel formed between two nonconductive plates that were separated into two halves by the electrode/PEM assembly, which produced a maximum power density 2.8 times higher than reported earlier.

In general, MFCs are typically designed as anode and cathode system separated by PEM. It is known that PEMs are quite expensive that limit the *applicability* of *MFC*. Liu and Logan [19] designed the MFC using carbon paper as anode and the cathode was carbon cloth bonded with PEM or rigid carbon paper without PEM (Fig. 4.3b) having a Coulombic efficiency of 28% and 20% with and without the PEM, respectively. This may be due to about 30-fold higher rate of diffusion of oxygen in MFC without PEM compared to that with PEM. However, the consequence of high oxygen flux in MFC without PEM is the loss of substrate due to the aerobic oxidation by bacteria in the anolyte. In another design up-flow mode without PEM (Fig. 4.3c) was employed by Jang et al. [20]. Despite of the relatively low Coulombic efficiencies in MFC design that lacks PEM, it could still be low a cost alternative for wastewater treatment. Using a rotating cathode (Fig. 4.3d) in river sediment MFC [21], the solubility of oxygen increased in the catholyte from 0.4 to 1.6 mg/L, leading to a higher power production (0.049 W/m^2) when compared to a non-rotating cathode (0.029 W/m^2).

So far, the voltage produced by a single MFC is not more than 0.5 V. The voltage can be enhanced by connecting the several MFC in series/parallel [22, 23]. Aelterman et al. [22] operated the stacked MFC in different modes and it was observed that MFC in parallel showed good performance. When multiple cells were stacked together, however, charge reversal could result in the reverse polarity of one or more cells and a loss of power generation [23]. The availability of low cost components needs to be increased to *make* it attractive for scale up.

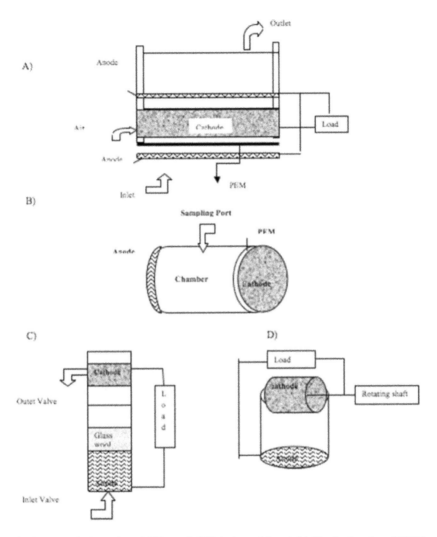

FIGURE 4.3 Schematics of different MFC designs: (a) and (b) Single chambered MFC; (c) Upflow mode MFC; (d) Rotating cathode MFC (Adapted from Liu et al. [17], Liu and Logan [19], Jang et al. [20] and He et al. [21], respectively).

4.2.3.2 Membranes

Microbial fuel cells make use of proton exchange membranes as the separating module between anode and the cathode compartments. The separator should be able to inhibit the transfer of substrates and oxygen, but allow the protons travel efficiently. Membranes used in MFCs include fluoropolymer-containing cation exchange materials such as Nafion [5, 25, 26], polystyrene and di-vinyl benzene with sulfonic acid groups [7] and dialysis membranes: 2000–14,000 Daltons [27]. The most commonly used membrane in MFC is Nafion, a *perfluorosulfonic acid polymer* developed by DuPont because of its highly selective permeability of protons and high ionic conductivity (10^{-2} S cm^{-1}). It consists of three regions (Fig. 4.4): (i) the Teflon-like, fluorocarbon backbone, (hundreds of repeating $-CF_2-CF-CF_2-$ units in length), (ii) the side chains, $-O-CF_2-CF-O-CF_2-CF_2-$, which connect the molecular backbone to the third region, (iii) the ion clusters consisting of sulfonic acid ions, SO_3^- H$^+$. The negative ion, SO_3^- is permanently attached to the side chain. However, when the membrane becomes hydrated by absorbing water, the hydrogen ion becomes mobile. Ion movement occurs by protons, bonded to water molecules, hopping from one SO_3^- site to another within the membrane. Thus, solid hydrated electrolyte is an excellent conductor of hydrogen ions. In general, proton conductivity increases as the hydration level of the polymer increases, with $H_2O:SO_3 \sim$ 15:1 being optimum [28]. However, the use of Nafion membrane may not be feasible for *large-scale* commercialization of MFC for waste water treatment due to its high cost and oxygen permeability (9.3×10^{-12} mol cm^{-1} s^{-1} [29].

Cation exchange membranes and anion exchange membranes are less expensive separators in MFC [30]. However, the transport of cation species other than protons (particularly of alkali and alkaline earth metals) through the CEM, causes a pH increase in the cathode chamber [26, 31].

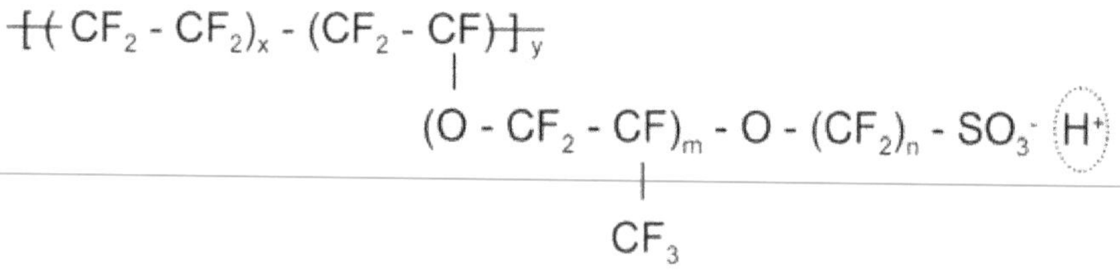

$$\left[\left(CF_2 - CF_2\right)_x - \left(CF_2 - CF\right)_y\right]$$
$$(O - CF_2 - CF)_m - O - (CF_2)_n - SO_3^- \; H^+$$
$$CF_3$$

| Nafion: | m = 1-3, n = 2, x = 5-13, y = 1000-1200 |
| Others: | m = 0, n = 1-5, x = 1-14, y = 800-1200 |

FIGURE 4.4   Proton membrane structure.

This pH increase negatively affects the MFC performance. Rozendal et al. [31] observed that for every increase in 3 pH units, a potential loss of 0.18 V occurs in the cathode.

A possible solution to the pH gradient associated potential losses is the application of an anion exchange membrane (AEM) instead of a CEM. In the case of an AEM, hydrogen at the cathode is not produced from the proton reduction, but from the reduction of water that diffuses through the membrane ($2H_2O + 2e^- \rightarrow 2OH^- + H_2$). The protons in the anode chamber are consumed by these hydroxyl ions transferred from anode to cathode and thus maintain *electroneutrality in the system*. However, the pH does increase more in the cathode chamber than with Nafion. In principle, only membranes that are truly 100% proton selective can prevent pH effects in MFC performance. But these types of membranes are not available at low costs.

Ultra filtration (UF) membranes, especially those developed for waste-water applications, may be suitable for use in MFCs. In this line of thought, Kim et al. [30] tested the UF with different molecular weight cut-off in MFC and found that these membranes had high internal resistances and thus produced low power output (5 ± 1 mW/m^2) when compared with CEM/AEM (35 ± 3 mW/m^2). On the other hand, membrane fouling always occurs in MFC, as biofilm with extracellular polymers can be formed on PEM during its long-term operation [32]. This would lead to deterioration in the MFC performance [33]. Therefore, *further research is required to develop* antifouling PEM with high proton diffusion coefficients at low costs.

4.2.4   ELECTRODES

Electron transfer in microbial fuel cells depends on the conductivity of the material, its surface characteristics and compatibility of its redox potential with the microbial metabolism. Among all *MFC* components, the *electrode* materials play a crucial *role* in electricity generation since the electrode cost will be a key factor deciding the implementation of the MFC technology at a large scale. The type of anode/cathode material used, desirable properties of anode/cathode, electrode performance, modification of electrode materials and costs are discussed in this section.

4.2.5 ANODIC MATERIAL

Anodic material must be a conductive surface whose redox behavior must be suited to microbial metabolism, conducive to biofilm formation, having high porosity with enhanced active surface area and non-corrosive. Various materials ranging from noncorrosive stainless steel to versatile carbons have been explored as anodes in various forms and shapes [13]. Some of the forms of carbon evaluated are carbon paper, cloth, foams, wires, reticulated vitreous carbon (RVC), graphite plate, thick graphite rod, foams, sheets, felt, graphite granules, fibers and brushes, conductive polymers, etc.

4.2.6 CARBON ELECTRODES

Carbon electrodes have some specific advantages, including a wide potential window, low residual current, long-term stability, excellent biocompatibility, high electrical conductivity, relatively high chemical stability, high corrosion resistance, low density, low thermal expansion and low cost [34, 35]. Homogeneous forms such as glassy carbon, graphite, vitreous carbon, screen printed, fullerenes, carbon nanotubes, carbon brushes and carbon fiber, etc., or heterogeneous forms like carbon paste electrodes, carbon fiber reinforced sheets, etc., have been evaluated as electrodes [36]. The vast range of carbon materials, both natural and synthetic, that have more disordered structures have traditionally been considered as variants of one or the other of the allotropic forms, such as graphite or diamond [36]. Thus, carbon materials differ in their physicochemical properties, electrochemical behavior and structural characteristics, and by deliberate modification of the surface or bulk matrix of the material it is possible to alter its electrochemical which would be suitable for MFC applications.

Carbon cloth is a woven material consisting of extremely thin fibers composed mainly of carbon atoms with great porosity, while carbon paper is nonwoven impregnation with a binder. Cheng and Logan [37] designed a single-chamber MFC that used carbon cloth as the electrode and domestic wastewater as the anolyte that produced a maximum power density of 1.640 Wm^{-2}. Wang et al. [38] built a single-chamber cell that produced a maximum power density of 0.483 Wm^{-2} when carbon cloth was the electrode and brewery wastewater was employed as the carbon source. The carbon cloth is currently too expensive (approximately $1,000 m^{-2}) to be used in large scale. Carbon paper is commonly available in plain and wet-proofed versions. It has high conductivity, but lacks durability and is fragile. Min and Logan [18] designed a flat plate MFC with carbon paper as an anode, achieving a maximum power density of 0.309 Wm^{-2} when using acetate as the substrate. In the dual chamber MFC constructed using carbon paper electrodes the power density reported by Kim et al. [30] was 0.040 Wm^{-2}. Carbon mesh is a possible alternative material for an MFC anode. It has a more open structure than carbon cloth due to a more open weave, which could help to reduce the biofouling. Pretreatment of carbon mesh might be required to get adequate performance in MFC. Wang et al. [39] reported power density upto 0.922 Wm^{-2} using heated carbon mesh.

Liu et al. [17] designed a single-chamber MFC that used eight graphite rod anodes produced a maximum power of 0.026 Wm^{-2}. Reimers et al. [40] used graphite rod as an anode and reported a maximum power density of 0.034 Wm^{-2}. Mohan et al. [41] reported a maximum power yield of 0.635 W/kg COD when a graphite sheet was used. However, the utility of the graphite rods and sheets were limited due to their low porosity and low surface area for biofilm formation. Graphite sheets are not porous and thus they produced less power per geometric (projected) surface area when compared with felts or foams. Chaudhuri and Lovley [6] found that the power output was three times larger when the graphite rod was replaced by graphite felt, indicating that increasing the surface area was beneficial to the performance of the MFC. Low electrochemical activity of commercial graphite felt is still one of the major hindrances that limit its applications in MFC.

A graphite fiber brush is made of graphite fiber that is wound around one or more conductive corrosion-resistant metal wires such as titanium. This method is attractive for its high surface area and low electrode resistance. Logan et al. [42] designed a flat-type MFC with graphite brush as the anode and 1.2 mg cm^{-2} cobalt tetramethylphenylporphyrin (CoTMPP) carbon cloth as the cathode that produced a power density of

2.4 Wm^{-2}, which was four times that obtained when using carbon paper as the anode (0.6 mWm^{-2}). Ahn and Logan [43] designed a single chamber air-cathode MFC in a continuous mode that used a graphite brush as the anode, generating a maximum power density of 0.422 Wm^{-2}. One limitation of the brush architecture is that the minimum electrode spacing is constrained by the brush size, which can leads to a high ohmic resistance.

Graphite granules can be of variable size, but only granules with a diameter of 1.5 to 5 mm have been used in MFCs. The granules were conductive (0.5–1.0Ω/granule). The first use of these materials was reported by Rabaey et al. [44]. You et al. [45] built a tube-type air cathode MFC that used graphite granules as the anode, and a graphite rod was used to collect electrons, achieving a maximum power density of 50.2 Wm^{-3} when using glucose as the substrate.

RVC has a low density and thermal expansion, high corrosion resistance and high thermal and electrical conductivities. RVC is an open-pore foam material of honeycomb structure composed solely of vitreous carbon. The conductivity of the material is 200 S cm^{-1}. He et al. [46] reported obtaining a power density of 0.170 Wm^{-2} when using sucrose as substrate in upflow MFC and RVC electrodes was used. Ringeisen et al. [47] constructed a miniature MFC catalyzed by *Shewanella oneidensis* producing a maximum power density of 0.024 W m^{-2}. The main disadvantage of the material is that it is quite brittle and hence short lived.

4.2.7 CARBON NANOTUBES AND CONDUCTIVE POLYMERS

Carbon nanotubes (CNT) have become one of the materials with very high potential because of its large specific surface area, high mechanical strength and ductility, and excellent stability and conductivity. However, it has been reported that CNT have a cellular toxicity that could lead to proliferation inhibition and cell death [48, 49]. Thus, they are not suitable for MFC unless modified to reduce the cellular toxicity. Recently, conductive polymer/CNT composites have received significant interest because by incorporating CNT in conductive polymers a synergistic effect may be obtained. Qiao et al. [50] reported the feasibility of an MFC using CNT/polyaniline composite as the anode material. Polyaniline as a conductive polymer, could not only protected the microorganisms but also improve their electro-catalytic activity [51]. Sahoo et al. [52] reported that polypyrrole (PPy)-coated carbon nanotubes had significantly higher electrical conductivity than that of carbon nanotubes alone. Among all the conducting polymers studied till date, PPy can be considered as one of the most attractive materials due to its excellent conductivity, stability and biocompatibility even in neutral pH. Yuan and Kim [53] modified the anode of the MFC with electropolymerized PPy and reported improvement in the power density of MFC greatly. Sharma et al. [54] developed MFC using an anode of carbon paper that was deposited with multiwalled CNT; to their great surprise, the power density was found to be approximately six times greater than that obtained with the graphite electrode. They have suggested that the carboxyl groups on the surface of multiwalled CNT could increase the chemical reactivity of metal nanoparticles.

There is no question that CNT can improve the performance of MFCs because they have very high surface areas (usually a few hundred to 1300 $m^2 g^{-1}$), and the groove openings that are formed between CNT bundles and the outside surface area of CNT bundles are supposed to be accessible by large sorbate species such as bacteria, which makes the increased surface area effective for microscale bacteria which is significantly larger than that of traditional microporous adsorbent media [55, 56]. The high cost of CNT manufacture, which was reported to be $80–100/kg is one of the major factors that limit its applications [57].

4.2.8 METALS

Dumas et al. [58] attempted sediment MFC using stainless steel as the electrode, but the maximum power density obtained was only 0.004 Wm^{-2}. Richter et al. [59] developed a *Geobacter sulfurreducens* catalyzed MFC that produced a steady current of 0.4–0.7 mA when gold was used as an anode. However, Heijne et al. [60] have reported that titanium as the anode was found unsuitable.

4.2.8.1   Cathodic Material

The cathode materials also have an equally important impact on the power generating capacity of MFCs. They must have a high redox potential to accept the electrons. The commonly used cathode materials are graphite, carbon cloth and carbon paper. However, it is difficult to obtain high cathodic potentials unless metal catalysts are used. Moon et al. [61] designed an MFC that used graphite felt containing platinum (Pt) as the cathode, and its power density reached 150 mWm^{-2}, which was three times higher than that for the graphite cathode. Alternatives to these catalysts include nonprecious metal catalysts linked with organics, such as Cobalt tetra methyl phenyl porphyrin (CoTMPP) and iron (II) phthalo cyanine (FePc) [62, 63]. These transition metal catalysts have excellent performance (Table 4.1), although higher loading rates than those used for Pt are needed to achieve similar performance [64].

4.2.8.2   Modification of Electrode Materials

Chemical modification methods can be effective for immobilizing metals, metal oxides or other active compounds on carriers, such as carbon materials or conductive polymers, to enhance the performance of MFCs.

TABLE 4.1   Metal Catalysts of Cathode in MFCs, from Zhou et al. [65]

Type of Catalyst	Cathode Materials	Power density (mW/m^2)
CoTMPP	Carbon cloth	369
FePc	Carbon paper	634
FePc	Graphite foils	0.01
PbO$_2$	Ti sheeting	485
β-MnO$_2$	Carbon cloth	0.003
Co-OMS	Carbon cloth	180
MnO$_x$	Carbon cloth	161
Co/Fe/N/CNT	Carbon cloth	751
Fe$_2$(SO$_4$)$_3$	Graphite composites	1.67

The modification methods, include ammonia or acid treatment, sintering, soaking method, chemical vapor deposition, carburization, etc.

4.2.9 CHEMICALLY MODIFIED CARBON ELECTRODES (CMCES)

Ammonia has been reported to be suitable for treatment of many forms of carbon including carbon cloth and carbon paper, and it has been regarded as one of the most effective ways to improve electrode performance. Ammonia gas treatment on carbon cloth was reported by Cheng and Logan [37]. In a nitrogen environment containing 5% NH_3, carbon cloth was heated at 700°C for 60 min, which resulted in an increase in the power density by 48% compared to untreated carbon cloth. This was due to the improvement in the surface charge from 0.38 to 3.99 mequiv.m^{-2}. Other CMCEs were prepared by heating the carbon cloth at 200°C or graphite sheet at 400°C for 48 h for carbonyl generation at the surface, followed by immobilization of neutral red (NR) or methylene blue (MB) into the matrix using N, N0-dicyclohexylcarbodiimide by the formation of a peptide bond between amine group of the redox dye and carbonyl group of electrode [66]. The best performing MFC with the highest power density was the MB modified carbon cloth anode, which gave a power output of 0.194 ± 0.020 mW/m^2 when compared to plain carbon cloth 0.133 ± 0.179 mW/m^2 [67]. The performance of the MFCs using carbon cloth showed a lot of variation in replicate experiments indicating that the material was very heterogeneous and the process of preparation of the carbon cloth did not result in producing a consistent product with respect to this application. However, in dyes immobilized in carbon cloth, the performance was more consistent. An attempt to immobilize the dyes to the graphite sheet was not successful because the material was damaged and became unstable after heat treatment.

Acid treatment, a very simple procedure, has also been evaluated for electrode modifications: the electrode material was placed into an acid solution, followed by washing with deionized water and drying. Erable et al. [68] modified the graphite granules with nitric acid. This process increased the cell voltage from 660 to 1050 mV, which was due to the formation of nitrogen containing superficial groups on the surface of graphite granules. Feng et al. [69] found that the combined process of the heat and acid treatment could result in improved power production to 1.370 Wm^{-2}, which was 34% greater than the untreated electrode.

4.2.10 SINTERING AND SOAKING

In this method the active components of catalysts were dispersed on the carriers or supports through impregnation. The steps followed included:
1. to increase the contact between the support and the active compo-nent precursor;
2. drying of the support impregnated with the active component pre-cursor;
3. activation of the catalyst by a thermal (sintering) or reactive (soak-ing) process.

Park and Zeikus [25] incorporated Mn^{4+} into the anode by the soak method, and the maximum power density reported was 0.788 Wm^{-2}. Lowy et al. [70] studied the performance of MFCs using several kinds of modified electrodes by the sintering process or the soak method. They prepared the graphite electrodes modified with anthraquinone-1,6-disul-fonic acid, 1,4-naphthoquinone, a graphite-ceramic composite containing Mn^{2+} and Ni^{2+} and a graphite paste containing Fe_3O_4 and Ni^{2+} and found that modified electrodes showed a power output 4–5 times that of the unmodified one.

Jayapriya et al. [66] reported the preparation of graphite/epoxy composite electrode (GECE) doped with different six types of metal ions (Zn^{2+}, Mg^{2+}, Mn^{2+}, Cu^{2+}, Co^{2+}, Fe^{3+}) enabling robustness through a wet impregnation procedure to facilitate electron transfer processes in biological applications (Fig. 4.5). It was observed that the Fe^{3+} doped GECE surfaces exhibited significantly high biofilm formation of 1.10 ($\pm$ 0.18)$\times 10^7$ CFU/cm^2 as compared to other dopants. Fe^{3+} doped GECE had the highest redox potential (0.11 V) compared to all the other electrodes tested, allowed better biofilm growth either because they allowed forma-tion of bacterial foci better or enhanced the growth of bacteria in the foci, or both. Their choice based on the closeness of their potential to the

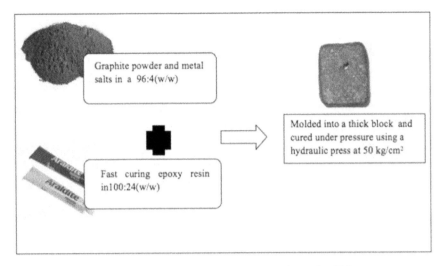

FIGURE 4.5 Preparation of Graphite epoxy composite electrodes [66].

oxidative phosphorylation chain of reactions in the bacterial membrane could have facilitated more efficient energy metabolism and thus better biofilm growth. Among the different combinations of the electrodes tested Fe^{3+} graphite cathode produced significantly higher power density $(1.680 \pm 0.098 \text{ mW/m}^2)$ [67]. These studies indicate the feasibility of the improvement in electrode design on the performance of MFC for wider applications.

4.2.11 CHEMICAL VAPOR DEPOSITION (CVD)

Chemical vapor deposition is often used in the semiconductor industry to produce thin films. In a typical CVD process, the substrate is exposed to one or more volatile precursors, which react and/or decompose on the substrate surface to produce a deposit. Liu and Logan [19] coated the anode with ferric oxide by a CVD technique and observed that the power density increased noticeably from 8 to 30 mWm^{-2} because of the enrichment of metal-reducing bacteria on the anode.

4.2.11.1   Biocatalysts in Microbial Fuel Cells

A major limitation to wider application of MFCs is the relatively low electron transfer between the bacterial catalysts and the anode, leading to high
internal resistance and reduced power generation. Hence, to improve MFC
performance understanding the metabolic impediments, dynamics of the
electrochemically active bacterial (EAB) communities in the biofilm and
how the bacterial species are capable to play a part in extracellular electron transfer. Although MFCs could be operated using pure cultures[5, 6,
47, 71–74], mixed cultures are more suitable for the use of complex fuels
such as wastewater [75]. MFCs that employ mixed bacterial cultures have
some important advantages over MFCs driven by axenic cultures: higher
resistance against process disturbances, higher substrate consumption rate,
wider substrate specificity and higher power output [9]. Mostly, the electrochemically active mixed cultures are enriched either from aquatic sediments (either marine or fresh water) or activated sludge from wastewater
treatment plants.

The mechanism of the electron transfer from the bacterial cell membrane or membrane organelle to anode will be addressed below.

4.2.11.2   Direct Electron Transfer

Many microorganisms possess the ability to transfer electrons derived
from their metabolism to the anode without a need of redox shuttle. These
microorganisms are ubiquitously present in places such as marine sediment, soil, wastewater, fresh water sediment and activated sludge [76,
77]. These microorganisms possess outer-membrane bound proteins such
as multiheme proteins (c type cytochromes) or nanowires that shuttle
the electron directly from the inside of the bacterial cell to solid electron
acceptor.

A real breakthrough was made in MFC when some dissimilatory metal
reducing bacteria (DMRB) were found to transfer the electrons directly
to an anode without the need for an electron shuttle. Several isolates *Shewanella putrefaciens* [7], *Geobacter metallireducens* [5], *Rhodoferax ferrireducens* [6] have been shown to generate electricity in mediator-less

MFC systems. In terms of performance, current densities of the order of 0.2–0.6 mA and a total power density of 1–17 mWm^{-2} graphite surface have been reported for *Shewanella putrefaciens* and *Geobacter sulfurreducens* when conventional (woven) graphite electrodes was used. Chaudhuri and Lovley [6] reported a Coulombic efficiency of 83% current production from glucose by *Rhodobacter ferrireducens*. In addition to membrane bound proteins, bacterial nanowires (also known as microbial nanowires) are electrically conductive appendages produced by a number of bacteria, most notably in the genera *Geobacter* and *Shewanella*, and reported to enhance the power output in MFCs [78]. Nanowires are about 3–5 nm wide and up to tens of micrometers long that facilitate the *long-range electron transfer* across thick biofilms. Conductive nanowires have also been confirmed in the oxygenic cyanobacterium *Synechocystis* PCC6803 and a thermophilic, methanogenic coculture consisting of *Pelotomaculum thermopropionicum* and *Methanothermobacter thermoautotrophicus* [79].

Recently, Malvankar et al. [80] reported that the electrical conductivity of microbial nanowires to be around 5 mS cm^{-1}, which is comparable to those of synthetic organic metallic nanostructures that are commonly used in the electronics industry. The wires were also seen to conduct over distances of centimeters, which is thousands of times the length of a bacterium itself. These findings could influence the design of energy-capture strategies, such as conversion of biomass and wastes to methane or electricity.

4.2.11.3   Mediated Electron Transfer

More generally, for electron transfer from a microbial electron donor to an electrode to occur, an electron mediator may be required. Wilkinson [81] has summarized the attributes of effective mediators as below: (i) the oxidized mediator should easily penetrate the bacterial membrane to reach the reductive species inside the bacteria; (ii) the redox potential of the mediator should match the potential of the reductive metabolite; (iii) the reduced mediator should easily escape from the cell; (iv) both the oxidized and reduced states of the mediator should be chemically stable in the electrolyte solution, should be easily soluble, and should not adsorb on the bacterial cells or electrode surface.

The mediators can be coupled to the MFC in two possible ways.

1. Exogenous Mediators: Many investigators [82–84] have reported that the metabolic reducing power produced by *Proteus vulgaris* or *Escherichia coli* can be converted to electricity by using electron mediators, such as thionin or 2-hydroxy-1,4-naphthoquinone (HNQ). Tanaka et al. [85] reported that light energy can be converted to electricity by *Anabaena variabilis* when HNQ was used as the electron mediator. Park et al. [86] reported that viologen dyes cross-linked with carbon polymers and absorbed on the cytoplasmic membranes of *Desulfovibro desulfuricans* mediated electron transfer from bacterial cells to electrodes or from electrodes to bacterial cells. Other organic dyes that have been tested include benzylviologen, 2,6-dichlorophenolindophenol, 2-hydroxy- 1,4-naphthoquinone, phenazines (phenazine ethosulfate, safranine), phenothiazines (alizarine brilliant blue, N,N dimethyl-disulfonated thionine, methylene blue, toluidine blue) and phenoxazines (brilliant cresyl blue, gallocyanine, resorufin) (Table 4.2). Among the dyes tested, phenoxazine, phenothiazine, phenazine, indophenol, bipyridilium derivatives, thionine and 2-hydroxy-1,4-naphthoquinone were found to be very efficient in maintaining relatively high cell voltage output when current was drawn from the biofuel cell [87]. Ferric chelate complexes (e.g., Fe (III) EDTA) were successfully used with *Lactobacillus plantarum*, *Streptococcus lactis* and *Erwinia dissolvens*, oxidizing glucose. A mixture of two mediators can be useful in optimizing the efficiency. Two mediators, namely thionine and Fe (III) EDTA were employed with *E.coli* for the oxidation of glucose. It was found that thionine was reduced over 100 times faster than Fe (III) EDTA [85].

2. Endogenous Shuttles: To facilitate the electron transfer from bacteria to the anode, addition of synthetic exogenous mediators could be helpful. However, such a solution can have some inherent disadvantage. Toxicity and instability of the added mediators can limit their applications in MFCs. One possible solution could be to use native (endogenous) electron shuttles produced

TABLE 4.2 Role of Mediators in the MFC Performance [12]

Microorganism	Substrate	Mediator	Anode	Power density (μA cm^{-2})
Pseudomonas methanica	Methane	1-Naphthol-2-sulpho-nate indo-2,6-dichloro-phenol.	Pt-black	2.8
Proteus vulgaris	Glucose	Thionine	RVC	4.3×10^{-4}
Proteus vulgaris	Glucose	HNQ	Graphite felt	0.095
Escherichia coli	Acetate	Neutral red	Graphite	1.4
Escherichia coli	Glucose	Neutral red	Graphite felt	37
Escherichia coli	Glucose	HNQ	Glassy carbon	180

by the microorganism itself. For MFC applications, the secondary metabolites which can work as redox mediators are of specific interest, as their *in situ* synthesis makes the electron transfer independent of the addition of exogenous redox shuttles. These metabolites can serve as reversible electron acceptors, carrying electrons from the bacterial cell either to an anode or into aerobic layers of biofilm, where they become reoxidized and are again available for subsequent redox processes.

Gamma proteobacteria such as *Pseudomonas aeruginosa* and *Shewanella oneidensis MR-1* have been found to produce their own electron shuttles [44, 88, 89]. The best-studied organism with respect to redox activity is *P. aeruginosa*, which is capable of producing pyocyanin and several other electron shuttling compounds [24, 90]. *Pseudomonasaeruginosa* is of interest due to its diverse metabolic activity, pathogenesis, ability to produce compounds such as siderophores and surfactants, applications in waste water treatment, etc. It is known to produce several phenazine derivatives, the regulation of which is partly determined by quorum sensing [91].

Phenazines comprise a large group of nitrogen-containing heterocyclic compounds that differ in their chemical and physical properties based on the type and position of functional groups present. More than 100 different phenazine structural derivatives have been identified in nature (Fig. 4.6), and over 6,000 compounds that contain phenazine as a central moiety

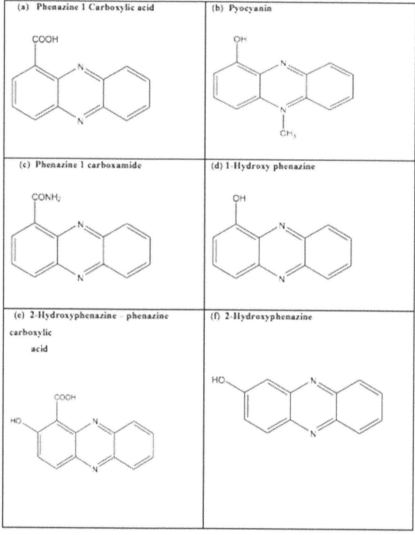

FIGURE 4.6. Some phenazine derivatives synthesized by *Pseudomonas.* (a) Phenazine 1 Carboxylic acid; (b) Pyocyanin; (c) Phenazine 1 carboxamide; (d) 1-Hydroxy phenazine; (e) 2-Hydroxyphenazine–phenazine carboxylic acid; (f) 2-Hydroxyphenazine.

have been synthesized [92]. The phenazine-based pyocyanin is of particular interest for its capability of producing reactive oxygen species [93]. It also has been reported to mediate the reoxidation of NADH under oxygen-limiting conditions, such as those found in mature biofilms [94–96]. Also several other intermediates in the pathway for pyocyanin and phenazine generation have been reported to hold electrochemical activity and could possibly exert a similar function.

   P. aeruginosa catalyzed MFC has been shown to provide a unique opportunity to combine the metabolic capability of the organism and to breakdown oxidizable pollutants and salvage energy from the process [24, 90]. MFCs with P. aeruginosa MTCC 2474 using graphite block electrodes generated a maximum power density of 392 $\mu W/m^2$ and it was observed that the 23 μM pyocyanin and 4.8 μM pyorubin were generated in the anolyte [24]. External supplementation of phenazines (pyocyanin, pyorubin, tubermycin or oxychloraphin) at a concentration of 3 mM to the anolyte resulted in different levels of power generation. Addition of tubermycin and oxychloraphin, which were not normally produced by *P. aeruginosa* MTCC 2474 under aerobic condition, enhanced the power density several fold. Particularly, the addition of oxychloraphin showed about 22.9-fold higher power density when using Cu^{2+} doped GECE anode as compared to unsupplemented MFC using graphite anode. However, the addition of pyocyanin or pyorubin, which are normally produced by the *P. aeruginosa* MTCC 2474, was not very helpful in improving the performance of the MFC. Also, the addition of these phenazines inhibited the growth of bacteria as well. The native phenazines that were generated as a response to stress, when added externally inhibited the organism, while the nonnative phenazines allowed a much more efficient metabolic conversion of the substrate available [24].

4.2.12   APPLICATIONS

The concept of MFCs has been known and demonstrated from the eighteenth century, the microorganisms working as catalysts in fuel cells to produce electricity [84, 97]. However, it is only recently that the thoughts of using MFC for practical applications have been entertained with increase in power output [68]. This was due to the paradigm shift in the discovery of microbial communities, such as dissimilatory metal reducing bacteria that directly transfer electrons without the need for exogenous mediators, development of system architecture and low cost components (electrodes, membranes, etc.) for MFCs. Some areas where MFCs offer immediate prospect of application are discussed in the following sub-sections.

4.2.12.1   Waste Water Treatment

The power consumption in the US for aerobic wastewater treatment, as estimated by Logan et al. [13] is 3% of total consumption. The cost of this process is becoming unacceptable since the immediate returns from the process are none. Domestic and industrial effluents contain a multitude of organic compounds that can fuel MFCs. Microorganisms in MFCs can discharge the dual duty of degrading the pollutants and generating power. Moreover, such a system can reduce solids generation by 50–90%, and reduce the load on their disposal [98]. Sanitary wastes, food processing wastewater, swine wastewater and corn stover are all great substrates for MFCs because their rich organic content [99–102]. About 80% of chemical oxygen demand (COD) of the domestic waste water was removed in a prototype single chambered MFC (SCMFC) reactor and generated 0.026 W/m^2 power density [19]. Using a single chamber MFC, a power density of 0.261 W/m^2, 0.205 W/m^2 and 0.672 W/m^2 had been achieved with wastewater streams of swine [101], brewery [103] and paper [104] industries. More than 80% COD removal was achieved in the combined anaerobic fluidized bed (AFB) and MFC with an electricity generation of 0.124 W/m^2 using distillery wastewater [105].

   MFCs were also explored to treat the azo dyes containing textile effluent and simultaneous recovery of power [106–108]. Azo dyes form the largest class of synthetic dyes with a variety of color and structure. The color of dyes is due to azo double bond and associated chromophores. Disposal of dyes into surface water not only affects the esthetics but can also lead to toxicity to the biota present. When azo dyes were used as catholyte of MFCs under anoxic conditions they could be employed as electron donors resulting in their decoloration. The reduction reactions in the cathode chamber are described by Goyal and Minocha [109], and Menek and Karaman [110] in which the –N=N–double bond was reduced to hydrazo (1) or amine (2) via the consumption of two or four electrons.

$$-N=N- + 2e + 2H^+ \rightarrow -NH-NH- \tag{8}$$

$$-N=N- + 4e + 4H^+ \rightarrow -NH_2-NH_2- \tag{9}$$

Different dyes were tested as catholytes in *Pseudomonas* catalyzed MFC by Jayapriya and Ramamurthy [67]. Compared to potassium ferricyanide and potassium permaganate, addition of methyl orange resulted in a higher power output.

Lovley [111] and Dumas et al., [58] designed MFCs using benthic organisms that oxidize the substrates in the sediment and transfer electrons to the anode either embedded in or rested on top of the sediment. The cathodes are suspended in the overlying seawater where oxygen reduction takes place. This MFC with *different industrial and domestic wastewater, tested at pilot scale, was found feasible, but would require higher efficiency and lower cost to be economically competitive.*

4.2.12.2 Biosensor

Another potential application of the MFC technology is to use it as a sensor for pollutant analysis and in situ process monitoring and control [112]. Although the five-day BOD test (BOD_5) is being widely used as the standard method to determine the concentration of biodegradable organics in wastewater [113], this is a time-consuming process. An MFC could be employed for real-time BOD monitoring by measuring current in an amperometric mode [114]. However, the signal from this MFC was reduced in the presence of high redox potential electron acceptors, such as nitrate and oxygen, and hence the analytes must be pretreated to remove the interfering compounds.

MFC based toxicity sensor for fast monitoring of acidic toxicity was developed by Shen et al. [115]. In principle, if a toxic event occurs, microbial activity is altered and thus the power output of the MFC is affected, therefore an MFC could serve as an early toxicity-warning device. Recently, MFC based biosensor array was designed by Kaur et al. [116] to measure the presence of different volatile fatty acids such as acetate, propionate and butyrate concentrations with sensitivity down to 5 mg l^{-1} and up to 40 mg l^{-1}.

4.2.12.3 Hydrogen Production

Hydrogen production by electrohydrogenesis in MFC is a new method for generating hydrogen from acetate and other fermentable products [31, 117–119]. In a microbial electrolysis cell (MEC), exoelectrogens oxidize a substrate and release electrons to the anode. When acetate is used as a substrate, a voltage of >0.2 V is applied for hydrogen evolution [120, 121]. This voltage is substantially less than the 1.8–2.0 V used in practice for hydrogen production via water electrolysis under alkaline conditions [122]. In a similar way, a DC Voltage (0.8 V) was supplied to the single chambered MEC using glucose, cumulative hydrogen production for a 5-day cycle of operation was 0.16 m^3 H_2/ day [123]. But the hydrogen production in MEC was instantaneously consumed by methanogens present in the system [117, 119], which is one of the major obstacles.

4.2.12.4 Powering Gadgets

MFCs are especially suitable for powering small telemetry systems and wireless sensors that have only low power requirements to transmit signals to receivers in remote locations [3, 124]. Realistic energetically autonomous robots could be equipped with MFCs that use different fuels like sugar, fruit, dead insects, grass and weed. The robot EcoBot-II solely powers itself by MFCs to perform some behavior including motion sensing, computing and communication [125–127]. Some scientists envision that in the future a miniature MFC can be implanted in a human body to power an implantable medical device with the nutrients supplied by the human body [128]. However, only after potential health and safety concerns related to the biocatalyst in the MFC are thoroughly understood, it could be widely deployed.

4.3 OUTLOOK

The potential of MFC has not been exploited for any specific application till date. A small but important subset of the parameters influences MFC performance: system architecture, the electrode and membrane characteristics, the roles of the microorganisms, and the need for mediators. From a practical viewpoint there have been a number of breakthroughs made in the recent years, some of which are listed below:

   a. High efficiency electrodes used in conventional fuel cells such as metal coatings, carbon-polymer composites and noble-metal catalyst although are commercially available are not being used in MFCs. These electrodes were not biocompatible and are susceptible to catalyst poisoning when industrial effluents/aquatic sediments were used as a fuel. This may be overcome by the development of anodes with enhanced active surface area, conductive surface and redox behavior suited to microbial metabolism and conducive to biofilm formation [1, 51, 66].

   b. EAB found in the diverse environments, understanding their metabolic impediments and physiology towards extracellular electron transfer can pave the way for significant performance improvement. The discovery of these electrochemically active bacteria/self-mediating bacteria may overcome requirements such as the need for synthetic mediator in MFC [5–7].

   c. Recent advances in MFC reactor design emphasize cost effectiveness and scalability. Some new designs are very unique such as the development of single chambered MFC without PEM (reducing the cost), stacked MFC module (gain of voltage) and rotating cathode (increasing the oxygen reduction rate) [19–22].

   d. Potassium ferricyanide is widely used as the electron acceptor in MFC. However, it requires the continuous replenishment over the cycle. Instead of potassium ferricyanide, some researchers claim that azo dyes posses high redox potential so that the azo dye could be used in the cathode with the dual objective of solving as facilitator for electron acceptor and the possibility of dye removal. These studies have suggested that the technology in the dual duty of degrading the azo dyes and simultaneous power production, by adapting the MFC for treating textile effluents could be exploited profitably. But, significant technical improvements are needed to scale up the process for addressing the industrial needs in a cost-effective way [67, 106–108].

e. The use of EAB to produce other products at cathode such as hydrogen, methane, hydrogen peroxide or ethanol by applying the voltage/ alter the design of MFC [119].

f. Some examples of commercial effort to employ MFCs are Trophos Energy (USA), Lebone (USA), IntAct Labs LLC, (USA), HySyEnce (USA), Plant-e (Netherlands) and Emefcy (Israel).

KEYWORDS

- Anion exchange membrane
- Carbon nanotubes
- Chemical vapor deposition
- Chemically modified carbon electrodes
- Cobalt tetramethylphenylporphyrin
- Dissimilatory metal reducing bacteria
- Electrochemically active bacterial
- Microbial fuel cell
- Proton exchange membrane
- Reticulated vitreous carbon
- Ultra filtration

REFERENCES

1. Park, D. H., Kim, S. K., Shin, I. H., Jeong, Y. J. (2000). Electricity production in biofuel cell using modified graphite electrode with Neutral Red.*Biotechnol.Lett. 22,* 1301–1304.

2. Tokuji, I., Kenji, K. (2003). Bioelectrocatalyzes-based application of quinoproteins and quinprotein- containing bacterial cells in biosensors and biofuel cells. *Biochimic. Biophys.Acta.* 1647, 121–126.

3. Ieropoulos, I. A., Greenman J., Melhuish C., Hart J. (2005). Comparative study of three types of microbial fuel cell. *Enzyme.Microb. Tech. 37,* 238–245.

4. Vega, C. A., Fernandez, I. (1987). Mediating effect of ferric chelate compounds in microbial fuel cells with *Lactobacillus plantarum, Streptococcus lactis and Erwiniadissolven. Bioelectrochem.Bioenerg.* 17, 217–222.

5. Bond, D. R., Lovley, D. R. (2003). Electricity production by *Geobacter sulfurreducens* attached to electrodes. *Appl. Environ. Microbiol. 69,* 1548–1555.

6. Chaudhuri, S. K., Lovley, D. R. (*2003)*.Electricity generation by direct oxidation of glucose in mediatorless microbial fuel cells. *Nat. Biotechnol.* 21, *1229–1232.*

7. Kim, H. J., Park, H. S., Hyun, M. S., Chang, I. S., Kim, M., Kim, B. H. (2002). A mediator-less microbial fuel cell using a metal reducing bacterium, *Shewanella putrefacians. Enzyme. Microbiol. Tech. 30,* 145–152.

8. Pham, C. A., Jung, S. J., Phung, N. T., Lee J., Chang, I. S., Kim, B. H., Yi, H., Chun, J. (2003). A novel electrochemically active and Fe (III)-reducing bacterium phylogenetically related to *Aeromonas hydrophila*, isolated from a microbial fuel cell. *FEMS Microbiol.Lett. 223,* 129–134.

9. Rabaey, K., Boon, N., Siciliano, S. D., Verhaege, M., Verstraete, W. (2004).Biofuel cells select for microbial consortia that self-mediate electron transfer. *App. Environ. Microbiol.* 70, 5373–5382.

10. Bard, A. J., Parsons, R., Jordan, J. Standard Potentials in Aqueous Solutions; Marcel Dekker: New York, 1985.

11. Alberty, R. A. Thermodynamics of Biochemical Reactions; John Wiley & Sons: New York, 2003.

12. Katz, E., Shipway, A. N., Willner, I. (2003). Biofuel cells: Functional design and operation. In Handbook of Fuel Cells – Fundamental, Technology, Applications; Vielstich, W., Gastieger, H., Lamin, A., Eds., Wiley: Chichester, Vol.1; 355–381.

13. Logan, B. E., Hamelers, B., Rozendal, R., Schroder, U., Keller, J., Freguia, S., Aelterman, P., Verstraete, W., Rabaey, K. (2006). Microbial fuel cells: Methodology and technology. *Environ. Sci. Technol.* 40, 5181–5192.

14. Larminie, J., Dicks, A. Fuel cell systems explained, John Wiley & Sons, Chichester. 2000.

15. Bockris, J. O. M., Reddy, A. K. N., Gamboa, A. M. Modern Electrochemistry: Fundamentals of Electrodics. Kluwer Academic Publishers: New York, 2000.

16. Hoogers, G. Ed., Fuel Cell Technology Handbook, CRC Press: Boca Raton, FL, 2003.

17. Liu, H., Ramnarayanan, R., Logan, B. E. (2004). Production of electricity during wastewater treatment using a single chamber microbial fuel cell. *Environ. Sci. Technol. 38,* 2281–2285.

18. Min, B., Logan, B. E. (2004). Continuous Electricity Generation from Domestic Wastewater and Organic Substrates in a Flat Plate Microbial Fuel Cell. *Environ. Sci. Technol. 38,* 5809–5814.

19. Liu, H., Logan, B. E. (2004). Electricity generation using an air-cathode single chamber microbial fuel cell in the presence and absence of a proton exchange membrane. *Environ. Sci. Technol. 38,* 4040–4046.

20. Jang, J. K., Pham, T. H., Chang, I. S., Kang, K. H., Moon, H., Cho, K. S., Kim, B. H. (2004). Construction and operation of a novel mediator-and membrane-less microbial fuel cell. *Process.Biochem. 39,* 1007–1012.

21. He, Z., Shao, H., Angenent, L. T. (2007). Increased power production from a sediment microbial fuel cell with a rotating cathode. *Biosens.Bioelectron. 22,* 3252–3255.

22. Aelterman, P., Rabaey, K., Pham, H. T., Boon, N., Verstraete, W. (2006). Continuous electricity generation at high voltages and currents using stacked microbial fuel cells. *Environ. Sci Technol. 40,* 3388–3394.

23. Oh, S. E., Logan, B. E. (2007). Voltage reversal during microbial fuel cell stack operation. *J.Power. Sources. 167,* 11–17.

24. Jayapriya, J., Ramamurthy, V. (2012). Use of nonnative phenazines improve performance of *Pseudomonas aeruginosa* MTCC 2474 catalyzed fuel cells. *Bioresour. Technol. 124,* 23–28.

25. Park, D. H., Zeikus, J. G. (2003). Improved fuel cell and electrode designs for producing electricity from microbial degradation. *Biotechnol.Bioeng. 81,* 348–355.

26. Gil, G. C., Chang, I. S., Kim, B. H., Kim, M., Jang, J. K., Park, H. S., Kim, H. J. (2003). Operational parameters affecting the performance of a mediator-less microbial fuel cell. *Biosens.Bioelectron. 18,* 327–334.

27. Kim, B. H., Kim, H. J., Hyun, M. S., Park, D. H. (1999). Direct electrode reaction of Fe (III)-reducing bacterium, *Shewanella putrefaciensJ. Microbiol.Biotechnol. 9,* 127–131.

28. Okada, T., Nakamura. N., Yuasa, M., SekineI. I. (1997). Ion and water transport characteristics in membranes for polymer electrolytefuel cells containing H^+ and Ca^{2+} cations. *J. Electrochem. Soc, 144,* 2744–2750.

29. Basura, V. I., Beattie, P. D., Holdcroft, S. (1998). Solid-state electrochemical oxygen reduction at Pt/Nafions 117 and Pt/BAM3GTM 407 interfaces. *J. Electroanal. Chem. 458,* 1–5.

30. Kim, J., Cheng, S. A., Oh, S., Logan, B. E. (2007). Power generation using different cation, anion, and ultrafiltration membranes in microbial fuel cells. *Environ. Sci. Technol. 41,* 1004–1009.

31. Rozendal, R. A., Hamelers, H. V. M., Buisman, C. J. N. (2006). Effects of membrane cation transport on pH and microbial fuel cell performance. *Environ. Sci. Technol. 40,* 5206–5211.

32. Chae, K. J., Choi, M., Ajayi, F. F., Park, W., Chang, I. S., Kim, I. S. (2008). Mass transport through a proton exchange membrane (Nafion) in microbial fuel cells. *Energy Fuels 22,* 169–176.

33. Xu, J., Sheng, G. P., Luob, H. W., Lib, W. W., Wang, L. F., Yub, H. Q. (2012). Fouling of proton exchange membrane (PEM) deteriorates the performance of microbial fuel cell, *Water Research 46,* 1817–1824.

34. Kalcher, K., Svancara, I., Metelka, R., Vytras K., Walcarius. A. (2006). Heterogeneous Carbon Electrochemical Sensors. In Encyclopedia of Sensors; Craig A. Grimes, Elizabeth C. Dickey, and Michael V. Pishko Eds., American Scientific Publishers, Stevenson Ranch, California, Vol. 4, pp. 283–430.

35. Uslu, B., Sibel, A. O. (2007). Electroanalytical application of carbon based electrodes to the Pharmaceuticals. *Anal.Lett. 40*, 817–853.

36. Greenville, W. A., Kintner, P. L. (1969). Carbon: observations on the new allotropic form. *Science 165*, 589–591.

37. Cheng, S., Logan, B. E. (2007). Ammonia treatment of carbon cloth anodes to enhance power generation of microbial fuel cells. *Electrochem.Commun. 9*, 492–496.

38. Wang, X., Feng, Y. J., Lee, H. (2008). Electricity production from beer brewery wastewater using single chamber microbial fuel cell. *Water Sci. Technol. 57*, 1117–1121.

39. Wang, X., Cheng, S., Feng, Y., Merrill, M. D., Saito, T., Logan, B. E. (2009). Use of carbon mesh anodes and the effect of different pretreatment methods on power production in microbial fuel cells. *Environ. Sci. Technol. 43*, 6870–6874.

40. Reimers, C. E., Tender, L. M., Fertig, S. J., Wang, W. (2001). Harvesting energy from the marine sediment-water interface. *Environ. Sci. Technol. 35*, 192–195.

41. Mohan, S. V., Saravanan, R., Raghavulu, S. V., Mohanakrishna, G., Sarma, P. N. (2008). Bioelectricity production from wastewater treatment in dual chambered microbial fuel cell (MFC) using selectively enriched mixed microflora: Effect of catholyte. *Bioresour Technol. 99*, 596–603.

42. Logan, B., Cheng, S., Watson, V., Estadt, G. (2007).Graphite fiber brush anodes for increased power production in air-cathode microbial fuel cells. *Environ. Sci. Technol. 41*, 3341–3346.

43. Ahn, Y., Logan, B. E. (2010). Effectiveness of domestic wastewater treatment using microbial fuel cells at ambient and mesophilic temperatures, *Bioresource. Technol. 101*, 469–475.

44. Rabaey, K., Ossieur, W., Verhaege, M., Verstraete, W. (2005). Continuous microbial fuel cells convert carbohydrates to electricity. *Water. Sci. Technol. 52*, 515–523.

45. You, S. J., Zhao, Q. L., Zhang, J. N., Jiang, J. Q., Wan, C. L., Du, M. A., Zhao, S. Q. (2007). A graphite-granule membrane-less tubular air-cathode microbial fuel cell for power generation under continuously operational conditions. *J. Power. Sources 173*, 172–177.

46. He, Z., Minteer, S. D., Angenent, L. T. (2005).Electricity generation from artificial wastewater using an upflow microbial fuel cell. *Environ. Sci. Technol. 39*, 5262–5267.

47. Ringeisen, B. R., Henderson, E., Wu, P. K., Pietron, J., Ray, R., Little, B., Biffinger, J. C., Jones-Meehan, J. M. (2006). High power density from a miniature microbial fuel cell using *Shewanella oneidensis* DSP10. *Environ Sci Technol. 40*, 2629–2634.

48. Magrez, A., Kasas, S., Salicio, V., Pasquier, N., Seo, J. W., Celio, M., Catsicas, S., Schwaller, B., Forro, L. (2006). Cellular toxicity of carbon-based nanomaterial. *Nano.Lett. 6*, 1121–1125.

49. Wu, Z., Feng, W., Feng, Y., Qiang Liu, Xu, X., Sekino, T., Fujii, A., Ozaki, M. (2007). Preparation and characterization of chitosan-grafted multiwalled carbon nanotubes and their electrochemical properties. *Carbon 45*, 1212–1218.

50. Qiao, Y., Li, C. M., Bao, S. J., Bao, Q. L. (2007). Carbon Nanotube/Polyaniline Composite as Anode Material for Microbial Fuel Cells. *J. Power. Sources 170*, 79–84.

51. Schroder, U., Niessen, J., Scholz, J. (2003). A generation of microbial fuel cells with current outputs boosted by more than one order of magnitude. *Angew. Chem. Int. Ed. 42*, 2880–2883.

52. Sahoo, N. G., Jung, Y. C., So, H. H., Cho, J. W. (2007). Polypyrrole coated carbon nanotubes: synthesis, characterization, and enhanced electrical properties. *Synthetic Metals 157*, 374–379.

53. Yuan, Y., Kim, S. (2008). Improved Performance of a Microbial Fuel Cell with Poly-pyrrole/Carbon Black Composite Coated Carbon Paper Anodes. *B. Kor. Chem. Soc. 29*, 1344–1348.

54. Sharma, T., Reddy, A. L. M., Chandra, T. S., Ramaprabhu, S. (2008). Devolopement of carbon nanotubes and nanofluids based microbial fuel cell. *Int. J. Hydrogen Energy 33*, 6749–6754.

55. Donaldson, K., Aitken, R., Tran, L., Stone, V., Duffin, R., Forrest, G., Alexander, A. (2006). Carbon nanotubes: A review of their properties in relation to pulmonary toxicology and workplace safety. *Toxicol. Sci. 92*, 5–22.

56. Upadhyayula, V. K. K., Gadhamshetty, V. (2010). Appreciating the role of carbon nanotubes in preventing biofouling and promoting biofilms on material surfaces in environmental engineering: review. *Biotechnol.Adv. 28*, 802–816.

57. Esawi, A. M. K., Farag, M. M. (2007). Carbon nanotube reinforced composites: Potential and current challenges. *Materials Design 28*, 2394–2401.

58. Dumas, C., Mollica, A., Féron, D., Basséguy, R., Etcheverry, L., Bergel, A. (2007). Marine microbial fuel cell: Use of stainless steel electrodes as anode and cathode materials. *Electrochim.Acta. 53*, 468–473.

59. Richter, H., McCarthy, K., Nevin, K. P., Johnson, J. P., Rotello, V. M., Lovley, D. R. (2008). Electricity generation by *Geobacter sulfurreducens* attached to gold electrodes, *Langmuir. 24*, 4376 – 4379.

60. Heijne, A. T., Hamelers, H. V. M., Saakes, M., Buisman, C. J. N. (2008). Performance of nonporous graphite and titanium-based anodes in microbial fuel cells. *Electrochim.Acta. 53*, 5697–5703.

61. Moon, H., Chang, I. S., Kim, B. H. (2006). Continuous electricity production from artificial wastewater using a mediator-less microbial fuel cell. *Bioresour. Technol. 97*, 621–627.

62. Zhao, F., Harnisch, F., Schroder, U., Scholz, F., Bogdanoff, P., Herrmann, I. (2005). Application of pyrolyzed iron (II) phthalocyanine and CoTMPP based oxygen reduction catalysts as cathode materials in microbial fuel cells, *Electrochem. Commun. 7*, 1405–1410.

63. Cheng, S., Liu, H., Logan, B. E. (2006). Power Densities Using Different Cathode Catalysts (Pt and CoTMPP) and Polymer Binders (Nafion and PTFE) in Single Chamber Microbial Fuel Cells. *Environ.Sci. Technol. 40*, 364–369.

64. Freguia, S., Rabaey, K., Yuan, Z., Keller. J. (2007). Non-catalyzed cathodic oxygen reduction at graphite granules in microbial fuel cells.*Electrochim.Acta. 53*, 598–603.

65. Zhou, M., Chi, M., Luo, J., He, H., Jin, T. (2011). An overview of electrode materials in microbial fuel cells.*J. Power Sources 196*, 4427–4435.

66. Jayapriya, J., Gopal. J., Ramamurthy, V., Kamachimudali, U., Raj, B. (2012). Preparation and characterization of biocompatible carbon electrodes. *Composites Part B-Eng. 43*, 1329–1335.

67. Jayapriya, J., Ramamurthy, V. (2014). The Role of Electrode Material in Capturing Power Generated in *Pseudomonas* catalyzed fuel Cells. *Can. J. Chem. Eng. 92,* 610–614.

68. Erable, B., Duteanu, N., Kumar, S. M. S., Feng, Y., Ghangrekar, M. M., Scott, K. (2009). Nitric acid activation of graphite granules to increase the performance of the noncatalyzed oxygen reduction reaction (ORR) for MFC applications. *Electrochem. Commun. 11,* 1547–1549.

69. Feng, Y., Yang, Q., Wang, X., Logan, B. E. (2010). Treatment of graphite fiber brush anodes for improving power generation in air-cathode microbial fuel cells. J. Power. Sources. *195,* 1841–1844.

70. Lowy, D. A., Tender, L. M., Zeikus, J. G., Park, D. H., Lovley, D. R. (2006). Harvesting energy from the marine sediment-water interface II Kinetic activity of anode materials. *Biosens.Bioelectron. 21,* 2058–2063.

71. Holmes, D. E., Bond, D. R., O'Neil, R. A., Reimers, C. E., Tender, L. R., Lovley, D. R. *(2004).Microbial communities associated with electrodes harvesting electricity from a variety of aquatic sediments.Microbial Ecol. 48,* 178–190.

72. Park, H. S., Kim, B. H., Kim, H. S., Kim, H. S., Kim, G. T., Kim, M., Chang, I. S., Park, Y. K., Chang, H. I. (2001). A novel electrochemically active and Fe(III)-reducing bacterium phylogenetically related to *Clostridium butyricum* isolated from a microbial fuel cell. *Anaerobe 7,* 297–306.

73. Prasad, D., Sivaram, T. K., Berchmans, S., Yegnaraman, V. (2006). Microbial fuel cell constructed with a microorganism isolated from sugar industry effluent. *J. Power. Sources 160,* 991–996.

74. Reguera, G., Nevin, K. P., Nicoll, J. S., Covalla, S. F., Woodard, T., L, Lovley, D. R. (2006). Biofilm and nanowire production leads to increased current in *Geobacter sulfurreducens* fuel cells. *Appl. Environ. Microbiol. 72,* 7345–7348.

75. Kim, J. R., Jung, S. H., Regan, J.M; Logan, B. E. (2007b). Electricity generation and microbial community analysis of alcohol powered microbial fuel cells. *Bioresour. Technol. 98,* 2568–2577.

76. Niessen, J., Harnisch, F., Rosenbaum, M., Schroder, U., Scholz, F. (2006). Heat treated soil as convenient and versatile source of bacterial communities for microbial electricity generation. *Electrochem.Commun. 8,* 869–873.

77. Zhang, E., Xu, W., Diao. G., Shuang, C. (2006). Electricity generation from acetate and glucose by sedimentary bacterium attached to electrode in microbial-anode fuel cells. *J. Power. Sources 161,* 820–825.

78. Reguera, G., McCarthy, K. D., Mehta, T., Nicoll, J. S., Tuominen. M. T., Lovley, D. R. (2005). Extracellular electron transfer via microbial nanowires. *Nature 435,* 1098–1101.

79. Gorby, Y. A., Yanina, S. McLean, J. S., Rosso, K. M., Moyles, D., Dohnalkova, A., Beveridge, T. J., Chang, I. S., Kim, B. H., Kim, K. S., Culley, D. E., Reed, S. B., Romine, M. F., Saffarini, D. A., Hill, E. A., Shi, L., Elias, D. A., Kennedy, D. W., Pinchuk, G., Watanabe, K., Ishii, S., Logan, B., Nealson, K.H; Fredrickson, J. K. (2006). Electrically conductive bacterial nanowires produced by *Shewanella oneidensis* strain MR-1 and other microorganisms. *Proc. Natl. Acad. Sci. USA 103,* 11358–11363.

80. Malvankar, N.S., Vargas, M., Nevin K. P., Franks, A. E., Leang, C., Kim, B. C., In-oue, K., Mester, T., Sean F. Covalla, S. F., Johnson, J. P., Rotello, V. M., Tuominen, M. T., Lovley, D. R. (2011). Tunable metallic-like conductivity in microbial nanow-ire networks. *Nature Nanotechnol. 6*, 573–579.

81. Wilkinson, S. (2000). Gastrobots – Benefits and challenges of microbial fuel cells in food powered robot applications. *Auton. Robot 9*, 99–111.

82. Dealney, G. M., Bennetto, H. P., Mason, J. R., Roller, S. D., Stirling, J. L., Thurston, C. F. (*1984*).Electron-transfer coupling in microbial fuel cells. 2. Performance of fuel cells containing selected microorganism-mediator-substrate combinations. *J. Chem. Technol. Biotechnol. 34*, 13–27.

83. Bennetto, H. P., Delaney, G. M., Mason, J. R., Roller, S. D., Stirling, J. L., Thurston, C. F. (*1985*).The sucrose fuel cell: Efficient biomass conversion using a microbial catalyst. *Biotechnol. Lett. 7*, 699–704.

84. Allen, R. M., Bennetto, H. P. (1993).Microbial fuel cells: Electricity production from carbohydrates. *Appl. Biochem. Biotechnol. 39*, 27–40.

85. Tanaka, K., Kashiwagi, N., Ogawa, T. (*1988*). Effects of light on the electrical output of bioelectrochemical fuel cells containing *Anabaena variabilis* M-2: mechanisms of the postillumination burst. *Chem. Technol. Biotechnol. 42*, 235–240.

86. Park, D. H., Kim, B. H., Moore, B., Hill, H. A. O., Song, M. K., Rhee, H. W. (1997). Electrode reaction of *Desulfovibrio desulfuricans* modified with organic conductive compounds. *Biotechnol.Tech. 11*, 145–148.

87. Kim, N., Choi, Y., Jung, S., Kim, S. (2000). Effect of initial carbon sources on the performance of microbial fuel cells containing *Proteus vulgaris. Biotechnol.Bioeng. 70*, 109–114.

88. Marsili, E., Baron, D, B., Shikhare, I. D., Coursolle, D., Gralnick, J. A., Bond, D. R. (2008). Shewanella secretes flavins that mediate extracellular electron transfer. *Proc. Natl. Acad. Sci. USA 105*, 3968–3973.

89. Von Canstein, H., Ogawa, J., Shimizu, S., Lloyd, J. R. (2008). Secretion of flavins by Shewanella species and their role in extracellular electron transfer. *Appl. Environ. Microbiol. 74*, 615–623.

90. Rabaey, K., Boon, N., Hofte, M., Verstraete, W. (2005a).Microbial phenazine pro-duction enhances electron transfer in biofuel cells. *Environ. Sci. Technol. 39*, 3401–3408.

91. Chin-A-W., Bloemberg, G. V., Lugtenberg, B. J. J. (*2003*). Phenazines and their role in biocontrol by *Pseudomonas* bacteria. *New.Phytol. 157*, 503–523.

92. Mavrodi, D. V., Blankenfeldt, W., Thomashow, L. S. (2006). Phenazine compounds in fluorescent *Pseudomonas* spp. Biosynthesis and regulation. *Annu. Rev. Phyto-pathol. 44*, 417–445.

93. Mavrodi, D. V., Bonsall, R. F., Delaney, S. M., Soule, M. J., Phillips, G., Thom-ashow, L. S. (2001). Functional analysis of genes for biosynthesis of pyocyanin and phenazine-1-carboxamide from *Pseudomonas aeruginosa* PAO1. *J. Bacteriol. 183*, 6454–6465.

94. Schroder, U. Anodic electron transfer mechanisms in microbial fuel cells and their energy efficiency, *Phys. Chem. Chem. Phys.* (2007). *9*, 2619–2629.

95. Dietrich, L. E. P., Teal, T. K., Whelan A. P., Newman, D. K. (2008). Redox-active antibiotics control gene expression and community behavior in divergent bacteria. *Science 321,* 1203–1206.

96. Maddula, V. S., Pierson, E. A., Pierson, L. S. (2008). Altering the ratio of phenazines in *pseudomonas chlororaphis* (aureofaciens) strain 30–84: effects on biofilm formation and pathogen inhibition. *J. Bacteriol. 190,* 2759–2766.

97. Habermann, W. and Pommer, E. H. (1991). Biological fuel cells with sulfide storage capacity. *Appl. Microbiol. Biotechnol. 35,* 128–133.

98. Holzman, D. C. (2005).Microbe power. *Environ. Health.Persp. 113,* A754–A757.

99. Suzuki, S., Karube, I., Matsunaga, T. (1978).Application of a biochemical fuel cell to wastewater. *Biotechnol.Bioeng.Symp. 8,* 501–511.

100. Oh, S. E., Logan, B. E. (2005). Hydrogen and electricity production from a food processing wastewater using fermentation and microbial fuel cell technologies. *Water Res. 39,* 4673–4682.

101. Min, B., Kim, J. R., Oh, S. E., Regan, J. M., Logan, B. E. (2005). Electricity generation from swine wastewater using microbial fuel cells. *Water Res. 39,* 4961–4968.

102. Zuo, Y., Maness, P. C., Logan, B. E. (2006). Electricity production from steam exploded corn stover biomass. *Energy Fuels 20,* 1716–1721.

103. Feng Y., Wang X., Logan B. E., Lee, H. (2008). Brewery wastewater treatment using aircathode microbial fuel cells. *Appl. Microbiol. Biotechnol. 78,* 873–880.

104. Huang, L., Logan B. E. (2008). Electricity generation and treatment of paper recycling wastewater using a microbial fuel cell. *Appl. Microbiol. Biotechnol. 80,* 349–355.

105. Huang, J. S., Yang, P., Guo, Y., Zhang, K. (2011). Electricity generation during wastewater treatment: An approach using an AFB-MFC for alcohol distilery wastewater. *Desalination 276,* 373–378.

106. Niu, C. G., Wang, Y., Zhang, X. G., Zeng, G. M., Huang, D. W., Ruan, M., Li, X. W. (2012). Decolorization of an azo dye Orange G in microbial fuel cells using Fe (II)-EDTA catalyzed persulfate. *Bioresource Technol. 126,* 101–106.

107. Hou, B., Sun, J., Hu, Y. (2011). Simultaneous Congo red decolorization and electricity generation in air-cathode single-chamber microbial fuel cell with different microfiltration, ultrafiltration and proton exchange membranes. *Bioresource Technol. 102,* 4433–4438.

108. Liu, L., Li, F. B., Feng, C. H., Li, X. Z. (2009). Microbial fuel cell with an azo-dye-feeding cathode. *Appl. Microbiol. Biotechnol. 85,* 175–183.

109. Goyal, R. N., Minocha, A. (1985). Electrochemical behavior of the bisazo dye, direct red-81. *J. Electroanal. Chem. Interfacial.Electrochem. 193,* 231–240.

110. Menek, N., Karaman, Y. (2005). Polarographic and voltammetric investigation of 8-hydroxy-7- (4-sulfo-1-naphthylazo) -5-quinoline sulfonic acid. *Dyes Pigments 67,* 9–14.

111. Lovely, D. R. Microbial fuel cells: novel microbial physiologies and engineering approaches. Biotechnol. (2006). *17,* 327–332.

112. Chang, I. S., Moon, H., Jang, J. K., Kim, B. H. (2005). Improvement of a microbial fuel cell performance as a BOD sensor using respiratory inhibitors. *Biosens. Bioelectron. 20,* 1856–1859.

113. APHA, Standard Methods for the Examination of Water and Wastewater, American Public Health Association, Washington, DC, USA, 1995.

114. Chang, I. S., Jang, J. K., Gil, G. C., Kim, M., Kim, H.J; Cho, B. W., Kim, B. H. (2004). Continuous determination of biochemical oxygen demand using microbial fuel cell type biosensor. *Biosens. Bioelectron.* 19, 607–613.

115. Shen, Y. J., Lefebvre, O., Tan, Z., Ng, H. Y. (2012). Microbial fuel-cell-based toxicity sensor for fast monitoring of acidic toxicity. *Water Sci. Technol.* 65, 1223–1228.

116. Kaur, A., Kim, J.R, ; Michie, I, ; Dinsdale, R.M, ; Guwy, A. J., Premier, G. C. Microbial fuel cell type biosensor for specific volatile fatty acids using acclimated bacterial communities. *Biosens.Bioelectron.* (2013). 47, 50–55.

117. Liu, H., Grot, S., Logan, B. E. (2005). Electrochemically assisted microbial production of hydrogen from acetate. *Environ. Sci. Technol.* 39, 4317–4320.

118. Rozendal, R. A. and Buisman, C. J. N. Bio-Electrochemical Process for Producing Hydrogen. Patent WO-2005–005981, 2005.

119. Call, D., Logan, B. E. (2008). Hydrogen production in a single chamber microbial electrolysis cell lacking a membrane. *Environ. Sci. Technol.* 42, 3401–3406.

120. Cheng, S., Logan, B. E. (2007). Sustainable and efficient biohydrogen production via electrohydrogenesis. *Proc. Natl. Acad. Sci. USA 104,* 18871–18873.

121. Hu, H., Fan, Y., Liu, H. (2008). Hydrogen Production Using Single-chamber Membrane-free Microbial Electrolysis Cells. *Water Res.* 42, 4172–4178.

122. Kinoshita, K. Electrochemical oxygen technology, Wiley, New York. 1992.

123. Gupta, P., Parkhey, P., Joshi, K., Mahilkar, A. (2013). Design of a microbial fuel cell and its transition to microbial electrolytic cell for hydrogen production by electrohydrogenisis. *Indian J Exp Biol. 51,* 860–865.

124. Shantaram, A., Beyenal, H., Veluchamy R. R. A., Lewandowski, Z. (2005). Wireless sensors powered by microbial fuel cells. *Environ. Sci. Technol.* 39, 5037–5042.

125. Ieropoulos, I., Melhuish, C., Greenman, J. (2005). EcoBot-II: An artificial agent with a natural metabolism. *Adv. Robot Syst. 2,* 295–300.

126. Ieropoulos, I., Melhuish, C., Greenman, J. (2004). Energetically autonomous robots. In: Intelligent autonomous systems, Groen F, et al., Eds. Amsterdam: IOS Press, Vol. 8, 128–135.

127. Melhuish, C., Ieropoulos, I., Greenman, J., Horsfield, I. (2006). Energetically autonomous robots: Food for thought. *Auton.Robot. 21,* 187–198.

128. Chiao, M. A miniaturized microbial fuel cell. In Technical digest of solid state sensors and actuators workshop.Hilton Head Island, 59–60, 2002.

CHAPTER 5

SYSTEMS BIOLOGY APPROACHES FOR MICROBIAL FUEL CELL APPLICATIONS

R. NAVANIETHA KRISHNARAJ,[1] and JONG-SUNG YU[2]

[1]R.Navanietha Krishnaraj, National Institute of Technology Durgapur, India

[2]Department of Advanced Materials Chemistry, Korea University, Korea

CONTENTS

ABSTRACT

Microbial fuel cells are ecofriendly electrochemical devices that help to transform the chemical energy from the organic materials into electricity with the help of the catalytic activity of the microorganisms. Numerous reports are available in the literature to improve the bioelectricity generation in Microbial fuel cells. Currently, few reports are available on the use of molecular biology and genetic engineering approaches for improving the catalytic activity and electron transfer characteristics of the microorganisms. However, most of these approaches were made based on the reductionist approach and hence they could not drastically enhance the yield of microbial fuel cells. System biology approaches seems to be a promising option for enhancing the power output of microbial fuel cells. Herein, the applications of system biology approaches for screening the electrochemically active bacteria are discussed. The chapter also discusses about the use of system biology techniques for understanding the mechanisms of electron transfer from the microorganism to the electrode. The chapter also throws light on the novel metabolic approaches for enhancing the power output of microbial fuel cells. The advantages of systems biology approaches for MFC applications and the limitations of the conventional reductionist approaches are also discussed in detail.

5.1   INTRODUCTION

Microbial fuel cells (MFCs) are devices that perform oxidative disintegration of the organic compounds with the inbuilt complex enzymatic machinery of the microorganisms thereby producing reduction equivalents for the generation of bioelectricity [1]. MFCs have several advantages over the conventional fuel cells such as its temperature of operation, pH conditions, ecofriendly nature, ease of operation and high conversion efficiency [2, 3]. Besides electricity generation, MFCs also aid in wastewater treatment, cellulose degradation, desalination, dye degradation, toxin removal and production of chemicals [4–8]. McCormick et al. [9] reported a photosynthetic MFC, which can generate electricity to run clock upon illumination with light. Efforts are also made widely with a goal of using MFCs to low power devices, charging mobile phones and light. MFCs find its application in powering autonomous robots,which are used to monitor or perform any function where humans cannot perform such as in deep sea or space [10]. MFCs can also be used to power sensors and other electronic devices [11]. MFCs can provide electricity to wireless sensors in remote locations while requiring less maintenance and reducing overall costs [12]. Biohydrogen production can also be coupled with the production of electricity in a MFC. MFCs can also be used to produce methane, and other valuable inorganic and organic chemicals besides electricity [7]. MFCs can also perform desalination process along with bioelectricity production [13]. MFCs can be integrated with other existing wastewater treatment processes. A novel up-flow anaerobic sludge blanket reactor-microbial fuel cell-biological aerated filter system to dispose the complex molasses wastewater was reported was also reported in the literature [14].MFCs are also used to reduce toxins such as microcystins. Though the MFCs have several advantages, they suffer from low power output[15]. Research activities throughout the world are underway to improve the yield. Several attempts are made by the researchers to improve the power output of MFCs. Increasing the size of the MFCs did not greatly aid in increasing the power output because of the scaling up limitations [16, 17]. Several electrode materials such as carbon felt, graphite, carbon foam, carbon paper, nickel foam have been attempted to improve the power outputs [18, 19]. As some of these materials are toxic to microorganisms, approaches were then made to modify the electrodes with biocompatible materials [20, 21]. Electrodes modified with biopolymers such as chitosan, alginate were shown to enhance the power output ofMFCs. Nanomaterials such as nanoparticles, carbon nanotubes, graphene were also shown to improve the electron transfer properties in MFC [22–24]. Reports were also documented on the use of conducting polymers such as polyaniline, polytoluene to improve the microbial electrocatalysis.Different configurations of fuel cells were also reported to improve the yield [25]. Increasing the conductivity of the electrolyte was reported to improve the yields to a certain levels [26]. Effect of the composition of anolyte and catholyte was also explored [27]. However, most of the methods rely on the physical parameters such as configuration of MFC, electrode materials to improve the power output of the MFCs. The microorganisms (electrigens/electroactive microorganisms) are the key players in the microbial fuel cells [28, 29]. Hence, improving the electrocatalytic activity and electron transfer characteristics of the microorganism plays a significant role in enhancing the power output of MFCs. Currently, the metabolic engineering of electroactive bacteria based on reductionist approach is attempted and they could not greatly aid in improving the power output of MFCs significantly. Systems biology approach is the ideal tool to understand the complete molecular mechanism behind the generation of bioelectricity using microorganisms and will aid to enhance the power output of the MFC.

5.2 SYSTEMS BIOLOGY

Systems biology is often hailed as the new interdisciplinary approach for studying the complex interactions in biological systems (Fig. 5.1). However, this is not always true. The classical biologists took the entire characteristics of the organisms for their experiment and made an integrative understanding of diverse biological systems. They did not consider the expression level of any particular gene or a protein or its mRNA transcript. *Gregor Johann Mendel*, the father of genetics performed his experiments by studying the complete system of *Pisum sativum*. Reductionist approaches such as proteomics, genomics or transcriptomics were not used for his investigations, instead the traits such as plant height, seed

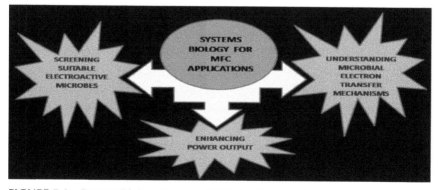

FIGURE 5.1   Systems biology for microbial fuel cell applications.

shape, pod shape, pod color, flower position, seed color and pea color was considered. Systems biology is the complete investigation on the behavior of complex biological organization and biochemical processes. It is just an opposite to that of reductionist approach. Reductionist approach studies the single/protein or a gene whereas the systems biology approach studies the expression different genes/proteins and the interactions between them. It considers the entirety of processes that happen in a biological system and helps to understand how exactly the system works. It is the detailed quantitative investigation on how the components of a biological system interact functionally with time. To understand the biology of the system, the features of the system as a whole have to be studied instead of study-ing the characteristics of the individual/isolated parts of the organisms. Systems biology aids in understanding the operation of the different com-ponents of the biological systems as a whole [30]. It is an emerging area and has the good potential to aid the researchers to understand the exact mechanisms of the biological systems. To enhance the yield of MFC or to identify a suitable microorganism for MFC application, one must under-stand the exact mechanism of oxidation/reduction of the substrate and the electron transfer characteristics of the microorganism. In the reductionist approach one take into account a single protein/gene. The biological sys-tem is very complex. Details about any one protein/gene might not give the complete picture of the electrocatalytic/electron transfer reactions. The mathematical modeling approaches and computation techniques will greatly aid in developing the models for different organisms/systems for diverse applications [31, 32]. Hence, an integrated approach such as sys-tems biology will help to clearly understand how the microorganisms are involved in oxidizing/reducing the substrates and how the microorganisms perform electron transfer reactions [33].

5.3   SYSTEMS BIOLOGY FOR UNDERSTANDING THE MECHANISM OF ELECTRON TRANSFER

Electrocatalytic activity and electron transfer characteristics are the key features of any electrochemically active microorganisms (Fig. 5.2). Hence understanding the complete mechanism of oxidation reaction in the case

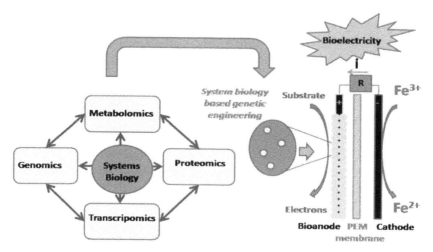

FIGURE 5.2 Developing genetic engineered microorganisms for microbial fuel cell applications.

of anodic biofilm or reduction reaction in the case of cathodic biofilm is essential. Microbial bioelectrocatalysis aids in the acceleration of the electrochemical reactions at the electrode-electrolyte interface with the aid of complex catalytic activity of the microorganisms. The electrochemical reaction aids in the transfer of electrons from the oxidation substrate to the electrode. Microorganisms differ greatly from the enzymes in their catalytic activity. Unlike enzymatic electrocatalysis that are very simple reactions, the microbial electrocatalytic reactions are very complex. The microorganisms can either perform the oxidation of substrates with a single enzyme or with the aid of a complex series of enzymes. For instance, let us look into how the microorganism catabolizes glucose. Some microorganisms are capable of producing the enzymes such as glucose oxidase or a glucose dehydrogenase. In such cases, whether the glucose oxidase or a glucose dehydrogenase enzyme are produced extracellularly or intracellularly and their catalytic rates must be investigated as these factors strongly influences the entire bioelectrochemical reactions. On the other hand, the microorganisms also have the inherent ability to perform the oxidation of glucose by the common glycolytic pathway, which involved ten different enzymes that works sequentially to finally form pyruvate.

Systems biology approaches using proteomic/genomic/transcriptomic/ metabolomic investigations might help to elucidate the mechanism of the microbial electrocatalytic reactions. There are certain enzymes such as PQQ glucose dehydrogenase that can perform both catalytic and electron transfer activities.

Investigations are also required to understand the mechanism electron transfer of the microorganisms. Microorganisms in the anodic chamber of MFCs on oxidation of the substrate transfer electrons to the anode either with the help of conductive proteins or with the help of mediators. MFCs operating with the direct electron transfer mechanism of the microorganisms are termed as mediator less MFC whereas MFCs where the shuttling of electrons happens with the aid of self-produced/externally supplemented mediators are termed as mediator based MFCs. Pham et al. [34] reported a mediatorless MFC with *Aeromonas hydrophila* for electrocatalytic activity. Karthikeyan et al. [35] reported the electrocatalytic activity of *Acetobacter aceti* and *Gluconobacter roseus*. However, in most cases, the mechanism of electron transfer from these microorganisms to the electrode is unrevealed. Microorganisms that cannot adhere to the electrode surface cannot perform direct electron transfer. To circumvent the issue, electrode modification techniques are used. However, modified electrodes did not drastically enhance the power output. Investigations on the genes involved with biofilm formation with system biology approaches will help to engineer microorganisms to perform direct electron transfer reactions. Numerous reports are available in the literature on the self produced mediators of the microorganisms used in MFC.

Freguia et al. [36] reported the use of soluble quinone excreted mediator in *Lactococcus lactis*. Fang et al. [37] documented the release of mediator such as 2,6-di-tertbutyl-p-benzoquinon (2,6-DTBBQ), as an excreted redox mediator in*K. pneumoniae strains* L17 based MFC. Mediator based MFC were reported with microorganisms such as *Erwinia dissolven, Escherichia coli, Klebsiella pneumonia, Lactobacillus plantarum* and *Proteus mirabilis* [38–40]. There are few reports in the literature on the mechanism of electron transfer mechanisms of certain well known electrochemically active microorganisms.Geobacter was reported to transfer electrons with the help of pili and cytochromes. Pili are short filamentous appendages popularly called bacterial nanowires and they transfer the electrons with the help of pili protein. Several reports were made on the electron transfer characteristics of pili independent of c-type cytochromes. Controversial report on this was made by Liu et al. [41]. Systems biology approaches might aid in understanding the exact mechanisms. Cytochrome mediated electron transfer is reported in the case of shewanella. The microorganism might also perform direct electron transfer with the help of two different direct electron transfer mechanisms. If the microorganisms transfer the electrons to the anode by more than one electron transfer mechanism, then identifying the contribution by each mechanism will be interesting. However, such investigations are more complex and cumbersome.

The systems biology tools such as computational techniques, graph theory approaches, modeling methods will aid in developing the complete genetic/proteomic networks of the electrochemically active microorganisms of interest for investigations related electrocatalysis/electron transfer reactions. Systems biology might help to get the true and a complete picture about the electron transfer mechanisms of any microorganism of interest. System biology will enable to elucidate whether the microbe performs direct electron transfer or mediator based electron transfer or both. If the microbe is able to transfer electron directly without the help of mediators, then systems biology can be used to get the exact picture on how the direct electron transfer reaction happens in the microorganism. System biology can vividly show the role of conductive proteins of the electrochemically active bacteria such as pili or any other membrane proteins, which are involved in direct electron, transfer reactions. Systems biology approach can help to understand whether the electron transfer is possible with the help of two different types of proteins. If the electron transfer reaction can take place with the help of more than one type of protein or more than one type of mechanism, then systems biology is the perfect tool to give the details about the contribution of each type of mechanism. Another interesting point to be investigated is that how the electrocatalytic mechanisms are correlated with electron transfer reactions. Theoretically, higher electron transfer kinetics should obviously enhance the bioelectrocatalysis. Genetic maps will aid to investigate the interaction of the two different groups of genes or proteins involved in the catalysis and electron transfer reactions. But currently reductionist approaches are practiced to understand the molecular mechanisms of electron transfer. Hence when we try to increase the one of the protein involved in electron transfer, the expression levels of the other type of electron transfer protein may go down. Hence, it is important to know complete details of the electron transfer characteristics of the microorganism. Such investigations are more essential before we can go in for a genetic engineering or a metabolic engineering of the microorganisms towards enhancing microbial electrocatalysis for microbial fuel cell applications.

5.4 SYSTEMS BIOLOGY APPROACH FOR IMPROVING THE POWER OUTPUT OF MICROBIAL FUEL CELLS

Systems biology approach can also be a suitable option for enhancing the power output of microbial fuel cells. There are a few metabolic engineering approaches for enhancing the microbial electrocatalysis for microbial fuel cell applications reported in the literature [42–44]. The use of antimycin to inhibit the ubiquinone and showed the enhancement in the power output. However, the effect of antimycin on the expression levels of different redox enzymes or conductive proteins is not explored.If such detailed investigation is made, it could really take MFC's to real time applications [43]. Hence systems biology approaches will aid to enhance either the electrocatalytic or electron transfer characteristics or both without any negative impacts on any one of the characteristics. Similarly a report was made on increasing the intracellular releasable electrons by deleting the lactate dehydrogenase gene for enhancing the power output of microbial fuel cells [45]. The increase the ratio of NADH/ NAD+ will theoretically lead to detrimental effects on the microorganisms. However, such investigations were not made/is not reported in the literature. The reductionist approaches are widely used to enhance the power output in MFC's.The molecular mechanism of bioelectricity from electroactive microorganisms is a very complex process and it involves numerous transcription and translation processes each of which are interrelated with each other in a complicated fashion as a web. Hence, current strategies based on one or few proteins cannot aid in enhancing the power output significantly.

5.5   SYSTEMS BIOLOGY APPROACHES FOR SCREENING MICROORGANISM

There are several reports in the literature on identifying the suitable micro-organisms for microbial fuel cells based on their oxidation/reduction ability and their electron transfer characteristics [35] (Fig. 5.3). Currently the electrocatalytic activities of the microorganisms are analyzed widely using the electrochemical techniques such as cyclic voltammetry, amperometry, polarization studies, coulombic efficiency and impedance spectroscopy. Cyclic voltammetry (CV) is one of the most commonly used electroanalytical techniques to analyze the redox reactions of the microorganisms by applying the electrochemical potentials. The bioelectrochemical investigation of the biofilm/microorganism with the cyclic voltammetry aids to identify the reversible, quasi-reversible and irreversible electrochemical reactions. Other electrochemical techniques such as polarization and amperometric also aids in screening microorganisms based on the power output. Electrochemical impedance spectroscopy is another interesting technique helpful for the analysis of ohmic internal resistance, mass and charge transfer impedances. Unlike OMICS techniques, which are based

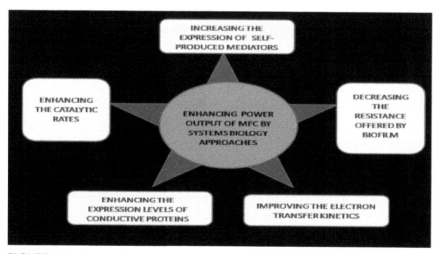

FIGURE 5.3   Routes for enhancing power output of microbial fuel cells using systems biology approaches.

on the bio-molecules such as protein, DNA, RNA or metabolites, the electrochemical techniques works based on the amount of electrons produced by the microorganisms. However, the electrochemical techniques cannot be categorized as systems biology technique or a part of it, as they do not involve molecular approaches such as genomics, proteomics, transcriptomics or metabolomics. The electrochemical techniques consider the electrocatalytic activity and electron transfer characteristics as a whole. Few genomic and proteomic investigations were also made to study the expression of certain genes/proteins from the anodic biofilm [1]. The levels of one or two proteins cannot be used as a tool for screening microorganisms because for a good electrocatalytic activity several proteins such as redox enzymes (for oxidation in the case of anodic biofilm/for reduction in the case of the reductive biofilm), conductive proteins (proteins for transferring the electron from the microorganism to the electrode such as pili, cytochromes, etc., in the case of mediator less MFC system) and the expression of mediators such as phenazine for shuttling electrons at the microbe-electrode interface. Hence systems biology approach such as a microarray by a complete investigation of several proteins/genes can be ideal tool to identify the suitable microorganism for MFC application.

5.6   FUTURE RESEARCH DIRECTIONS

Lack of proper approach for research is the key factor hindering the improvement of microbial fuel cells. Unlike other technologies, Microbial fuel cell is an interdisciplinary area where the role of physicists, materials scientists, engineers and molecular biologists is involved. But currently, more of biological approaches were not attempted. Systems biology approach is very much essential to increase the power output of MFC significantly and to improve it for commercialization. Bioinformatics techniques are now developed; however, building the models for such huge biological pathways is difficult. Hence, systems biology approaches like microarrays can be used to understand the mechanisms. Based on the data, the computational or mathematical techniques can be used to develop models for the pathways involved in electron transfer reactions.

5.7   CONCLUSION

The complete quantitative and qualitative study on the expression of all the proteins related to electrocatalytic activity and extracellular electron transfer is essential to conclude the exact electron transfer of any microorganism. Such studies can be possible only with the help of system biology approach and not by the conventional reductionist approaches. Although processing the huge volume of biological data computational techniques remains a challenging task, this can be resolved by the use of upcoming information technologies.

KEYWORDS

- **Electrode**
- **Electron transfer**
- **Microbial fuel cells**
- **Power output**

REFERENCES

1. Schröder, U. (2007). Anodic electron transfer mechanisms in microbial fuel cells and their energy efficiency. *Physical Chemistry Chemical Physics. 9,* 2619–2629.
2. Lovley, D. R. (2008). The microbe electric, conversion of organic matter to electricity. Current Opinion in Biotechnology. *19(6)*, 564–71.
3. Logan, B. E., Hamelers, B., Rozendal, R., Schröder, U., Keller, J., Freguia, S., Aelterman, P., Verstraete, W., Rabaey, K. (2006). Microbial fuel cells, methodology and technology. Environmental Science &Technology. *40*(17), 5181–5192.
4. Ahn, Y., Logan, B. E. (2013). Domestic wastewater treatment using multielectrode continuous flow MFCs with a separator electrode assembly design. *Applied Microbiology* and *Biotechnology. 97(1),* 409–16.
5. Bhuvaneswari, A., Navanietha Krishnaraj, R., Sheela Berchmans. (2013). Metamorphosis of pathogen to electrigen at the electrode/electrolyte interface, Direct electron transfer of *Staphylococcus aureus* leading to superior electrocatalytic activity. Electrochemistry Communications. *34,* 25–28.
6. Hou, B., Sun, J., Hu, Y. (2011). Effect of enrichment procedures on performance and microbial diversity of microbial fuel cell for Congo red decolorization and electricity generation. Applied Microbiology & Biotechnology. *90(4),* 1563–72.

7. Logan, B. E., Rabaey, K. (2012). Conversion of Wastes into Bioelectricity and Chemicals by using Microbial Electrochemical Technologies. Science. 337 (6095), 686–90.

8. Yuan, Y., Chen, Q., Zhou, S., Zhuang, L., Hu, P. (2011). Bioelectricity generation and microcystins removal in a blue-green algae powered microbial fuel cell. Journal of Hazardous Materials. *187*, 591–595.

9. McCormick, A. J., Bombelli, P., Scott, A. M., Philips, A. J., Smith, A. G., Fisher, A. C., Howe, C. J. (2011). Photosynthetic biofilms in pure culture harness solar energy in a mediatorless bio-photovoltaic cell (BPV) system. Energy & Environiromental Science. *4*, 4699–4709.

10. Wilkinson, S. (2000). "Gastrobots" Benefits and challenges of microbial fuel cells in food powered robot applications. Autonomous Robots. *9*, 99–111.

11. Bond, D. R., Holmes, D. E., Tender, L. M., Lovley, D. R. Electrode-reducing micro-organisms that harvest energy from marine sediments. Science. (2002). *295*, 483–485.

12. Zhang, J., Zhang, B., Tian, C., Ye, Z., Liu, Y., Lei, Z., Huang, W., Feng, C. (2013). Simultaneous sulfide removal and electricity generation with corn stover biomass as cosubstrate in microbial fuel cells. Bioresource Technology. *138*, 198–203.

13. Luo, H., Xu, P., Roane, T. M., Jenkins, P. E., Ren, Z. (2012). Microbial desalination cells for improved performance in wastewater treatment, electricity production, and desalination. Bioresource Technology. *105*, 60–66.

14. Zhang, B. G., Zhao, H. Z., Zhou, S. G., Shi, C. H., Wang, C., Ni, J. R. (2009). A novel UASB-MFC-BAF integrated system for high strength molasses wastewater treatment and bioelectricity generation. Bioresource Technology. *100*, 5687–5693.

15. Dewan, A., Donovan, C., Heo, D., Beyenal, H. (2010). Evaluating the performance of microbial fuel cells powering electronic devices, Journal of Power Sources. *195*, 90–96.

16. Aelterman, P., Rabaey, K., Pam, H. T., Boon, N., Verstaraete, W. (2006). Continuous Electricity Generation at High Voltages and Currents Using Stacked Microbial Fuel Cells, Environmental Science &Technology. *40 (10)*, 3388–3394.

17. Dewan, A., Beyenal, H., Liwandowski, Z. (2008). Scaling up Microbial Fuel Cells, Environmental Science & Technology. *42 (20)*, 7643–7648.

18. Karthikeyan, R., Ganesh, V., Berchmans, S. (2012). Bio-electrocatalysis of *Acetobacter aceti* through direct electron transfer using a template deposited nickel anode. Catalysis Science Technology. *2(6)*, 1234–1241.

19. Wei, J., Liang, P., Huang, X. (2011). Recent progress in electrodes for microbial fuel cells. Bioresource Technology. *102 (20)*, 9335–9344.

20. Navanietha Krishnaraj, R., Karthikeyan, R., Berchmans, S., Saravanan Chandran, Parimal Pal. (2013). Functionalization of electrochemically deposited chitosan films with alginate and Prussian blue for enhanced performance of Microbial fuel cells. Electrochimica Acta. *112*, 465–472.

21. Liu, X. W., Sun, X. F., Huang, Y. X., Sheng, G. P., Wangand, S. G., Yu, H. Q. (2011). Carbon nanotube/chitosan nanocomposite as a biocompatible biocathode material to enhance the electricity generation of a microbial fuel cell, Environmental Science & Technology. *4*, 1422–1427.

22. Higginsa, S. R., Foerstera, D., Cheunga, A., Laua, C., Bretschgera, O., Minteera, S. D., Nealsona, K., Atanassova, P., Cooney, M. J. (2011). Fabrication of macroporous chitosan scaffolds doped with carbon nanotubes and their characterization in microbial fuel cell operation. Enzyme and Microbial Technology. *48 (6–7),* 458–65.

23. Wen, Z., Ci, S., Mao, S., Cui, S., Lu, G., Yu, K., Luo, S., He, Z., Chen, J. (2013). TiO_2 nanoparticles-decorated carbon nanotubes for significantly improved bioelectricity generation in microbial fuel cells, Journal of Power Sources. *234,* 100–106.

24. Xiao, L., Damien, J., Luo, J., Jang, H. D., Huang, J., He, Z. (2012). Crumpled graphene particles for microbial fuel cell electrodes, Journal of Power Sources. *208,* 187–192.

25. Lai, B., Tang, X., Li, H., Du, Z., Liu, X., Zhang, Q. (2011). Power production enhancement with a polyaniline modified anode in microbial fuel cells. Biosensors and Bioelectronics. *28,* 373–377.

26. Huang, L., Logan, B. E. (2008). Electricity generation and treatment of paper recycling wastewater using a microbial fuel cell. Applied Microbiology Biotechnology. *80,* 349–55.

27. Amutha, R., Josiah, J. J. M., Adriel Jebin, J., Jagannathan, P., Sheela Berchmans. (2010). Chromium hexacyanoferrate as a cathode material in microbial fuel cells, Journal of Applied Electrochemistry. *40,* 1985–1990.

28. Logan, B. E. (2009). Exoelectrogenic bacteria that power microbial fuel cells. Nature Reviews Microbiology. *7,* 375–381.

29. Liang, S., Wenzhao, J., Changjun, H., Yu, L. (2011). Microbial biosensors, A review, Biosensors and Bioelectronics. *26(5),* 1788–99.

30. Dhar, P. K., Zhu, H., Mishra, S. K. (2004). Computational approach to systems biology, from fraction to integration and beyond. IEEE Trans Nanobioscience. *3(3),* 144–52.

31. Friboulet, A., Thomas, D. (2005). Systems Biology-an interdisciplinary approach. Biosensors and Bioelectronics. *20 (12),* 2404–2407.

32. Johnson, C. G., Goldman, J. P., Gullick, W. J. (2004). Simulating complex intracellular processes using object-oriented computational modeling. Progress in *Biophysics & Molecular Biology. 86 (3),* 379–406.

33. Kholodenko, B. N., Bruggeman, F. J, Sauro, H. M. (2005). Mechanistic and modular approaches to modeling and inference of cellular regulatory networks. Systems BiologyTopics in Current Genetics. *13,* 143–159.

34. Pham, C. A., Jung, S. J., Phung, N. T., Lee, J., Chang, I. S., Kim, B. H., Yi, H., Chun, J. (2003). A novel electrochemically active and Fe (III) -reducing bacterium phylogenetically related to *Aeromonas hydrophila*, isolated from a microbial fuel cell. FEMS Microbiology Letters. *223,* 129–134.

35. Karthikeyan, R., Sathish kumar, K., Murugesan, M., Sheela Berchmans; Yegnaraman, V. (2009). Bioelectrocatalysis of *Acetobacteraceti* and *Gluconobacter roseus*for Current Generation, Environmental Science & Technology. *43 (22),* 8684–8689.

36. Freguia, S., Masuda, M., Tsujimura, S., Kano, K. (2009). *Lactococcus lactis* catalyzes electricity generation at microbial fuel cell anodes via excretion of a soluble quinone. Bioelectrochemistry. *76,* 14–18.

37. Fang, D. L., Bai, L. F., Gui, Z. S., Yin, H. D., Ren, N. J. (2010). A study of electron-shuttle mechanism in Klebsiella pneumoniae based-microbial fuel cells, Chinese science bulletin. *55*, 99–104.

38. Vega, C. A., Fernandez, I. (1987). Mediating effect of ferric chelate compounds in microbial fuel cells with *Lactobacillus plantarum*, *Streptococcus lactis*, and *Erwinia dissolvens*. Bioelectrochemistry & Bioenergetics. *17*, 217–222.

39. Fang, D. L., Bai, L. F., Gui, Z. S., Yin, H. D., Ren, N. J. (2010). A study of electron-shuttle mechanism in *Klebsiella pneumoniae* based-microbial fuel cells, Chinese science bulletin. *55*, 99–104.

40. Choi, Y., Jung, E., Kim, S., Jung, S. (2003). Membrane fluidity sensing microbial fuel cell. Bioelectrochemistry. *59*, 121–127.

41. Liu. X., Tremblay. P. L., Malvankar. N. S., Nevin. K. P., Lovley. D. R., Vargas, M. (2013). *Geobacter sulfurreducens* Strain Expressing *Pseudomonas aeruginosa* Type IV Pili Localizes OmcS on Pili but is Deficient in Fe (III) Oxide Reduction and Current Production. Doi: 10. 1128/AEM. 02938-13.

42. Fishilevich, S., Amir, L., Fridman, Y., Aharoni, A., Alfonta, L. (2009). Surface Display of Redox Enzymes in Microbial Fuel Cells, Journal of American Chemical Soceity. *131*(34), 12052–12053.

43. Szczupak, A., Kol-Kalman, D., Alfonta, L. (2012). A hybrid biocathode, surface display of O_2-reducing enzymes for microbial fuel cell applications. Chemical Communications. *48,* 49.

44. Alfonta. L. (2010). Genetically Engineered Microbial Fuel Cells, Electroanalysis. *22,* 822–831.

45. Yong, Y.-C., Yu, Y.-Y., Yang, Y., Li, C. M., Jiang, R., Wang, X., Wang, J. Y., Song, H. (2012). Increasing intracellular releasable electrons dramatically enhances bioelectricity output in microbial fuel cells. Electrochemistry Communications. *49 (91),* 10754–10756.

PART 3

BIOETHANOL PRODUCTION

CHAPTER 6

POTENTIALS OF *OSCILLATORIA ANNAE* IN PRODUCING BIOETHANOL BY DEGRADATION OF SELECTED LIGNOCELLULOSICS

P. MALLIGA*, V. VISWAJITH, D. MALINI, and R. SABITHA

D. Malini, Research Scholar, Dept. of Marine Biotechnology and R. Sabitha, Research Scholar, Dept. of Biomedical Sciences,

*Professor, Department of Marine Biotechnology,
Bharathidasan University, Tiruchirappalli–620024, Tamil Nadu, India

CONTENTS

ABSTRACT

Globally, lignocellulosic materials are considered to be the most abundant renewable source available throughout the world in large quantities for the production of ethanol. The agroeconomic perspective, the present study targeted to exploit the ability of cyanobacteria to convert lignocellulosics for ethanol production. Lignocellulosic materials, coir pith, *Prosopis juliflora* and *Lantana camara L.* were allowed to treat with fresh water cyanobacterium (*Oscillatoria annae*). Bioconversion of lignocellulosics to ethanol requires initial dilute acid hydrolysis and fermentation of yeast (*Pichia stipitis*). At 36th hour, the *O. annae* treated *L. camara* produced maximum ethanol yield of 0.41 g/g (9.36 gL^{-1}). This was succeeded by the *O. annae* treated *P. juliflora* ethanol yield of 0.42 g/g (9.022 gL^{-1}) and least amount of 0.474 g/g (4.32 gL^{-1}) was obtained with *O. annae* treated coir pith. Hence, the *O. annae* treated lignocellulosics rendered higher amount of ethanol as compared with the fermentation of yeast (*P. stipitis*) alone. Thus, the study clearly evidenced that the *O. annae* degrades lignocellulosic and making it a better source for ethanol production by yeast (*P. stipitis*) fermentation.

6.1 INTRODUCTION

Energy consumption has increased steadily over the last century as the world population has grown and more countries have become industrialized. Bioethanol can also be important in helping meet the growing demand for energy in the developing world as these countries improve the living standards of more and more people [1]. In spite of, ethanol production only uses energy from renewable sources and it plays an important role in the reduction of greenhouse gas emissions. Hence, no net carbon dioxide is added to the atmosphere, making ethanol an environmentally beneficial energy source [2, 3]. Nigam [4] scrutinized the ethanol production from wheat straw hemicellulose hydrolysate by *Pichia stipitis* and reported a maximum yield of 0.41 ± 0.01 g$_p$, g$_s^{-1}$ (g$_{ethanol}$, g$_{sugar}$), equivalent to $80.4 \pm 0.55\%$ theoretical conversion efficiency. The route of bioethanol production from lignocellulosic material such as cereal

straw and corn stover has also been reported [5, 6]. Bioethanol production has been running for several years, improvements are required to
increase process performance [7]. The choice of the best technology
for lignocellulose to bioethanol conversion should be decided on the
basis of overall economics (lowest cost), environmental (lowest pollutants) and energy (higher efficiencies). Deplorably, much lignocellulose
waste is often disposed of by biomass burning, which is not restricted
to underdeveloped countries alone, but is considered a global phenomenon [8]. In India, lignocellulosic biomass (crop residues, forestry and
fruit and vegetable waste and weeds) is available in plenty. Renewable
fuels particularly ethanol should get more and more attention all over
the world. *Lantana camara* is an important non-edible lignocellulosic
biomass, which grows widely throughout India. It contains high cellulose and no competition with the food chain makes it an ideal substrate
for bioethanol production [9, 10]. *Prosopis juliflora* or mesquite is a
drought resistant, grows in semiarid and arid tracts of tropical and sub-
tropical regions of the world and is spreading fast because the leaves
are unpalatable and animals do not digest its seed [11]. Aqeel [12]
investigated the antidermatophytic activity of juliflorine and a benzene
insoluble alkaloidal fraction obtained from *P. juliflora* against *Tricho-
phyton mentagrophytes* infection in rabbits. The waste products of coir
yarn industry are coir dust and coir pith or coco peat which is the waste
obtained from the coconut (*Coccus nucifera* Linn.), which constitute
about 70% of the husk. In spite of their limited use as soil conditioners,
the quantity of coir dust produced is so enormous making its disposal
difficult because of its lignocellulosic nature and slow degradation in
the natural environment [13]. Degradation of coir pith can be effectively done with suitable species of Basidiomycetes fungus (*Pleurotus
sajor caju*) and in combination with nitrogen fixing bacteria. Bacterial
isolates from compost or soil, namely *Azotobacter, Bacillus megatar-
ium* and *Serratia marcescens*, were capable of decolorizing or solubi-
lizing industrial lignin [14, 15]. Ligninolytic fungi are well known, not
only as decomposers of lignin but also for their ability to degrade a wide
variety of organopollutants [16]. Cyanobacteria are capable of abating
various kinds of pollutants and have been used in the production of
energy, fertilizer [17]. Moreover, technologies for producing bioethanol

from seaweed or marine cyanobacteria have been reported, a method for producing ethanol using freshwater cyanobacteria has yet not to be reported [18]. Therefore, the intention of this study aimed to use fresh water cyanobacterium (*O. annae*) and yeast (*P. stipitis*) for the production of bioethanol by using three-lignoicellulosic materials coir pith, *L. camara* and *P. juliflora*. This study also delineates the comparative effect of bioethanol production with yeast (*P. stipitis*).

6.2 MATERIALS AND METHODS

6.2.1 LIGNOCELLULOSIC MATERIALS

Coir pith, a waste by-product of coir rope industry was collected from coir industries near Srirangam, Tiruchirappalli, Tamilnadu, India. Woody stems of *Prosopis juliflora* and *Lantana camara*, were obtained from Bharathidasan University campus, dried and ground to pass a 0.5 mm screen.

6.2.2 MICROORGANISM AND CULTURE CONDITIONS

Fresh water cyanobacterium was grown and maintained in BG11 medium [19] under white fluorescent light of 13.8 μE m^{-1} s^{-1} at 25 $\pm$2°C with 14/10 D/L cycle. It was grown along with lignocellulosics (coir pith, *P. juliflora* and *L. camara*) separately at a dry weight ratio of 1:10 (cyanobacteria: lignocellulosics) in BG11 medium for 15 days. Large-scale cultivation of *O. annae* was carried out initially in boxes, tanks and further increased to pits dug on ground and covered with polythene sheets. Cyanobacteria with coir pith, *P. juliflora* and *L. camara* was inoculated in separate pits in urea media. After 15 days of incubation, the degraded wood materials were collected and separated from cyanobacterial mat and air-dried for further use as source for bioethanol.

Pichia stipitis NCIM 3498 was obtained from the collection of National Chemical Laboratory, Pune, India. *P. stipitis* culture was maintained in a medium containing peptone 2.0 gL^{-1}, yeast extract 2.0 gL^{-1}, (NH)$_2$SO$_4$ 3.0 gL^{-1}, KH$_2$PO$_4$ 2.0 gL^{-1}, MgSO$_4$ 1.0 gL^{-1} and xylose 10.0 gL^{-1}.

6.2.3 BIOETHANOL PRODUCTION

6.2.3.1 Acid Hydrolysis

Dilute acid hydrolysis method by using H_2SO_4 was used for the saccharification process. Varying concentration of sulfuric acid (0.75, 1.5, 2.25, 3.0 and 3.75% v/v) was tested for the maximum release of sugars from the wood materials. Dilute acid hydrolysis was combined with heat for enhancing the release of sugars from the wood. Each concentration of dilute acid hydrolysis was combined with three different temperatures (100, 120 and 140°C) for 1 h using a special Russian autoclave.

6.2.3.2 Detoxification Methods

In order to decrease the toxicity for *P. stipitis*, considerable efforts have been focused on detoxification procedure prior to fermentation, including neutralization, overliming and overliming combined with activated charcoal.

6.2.3.3 Neutralization

Calcium hydroxide was added to the dilute acid hydrolysate to increase the pH to 5.8 and the mixture was stirred for 30 min. It was then filtered to remove any precipitate formed before using as substrate for fermentation [20].

6.2.3.4 Overlimiting

Calcium hydroxide was added to the dilute acid hydrolysate to increase the pH to 10.5 and the whole mixture was stirred for 30 min. It was then filtered and the pH was adjusted to 5.8 using 2N H_2SO_4. The sample was once again filtered and the pH was brought to 5.8 using 2N H_2SO_4 [21].

6.2.3.5   Overlimiting Combined With Activated Charcoal

Calcium hydroxide was added to the dilute acid hydrolysate to increase the pH to 10.5 and the whole mixture was stirred for 30 min. It was then filtered and 1% activated charcoal was added and stirred for 20 min. It was once again subjected to filtration [22].

6.2.3.6   Analytical Methods

Samples were withdrawn periodically to determine biomass of yeast (*P. stipitis*) cell mass, for reducing sugars and ethanol in the broth.

6.2.3.7   Reducing Sugar Estimation

According to Miller [23], 1 mL of sample (*O. annae* untreated, treated lignocellulosics), 1 mL of reagent (100 mL distilled water and 1 g Dinitrosalicylic acid (DNSA), 20 mL 2N NaOH, 30 g of Rochella (NaK tartrate)) was added. The tubes were kept in boiling water bath for 10 min. After cooling it was made up to 10 mL by adding distilled water. Absorbance was measured spectrophotometrically at 540 nm. Concentration of sugar was determined by computing optical density against the standard curve, prepared using sugar (glucose) concentration from 10–100 µg/mL.

6.2.3.8   Estimation of Total Phenolics

Two hundred and 50 µL of Folin-Ciocalteu reagent was added to 50 µL of test sample (*O. annae* untreated, treated lignocellulosics). The mixture was vortexed well and 750 µL of 20% sodium carbonate was added and the final volume of the mixture was made to 5 mL using distilled water. The mixture was incubated in dark at room temperature for 1 hr. The absorbance was read at 760 nm against blank containing distilled water instead of test sample. The standard curve was prepared using vanillin (50–500 µg/mL) [24].

6.2.3.9   Ethanol Production and Yeast Fermentations

Fermentation studies started with 3.5% (v/v) of *P. stipitis* inoculum. Ethanol fermentations by *P. stipitis* were evaluated at 30 °C in 500 mL Erlenmeyer flasks containing 100 mL appropriate media in a shaker at 150 rpm. To determine the dry weight of yeast cells, the homogenized cell suspension (1 mL) was centrifuged (10,000rpm.) in preweighed eppendorf. The cell pellets were rinsed with sterilized distilled water, recentrifuged at 10,000rpm and dried till became, constant weight at 60 °C. All the experiments were carried out in triplicates.

6.2.3.10   Ethanol Estimation

Ethanol was estimated by gas chromatography (GC) (Perkin Elmer, Clarus 500) with an elite-wax (cross bond-Poly ethylene glycol) column (30 m × 0.25 mm) at 120°C, flame ionization detector at 210°C and injector at 180°C using isopropanol as standard. The carrier gas was nitrogen (N_2). Followingly, 0.2 µL of sample was injected into the column for estimation.

6.3   RESULTS AND DISCUSSION

Lignocellulosic waste technological developments can help to move the price of bioethanol closer to or even less than that of petroleum- based fuel. The use of ethanol as an alternative motor fuel has been steadily increasing around the world for a number of reasons. Domestic production and use of ethanol for fuel can increase dependency on foreign oil, reduce trade deficits, create jobs in rural areas, and reduce global climate change and carbon dioxide buildup [25].

6.3.1   *OPTIMIZATION OF ACID HYDROLYSIS AND TEMPERATURE*

Several inhibitory compounds are formed during acid hydrolysis of lignocellulosic materials. Such compounds are divided into four groups:

(a) sugar degradation products including furfural and hydroxymethyl-furfural; (b) lignin degradation products including several aromatic and polyaromatic compounds; (c) substances released from the hemicellu-losic structure, including acetic acid, terpenes and tannins; and (d) met-als released from equipment like chromium, copper, iron and nickel [26]. Optimization of acid hydrolysis for the maximal release of reducing sug-ars was carried out using sulfuric acid (H_2SO_4) at varying concentrations (0.75, 1.5, 2.25, 3.25 and 3.75% v/v) and at variable temperatures (100, 120 and 140°C) for 1 h. The effect of operation variables namely tem-perature and acid concentration were studied for the maximum sugar recovery from the lignocellulosic substrates (both untreated and *O. annae* treated coir pith, *P. juliflora* and *L. camara*). The concentration of total reducing sugars varied systematically for all the hydrolysis condi-tions carried out.

Acid hydrolysis leads to the higher amount of sugars in shorter time and it also commercially viable process. During acid hydrolysis of lig-nocellulosics, aliphatic acids (acetic, formic and levulinic acid), furan derivatives [furfural and 5-hydroxymethylfurfural (HMF)] and phenolic compounds are formed in addition to the sugars. These compounds are known to inhibit ethanol fermentation performance [27, 20]. Therefore, in order to improve the fermentability and quality of the hydrolysates, it is necessary to remove the toxic compounds prior to fermentation. Various methods of detoxification such as neutralization, overliming with calcium hydroxide ($Ca(OH)_2$) and overliming combined with activated charcoal were employed for this purpose. Irrespective of the temperature chosen all the three substrates showed an increase amount of sugar and were con-centration dependent, of the three substrates subjected to pretreatments *O. annae* treated *P. juliflora* could release higher amount of sugars whereas the least was noticed with coir pith (Figs. 6.1A–6.1C).

The release of phenols with the three substrates noticed were with refer-ence to temperature, though the phenol release was concentration depen-dent as far as the acid treatment is considered the maximum phenolic release observed at 140°C with *O. annae* treated coir pith (Figs. 6.2A–6.2C).

Figures 6.1–6.3 depict the phenolics released from untreated and *O. annae* treated lignocellulosics at various temperatures when treated with

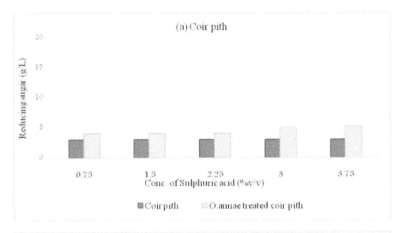

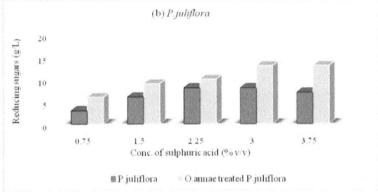

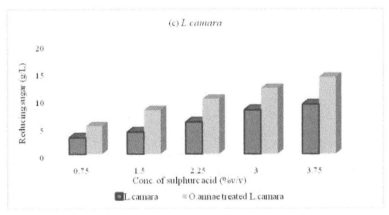

FIGURE 6.1A   Effect of thermochemical pretreatment (H_2SO_4, 100°C, 1 hr) on release of reducing sugars from selected lignocellulosics.

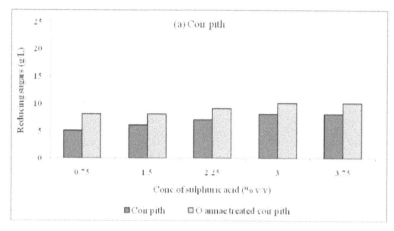

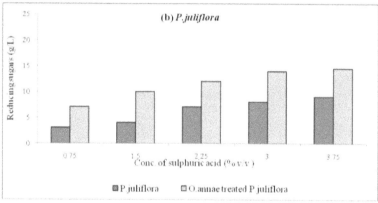

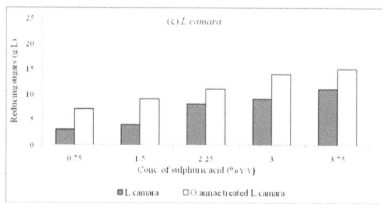

FIGURE 6.1B   Effect of thermochemical pretreatment (H_2SO_4, 120°C, 1 hr) on release of reducing sugars from selected lignocellulosics.

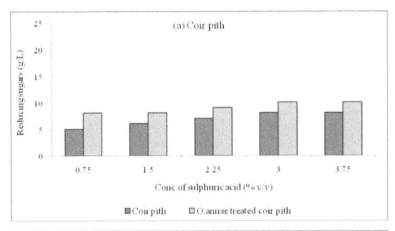

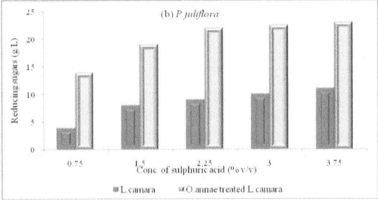

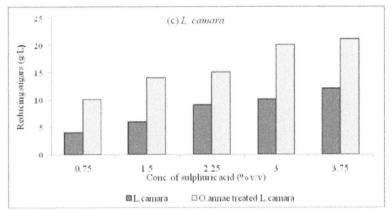

FIGURE 6.1C Effect of thermochemical pretreatment (H_2SO_4, 140°C, 1 hr) on release of reducing sugars from selected lignocellulosics.

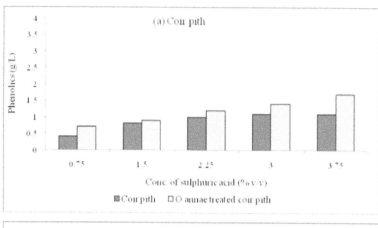

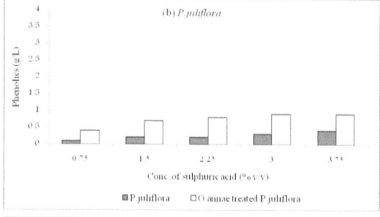

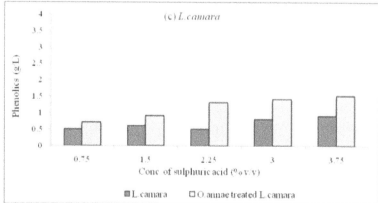

FIGURE 6.2A   Effect of thermochemical pretreatment (H_2SO_4, 100°C, 1 hr) on release of phenolics from selected lignocellulosics.

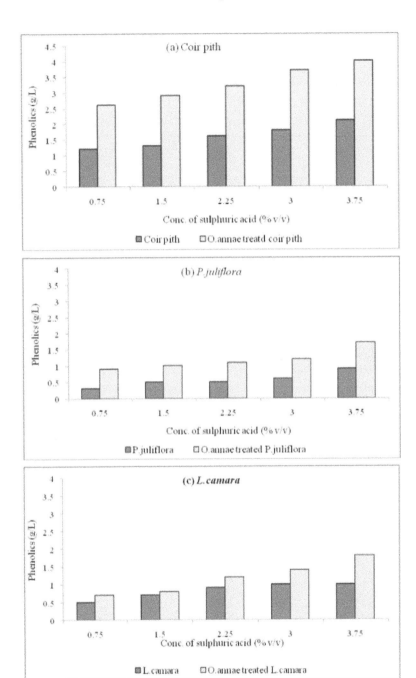

FIGURE 6.2B   Effect of thermochemical pretreatment (H_2SO_4, 120°C, 1 hr) on release of phenolics from selected lignocellulosics.

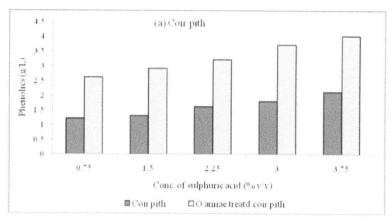

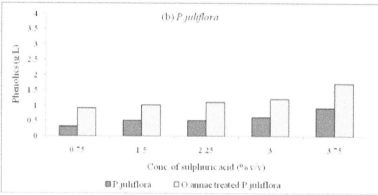

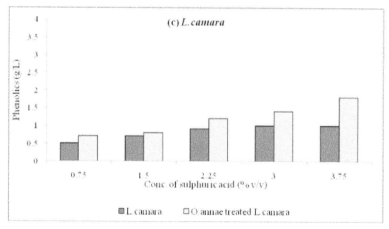

FIGURE 6.2C   Effect of thermochemical pretreatment (H_2SO_4, 140°C, 1 hr) on release of phenolics from selected lignocellulosics.

varied concentrations of H_2SO_4. Primarily, all the following lignocellu-losic materials were treated with O. annae. The maximum phenolics yield was obtained from *O. annae* treated coir pith 3.9 ± 0.2 gL^{-1} followed by *O. annae* treated *L. camara* 1.67 ± 0.04 g L-1 and *P. juliflora* 1.6 ± 0.08 gL^{-1} with 3.75% v/v H_2SO_4 at 140 °C for 1 hr. Among the control untreated lignocellulosics coir pith produced maximum reducing sugar of 2.1 ± 0.08 g L^{-1} followed by *L. camara* 0.97 ± 0.03 gL^{-1} and *P. juliflora* 0.84 ± 0.03 gL^{-1} with 3.75% v/v H_2SO_4 in 1 hr at 140 °C.

Though concentrated acids such as H_2SO_4 and HCl have been used to treat lignocellulosic materials dilute acid hydrolysis has been success-fully developed for the hydrolysis of lignocellulosic materials. At moder-ate temperature, direct saccharification offered from low yields because of sugar decomposition. High temperature (300 °C) in dilute acid treat-ment is favorable for cellulose hydrolysis [28]. Recently developed dilute acid hydrolysis processes use less severe conditions and achieve high xylan to xylose conversion yields. Achieving high xylan to xylose conver-sion yields is necessary to achieve favorable overall process economics because xylan accounts for up to a third of the total carbohydrate in many lignocellulosic materials [29].

Dilute acid hydrolysis is employed for depolymerization of hemicel-lulosic polymer into its monomeric constituents mainly in the form of xylose. The maximum sugar yield (25.0 gL^{-1}) was achieved at 140°C, 3.5% (v/v) H_2SO_4 for 1h in the present investigation. Previous reports of recovered sugars yield from hemicellulose hydrolysis of birch and spruce [30]. Also, closely agree with those found in the present study. The acid hydrolysis using 3.75% H_2SO_4 released maximum amount of reducing sugars in both control and test samples at 140°C for 1 hr. Hence, 3.75% H_2SO_4 for 1 hr at 140°C was considered as optimum for maximum release of reducing sugars from the selected lignocellulosic materials though there have been significant differences between the release of sugars between lignocellulosic materials and the concentration of acid treatment within the lignocellulosic materials. It has been reported that the cell wall struc-ture and components may be significantly different in plants, which may influence the biomass digestibility resulting in variation in the release of reducing sugar [31, 32].

6.3.2 DETOXIFICATION

The neutralization of acid hydrolysate by increasing the hydrolysate pH up to fermentable pH (5.8) required for yeast growth caused reduction in total phenolics (28.22%).

Overliming of acid hydrolyzates using calcium hydroxide resulted in decrease of total phenolics up to 58.13%. Increasing the pH to 10.5 by $Ca(OH)_2$ and readjustment to 6.5 with H_2SO_4 caused detoxification of lignocellulosic hydrolysate. The detoxifying effect of overliming is due to, both (a) precipitation of toxic components and (b) instability of some inhibitors at high pH. A critical issue in the conversion of dilute acid hydrolysates has been the ability to withstand inhibitors [26] and most often a detoxification step is needed to improve the fermentation efficiency. However, the introduction of a detoxification step increases the process complexity and may give precipitation problems [4].

More recently, Chandel [20] also observed similar results while detoxifying sugarcane baggase hemicellulose hydrolysate. According to Roberto [33], pH adjustment with a combination of bases and acids is a low-cost treatment that gives good results. By adjusting the pH of sugarcane bagasse hemicellulosic hydrolyzate first to 10.0 with $Ca(OH)_2$ and then to 6.5 with H_2SO_4, these authors obtained a partial removal of phenolic and other compounds and a xylitol yield of 0.48 g/g. Van Zyl [34] adjusted the pH of sugarcane bagasse hemicellulose hydrolyzate first with $Ca(OH)_2$ and then with NaOH and found that the former treatment enhanced the hydrolyzate fermentability, whereas the latter increased the precipitation of toxic compounds. Martinez [21] reported that using $Ca(OH)_2$ to adjust the pH of sugarcane bagasse hemicellulose hydrolysate (overliming treatment) to 9.0, proved to be a very efficient detoxification method. The fermentability of a corncob, acid-hydrolysates hemicellulose was considerably increased from an ethanol yield of 0.21 g/g to 0.32 g/g [35].

Among the known detoxification methods activated charcoal treatment reduces all class of fermentation inhibitors from hydrolysate because it has high capacity to adsorb toxic compounds. Thus, the combined treatment brought about reduction of 75.28% total phenolics from acid hydrolysates.

6.3.3   FERMENTATION OF ACID HYDROLYSATES

A typical fermentation profile of *P. stipitis* NCIM 3498 in batch culture using sugars from acid hydrolysate of untreated and *O. annae* treated coir pith, *P. juliflora* and *L. camara* (Figs. 6.3A–C and 6.4A–C) has been shown. The ethanol production level reached a maximum at 36th hour of incubation. The ethanol yield and productivity of control and test samples of the respective lignocellulosics were compared.

The untreated and *O. annae* pretreated (acid) hydrolysates were used for ethanol production separately in batch mode of cultivation. The fermentation of different types of hydrolysate showed distinct characteristics due to the differences in the initial concentration of sugars and the level of different fermentation inhibitors present in hydrolysates. With respect to the hydrolysate fermentation, it can be observed that the sugar consumption rate ranged from 80 to 94%, irrespective of initial concentration of sugars in medium. However, the amount of ethanol production varied significantly. The reason could be the difference in level of fermentation inhibitors present in hydrolysate even after the detoxifications. These inhibitors affect the ethanol production efficiency of yeast. Data indicates that the rate of sugar consumption, biomass production rate and ethanol production rates was much higher upon fermentation of *O. annae* pretreated and acid hydrolyzed (Table 6.1).

Variety of yeasts (*Brettanomyces, Candida, Clavispora, Kluvyvero-myces, Pachysolen, Pichia* and *Schizosaccharmoyces*) has been used for ethanol production using xylose as carbon source. Among these, *Pichia stipitis* has shown promise for industrial application as it ferments xylose rapidly with a high ethanol yield (0.47 g/g) and production (22.3 g L^{-1}) [36] with insignificant amount of xylitol (0.03 gL^{-1}) [37]. Furthermore, *P. stipitis* has no absolute vitamin requirement for xylose fermentation [38, 39].

A typical fermentation profile of *P. stipitis* NCIM 3498 in batch culture is shown in Figs. 6.7 and 6.8. Bioethanol production from *O. annae* treated lignocellulosics was distinctly higher when compared to the untreated lignocellulosics. Among the lignocellulosics studied a maximum ethanol production of 9.36 gL^{-1} with 0.41 g/g ethanol yield and

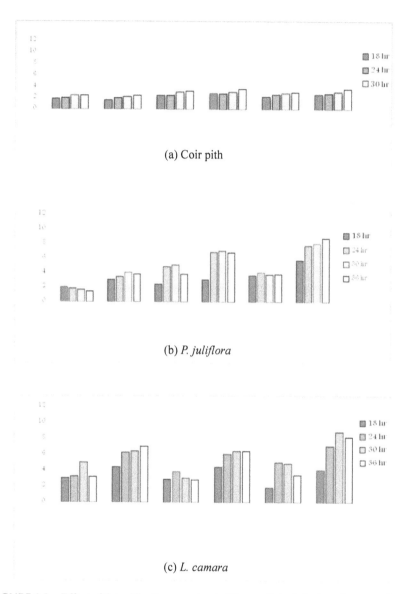

(a) Coir pith

(b) *P. juliflora*

(c) *L. camara*

FIGURE 6.3   Effect of detoxification on untreated lignocellulosic hydrosylates on ethanol production by *P. stipitis* NCIM 3498.

0.26 $gL^{-1}hr^{-1}$ ethanol productivity was obtained from *O. annae* treated *L. camara* hydrolysate at 36th hour. This was followed by the *O. annae* treated *P. juliflora* bioethanol production of 9.022 gL^{-1} with 0.42 etha-

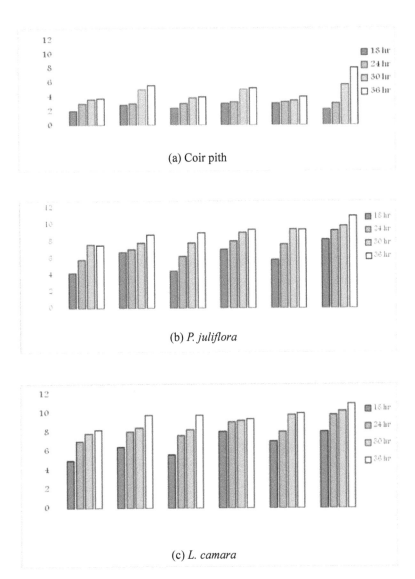

(a) Coir pith

(b) *P. juliflora*

(c) *L. camara*

FIGURE 6.4   Effect of detoxification on *O.annae* treated lignocellulosic hydrolysates on ethanol production by *P. stipitis* NCIM 3498.

nol yield and ethanol productivity of 0.25 g $L^{-1}hr^{-1}$ at 36th hour. Among the test samples least amount of bioethanol production was obtained from *O. annae* treated coir pith 4.32 gL^{-1} and ethanol yield of 0.474 gL^{-1} and productivity of 0.114 $gL^{-1}hr^{-1}$. Statistical analysis showed that the etha-

TABLE 6.1   Effect of Detoxification on Untreated and *O.annae* Treated Lignocellulosic Hydrolysates for Ethanol and Biomass Yield

Samples	Method	Ethanol yield (g/g)	Biomass yield (g/g)
Untreated coir pith	Neutralization,	0.3823	0.147
	Overliming,	0.444	0.5455
	Overliming + Activated Charcoal (AC)	0.456	0.585
O.annae treated Coir pith	Neutralization,	0.416	0.591
	Overliming,	0.428	0.616
	Overliming + Activated Charcoal (AC)	0.474	0.675
Untreated *P. juliflora*	Neutralization,	0.25	0.396
	Overliming,	0.45	0.42
	Overliming + Activated Charcoal (AC)	0.39	0.512
O.annae treated *P. juliflora*	Neutralization,	0.323 0.41	0.375
	Overliming,	0.42	0.812
	Overliming + Activated Charcoal (AC)		0.875
Untreated *L. camara*	Neutralization,	0.313	0.41
	Overliming,	0.262	0.41
	Overliming + Activated Charcoal (AC)	0.37	0.48
O.annae treated *L. camara*	Neutralization,	0.35	0.71
	Overliming,	0.4	0.71
	Overliming + Activated Charcoal (AC)	0.41	0.8

nol production from test samples was statistically significant compared to respective control samples.

The higher ethanol and biomass yield and ethanol productivity of cyanobacteria pretreated lignocellulosics is due to higher release of reducing sugars and absence of fermentation inhibitors due to detoxification (Fig. 6.5 and Table 6.1).

Cp – Coir pith;   Oa + Cp-*O. annae* + Coir pith;

Pj – P. juliflora;   Oa +Pj-*O. annae* + *P. juliflora;*

Lc – L. camara;   Oa + Lc-*O. annae* + *L. camara.*

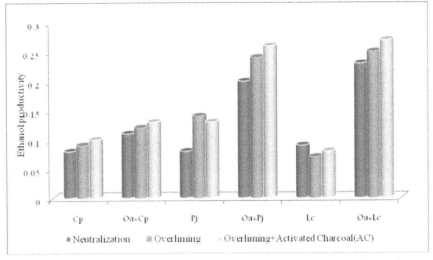

FIGURE 6.5   Effect of detoxification on untreated and *O.annae* treated lignocellulosic hydrolysates for ethanol productivity (g/l/h).

This could also be well supported by the previous studies of several authors [40–42]. Saha [43] found the better ethanol yield in acid pretreated enzymic hydrolysate fermented with recombinant *Escherichia coli* strain FBR5 giving 19 g L^{-1} ethanol with a yield of 0.24 g/g. Gupta [15] reported an ethanol production of 5.0 gL^{-1} in batch fermentation and 26.2 gL^{-1} in fed batch fermentation using *P. stipitis* NCIM-3498 from *Saccharum spontaneum*. For the detoxification three processes were undertaken namely, neutralization, overliming, overliming with activated charcoal. Both the untreated and *O. annae* treated lignocellulosic hydrolysates were able to produce ethanol with *P. stipitis*. The ethanol production was noticed to be high in combined treatment (overliming with activated charcoal). This is mainly due to the better removal of inhibitory substance [42]. The ethanol production was analyzed and estimated for 18, 24, 30, and 36 hours. Both the untreated and *O. annae* treated substrates (coir pith, *P. juliflora* and *L. camara*) showed maximum production by 36th hour. As expected when compared to neutralization process alone because of the removal of other ethanol producing inhibitory substances overliming and the combines process of overliming with activated charcoal yielded almost the double the quantity of ethanol.

Tables 6.2–6.5 depicts the ANOVA for reducing sugar released by *O. annae* with selected lignocellulosics, phenolics released by *O. annae* with

TABLE 6.2   ANOVA of Reducing Sugar Released by *O. annae* with Selected Lignocellulosics

S. no.	Parameters	99% confidence limits		Null hypoth-esis	Alternate hypoth-esis
		F	Fcrit		
1	Reducing sugar at 100°C and acid concentration of coir pith hydrolysates	28. 31301	4.45897	Rejected	Accepted
2	Reducing sugar at 120°C and acid concentration of coir pith hydrolysate	147.2966	4.45897		
3	Reducing sugar at 140°C and acid concentration of coir pith hydrolysate	861.5414	4.45897		
4	Reducing sugar at 100°C and acid concentration of *P.juliflora* hydrolysate	49.21591	4.45897		
5	Reducing sugar at 120°C and acid concentration of *P.juliflora* hydrolysate	73.43046	4.45897		
6	Reducing sugar at 140°C and acid concentration of *P.juliflora* hydrolysate	193.7169	4.45897		
7	Reducing sugar at 100°C and acid concentration of *L. camara* hydrolysate	61.82457	4.45897		
8	Reducing sugar at 120°C and acid concentration of *L. camara* hydrolysate	54.0685	4.45897		
9	Reducing sugar at 140°C and acid concentration of *L. camara* hydrolysate	66.78603	4.45897		

F = Experimental value; Fcrit = Tabulated value (significant value).

selected lignocellulosics, effect of detoxification of selected lignocellulosics hydrosylates by *O. annae* for improves bioethanol production and bioethanol yield by *O. annae* treated lignocellulosics hydrosylates. Thus, from the results it can be concluded that *O. annae* pretreatment of lignocellulosics before thermochemical treatment can be beneficially used as an effective method to enhance the release of reducing sugar from the lignocellulosic materials. Among the lignocellulosics studied *P. juliflora* and

L. camara can be considered as a better source of reducing sugar for bio-ethanol production due to its high sugar and low phenolics content when compared to coir pith. As far as our study concerned both *P. juliflora* and *L. camara* were adjudged to be the candidates for ethanol production after cyanobacterial growth followed by detoxification by overliming. In line with the findings of bacteria and fungi [45] fresh water cyanobacterium *O. annae*'s ability to degrade lignocellulosics and making it a better substrate for ethanol production.

TABLE 6.3 ANOVA of Phenolics Released by *O. annae* with Selected Lignocellulosics

S. no.	Parameters	99% confidence limits		Null hypothesis	Alternate hypothesis
		F	Fcrit		
1	Phenolics at 100°C and acid concentration of coir pith hydrolysates	11.30519	4.45897		
2	Phenolics at 120°C and acid concentration of coir pith hydrolysate	12.14238	4.45897		
3	Phenolics at 140°C and acid concentration of coir pith hydrolysate	19.45638	4.45897		
4	Phenolics at 100°C and acid concentration of *P.juliflora* hydrolysate	16.36974	4.45897	Rejected	Accepted
5	Phenolics at 120°C and acid concentration of *P.juliflora* hydrolysate	15.45043	4.45897		
6	Phenolics at 140°C and acid concentration of *P.juliflora* hydrolysate	14.5373	4.45897		
7	Phenolics at 100°C and acid concentration of *L. camara* hydrolysate	17.06229	4.45897		
8	Phenolics at 120°C and acid concentration of *L. camara* hydrolysate	13.67615	4.45897		
9	Phenolics at 140°C and acid concentration of *L. camara* hydrolysate	11.73684	4.45897		

F = Experimental value; Fcrit = Tabulated value (significant value).

TABLE 6.4   ANOVA on Effect of Detoxification of Selected Lignocellulosics hydrolysates by *O. annae* for Improves Bioethanol Production

S. no.	Parameters	99% confidence limits		Null hypoth-esis	Alternate hypoth-esis
		F	Fcrit		
1	Untreated coir pith	2.001896	3.885294	Rejected	Accepted
2	Treated coir pith	4.478342	3.885294		
3	Untreated *P. juliflora*	0.992912	3.159908		
4	Ttreated *P. juliflora*	6.775324	3.159908		
5	Untreated *L. camara*	3.3427673	3.159907		
6	Treated *L. camara*	6.361262	3.159908		

TABLE 6.5   ANOVA on Bioethanol Yield by *O. annae* Treated Lignocellulosics hydrolysates

S. no.	Parameters	99% confidence limits		Null hypoth-esis	Alternate hypothesis
		F	Fcrit		
1	Untreated coir pith	1.65313	6.94427	Rejected	Accepted
2	Treated coir pith	8.987561	2.99612		
3	Untreated *P. juliflora*	8.478342	6.944272		
4	Ttreated *P. juliflora*	8.987561	6.944272		
5	Untreated *L. camara*	4.100723	6.944272		
6	Treated *L. camara*	7.0	6.944272		

6.4   CONCLUSION

Lignocellulose (lignin, cellulose and hemicellulose) is the major structural component of woody and non woody plants make them a substrate of enormous biotechnological value. These huge amounts of residual plant biomass considered as waste can be potentially converted into various value added products namely biofuels, chemicals, and cheap energy source for fermentation. Lignocellulosic wastes such as *L.camara, P. juliflora* and coir pith were used in an efficient way for producing bioethanol.

Cyanobacteria are known for their ability to degrade various pollutants such as lignin, pesticides and phenol. In view of producing bioethanol, the fresh water cyanobacterium *O. annae* treated lignocellulosics (15th day) were subjected to thermochemical hydrolysis. The efficiency of bioethanol production by *Pichia stipitis* NCIM 3498 after detoxification by neutralization, overliming and overliming combined with activated charcoal were analyzed in untreated and *O. annae* treated lignocellulosics, respectively. Ethanol produced using *Pichia stipitis* NCIM 3498 were analyzed by gas chromatography and the fermentation efficacy was calculated. The above findings demonstrate that freshwater cyanobacterium *O. annae* was found to be as efficient degrader of all the three lignocellulosics namely coirpith, *P. juliflora* and *L. camara* as substrates and release fermentable sugars followed fermentation of yeast (*P. stipitis*) for bioethanol production. Hence, this is cost effective, ecofriendly, zero pollution lignocellulosic-based technology could be another lead for agro-economy development to India.

KEYWORDS

- **Bioethanol**
- **Coir pith**
- *Lantana camara*
- **Lignocellulosic**
- *Oscillatoria annae*
- *Pichia stipitis*
- *Prosopis juliflora*

REFERENCES

1. Beck, R. J. Worldwide Petroleum Industry Outlook: 1998–2002 Projections to 2007, 14th ed., Pennwell, Tulsa, OK. 1997.
2. Bull, S. R., Riley, C. J., Tyson, K. S., Costella, R. (1992). Total fuel cycleand emission analyze of biomass to ethanol. *In:* Energy fromBiomass and wastes vol. XVI (Klass DL ed.) Institute of gas technology, Chicago, 1–14.

3. Kheshgi H. S., Prince, R. C., Marland, G. (2000). The potential of biomass fuels in the context of global climate change; Focus on transportation fuels. *Annual Rev. Energy Environ. 25*, 199–244.

4. Nigam, J. N. (2001). Ethanol production from wheat straw hemicellulose hydrolysate by *Pichia stipitis*. *J. Biotechnol. 87*, 17–27.

5. Bjerre, A. B., Olesen, A. B., Fernqvist, T. (1996). Pretreatment of wheat straw using combined wet oxidation and alkaline hydrolysis resulting in convertible cellulose and hemicellulose. *J. Biotechnol. Bioeng. 49*, 568–577.

6. Varga, E., Schmidt, A. S., Reczey, K., Thomsen, A. B. (2003). Pretreatment of corn stover using wet oxidation to enhance enzymatic digestibility. *J. Appl. Biochem. Biotechnol. 10*, 37–50.

7. Elmer, C. R., Costa, A. C., Lunelli, B. H., Maciel, M. R. W., Filho, R. M. (2007). Kinetic Modeling and Parameter Estimation in a Tower Bioreactor for Bioethanol Production. *J. Appl. Biochem. Biotechnol. 148*, 163–173.

8. Levine, J. S. (1996). Biomass burning and global change. Remote sensing and inventory development and biomass burning in Africa. *In:* Levine JS (Ed). The MIT Press, Cambridge, Massachusetts, USA. *1*, 35.

9. Hahn-Hägerdal, B., Galbe, M., Gorwa- Grauslund, M. F., Lidén, G., Zacchi, G. (2006). Bioethanol the fuel of tomorrow from the residues of today. *Trends Biotechnol. 24*, 549–56.

10. Hiremath, A. J., Bharath, S. (2005). The fire-lantana cycle hypothesis in Indian forests. *Conservat Soc. 3*, 26–42.

11. Sawal, R. K., Ratan, R., Yadav, S. B. S. (2004). Mesquite (*Prosopis juliflora*) pods as a feed resource for livestock. *Asian Australasian J. Animal Sci. 17 (5)*, 719–725.

12. Aqeel, A., Viquaruddin, A., Khalid, S. M., Ansari, F. A., Khan, K. A. (1997). Study on the antifungal efficacy of juliflorine and a benzene-insoluble alkaloid fraction of *Prosopis juliflora.Philippine J. Sci. 126(2)*, 175–182.

13. Malliga P., Uma, L., Subramanian, G. (1996). Lignolytic activity of the cyanobacterium *Anabaena azollae* ML2 and the value of coir waste as a carrier for biofertilizer. *Microbios. 86*, 175–183.

14. Perestelo, F., Falcon, M. A., Carnicero, A., Rodriguez, A., Feunte, D. L. G. (1994). Limited degradation of industrial, synthetic and natural lignin by *Serratia marcescens*. *Biotechnol. Lett. 16*, 299–302.

15. Morii, H., Nakamiya, K., Kinnoshita, S. (1995). Isolation of lignin decolorizing bacterium. *J. Ferment. Bioeng. 80*, 296–299.

16. Bumpus, J. A., Aust, S. D. Biodegradation of DDT (1,1,1-tri-chloro-2,2-bis-(4-chlorophenyl) -ethane) by the white-rot fungus *Phanerochaete chrysosporium*. *Appl. Environ Microbiol.* 53, 2001–2008.

17. Hall, D. O., Markov, S. R., Watanable, Y., Rao, K. K. (1995).The potential applications of cyanobacterial photosynthesis in clean technologies. *Photosynthesis Research. 46*, 159–167.

18. Dongjin Pyo, ; Taemin Kim, ; Jisun Yoo. (2013). Efficient Extraction of Bioethanol from Freshwater Cyanobacteria Using Supercritical Fluid Pretreatment. *Bull. Korean Chem.Soc. 34*, 2, 379.

19. Rippka, R., Deruelles, J., Waterbury, J. B., Herdman, M., Stainer, R. Y. (1979). Genetic assignments, strain histories and properties of pure cultures of cyanobacteria. *J. Gen. Microbiol. 111*, 1–61.

20. Chandel, A. K., Kapoor, R. K., Singh, A., Kuhad, R. C. (2007). Detoxification of sugarcane bagasse hydrolysate improves ethanol production by *Candida shehatae* NCIM 3501. *Biores. Technol. 98, 2*, 1947–1950.

21. Martinez, A., Rodriguez, M. E., Wells, M. L., York, S. W., Preston, J. F., Ingram, L. O. (2001). Detoxification of dilute acid hydrolysates of lignocellulose with lime. *Biotechnol. Progress. 17*, 287–293.

22. Gong, C. S., Chen, C. S., Chen, L. F. (1993). Pretreatment of sugarcane bagasse hemicellulose hydrolyzate for ethanol production by yeast. *Appl. Biochem. Biotechnol. 39–40*, 83–88.

23. Miller, G. L. (1959). Use of dinitrosalicylic acid reagent for determination of reducing sugar. *Analytical Chem. 31*, 426–428.

24. Wildenradt, H. L., Singleton, V. L. (1974). The production of aldehydes as a result of oxidation of polyphenolic compounds and its relation to wine aging. *Am. J. Enol. Vitic. 25*, 119–126.

25. Badger, P. C. (2002). Ethanol from cellulose: a general review. *In:* Janick J, Whipkey A, editors. Trends in new crops and new uses. Alexandria, VA: ASHS Press; 17–21.

26. Olsson, L., Hahn-Hagerdal, B. (1996). Fermentation of lignocellulosic hydrolysates for ethanol production. *Enz. Microb. Technol. 18*, 312–331.

27. Sun, Y., Cheng, J. (2002). Hydrolysis of lignocellulosic materials for ethanol production. *Biores. Technol. 83*, 1–11.

28. McMillan, J. D. (1994). Pretreatment of lignocellulosic biomass. *In*: Conversion of Hemicellulose Hydrolyzates to Ethanol. Himmel, M. E., Baker, J. O., Overend, R. P. (editors), American Chemical Society Symposium, Washington, DC. 292–324.

29. Hinman, N. D., Schell, D. J., Riley, C. J., Bergeron, P. W., Walter, P. J. Preliminary estimate of the cost of ethanol production for SSF technology. *J. Appl. Biochem. Biotechnol.* (1992). *34–35*, 639–649.

30. Taherzadeh, M. J., Niklasson, C., Lidén, G. (1997). Acetic acid – friend or foe in anaerobic batch conversion of glucose to ethanol by *Saccharomyces cerevisiae*? *Chem. Eng. Sci. 52, 15*, 2653–2659.

31. Hartley, R. D., Jones, E. C. (1977). Phenolic components and degradability of cell walls of grass and legume species. *Phytochem. 16*, 1531–1534.

32. Hopkins, W. G. (1999). Introduction to Plant Physiology. John Wiley and Sons, Inc., New York. *12*, 1–528.

33. Roberto, C., Lacis, L. S., Barbosa, M., Mancilha, I. M. D. (1991). Utilization of sugarcane bagasse hemicellulosic hydrolysate by *Pichia stipitis* for the production of ethanol. *Process Biochem. 26*, 15–21.

34. Van Zyl, C., Prior, B. A., Preez, J. C. D. (1998). Production of ethanol from sugar cane bagasse hemicellulose hydrolysate by *Pichia stipitis*. *J. Appl. Biochem. Biotechnol. 17*, 357–369.

35. Amartey, S., Jeffries, T. (1996). An improvement in *Pichia stipitis* fermentation of acid hydrolysates hemicellulose achieved by overliming (calcium hydroxide treatment) and strain adaptation. *World J. Microbiol. Biotechnol. 12*, 281–283.

36. Kruse, B., Schugrel, K. (1996). Investigation of ethanol formation by *Pachysolen tannophilus* from xylose and glucose-xylose substrates. *Process Biochem. 31,* 389–408.

37. Dominguez, J. M., Gong, C. S., Tsao, G. T. (1996). Pretreatment of sugarcane bagasse hemicellulose hydrolyzate for xylitol production by yeast. *Appl. Biochem. Biotechnol. 57–58,* 49–56.

38. Dellweg, H., Rizzi, M., Methner, H; Debus, D. (1984). Xylose fermentation by yeasts. *Biotechnol. Lett. 6,* 354–395.

39. Du Preez, J. C., Bosch, M., Prior, B. A. (1986). The fermentation of hexose and pentose sugars by *Candida shehatae* and *Pichia stipitis*. *Appl. Microbiol. Biotechnol. 23,* 228–233.

40. Chung, I. S., Lee, Y. Y. (1985). Ethanol production of crude acid hydrolysate of cellulose using high-level yeast inocula. *Biotechnol. Bioeng. 27,* 308–315.

41. Singh, A., Das, K., Sharma, D. K. (1984). Production of xylose, furfural, fermentable sugars and ethanol from agricultural residues. *J. Chem. Technol. Biotechnol. 34,* 51–61.

42. Mussatto, S. I., Roberto, I. C. (2004). Alternatives for detoxification of diluted-acid lignocellulosic hydrolyzates for use in fermentative processes: a review. *Biores. Technol. 93,* 1–10.

43. Saha, B. C., Iten, L. B., Cotta, M. A., Wu, Y. V. Dilute acid pretreatment, enzymatic saccharification and fermentation of wheat straw to ethanol. *Process Biochem.* 2005.

44. Gupta, P. Bioethanol from *Saccharum spontaneum*. M.Sc thesis. Delhi University, New Delhi, India. 2006.

45. Zaccaro, M. C., Kato, A., Zulpa, G., Storni, M. M., Steyerthal, N., Lobasso, K., Stella, A. M. (2006). Bioactivity of *Scytonema hofmanni* (Cyanobacteria) in *Lilium alexandrae* in vitropropagation. *Electronic J. Biotechnol. 9, 3,* 210–214.

CHAPTER 7

CHALLENGES IN HARNESSING THE POTENTIAL OF LIGNOCELLULOSIC BIOFUELS AND THE PROBABLE COMBATING STRATEGIES

LAKSHMI SHRI ROY, VIJAY KUMAR GARLAPATI, and RINTU BANERJEE

Microbial Biotechnology and DSP Laboratory, Agricultural and Food Engineering Department, IIT Kharagpur, Kharagpur, West Bengal – 721302, India

CONTENTS

ABSTRACT

Thrust towards embarking on building a new economy based on non-fossil fuel energy has imparted impetus to research on lignocellulosic biofuel industry. Lignocellulosics form a large class of renewable feed-stocks, which include agricultural residues, municipal solid waste and food processing and industrial wastes. Although lignocellulosics are very energy rich substrates, unlocking its potential and extending it to practical large scales has a number of bottlenecks. The challenges in instituting and exploiting these resources for biofuel production compel us to rethink current paradigm of supply and usage. Hence there is an urgent need to combat these issues to scale it up and execute at practical scales. Technological constraints do not support cost-effective and competitive production of lignocellulosic bioethanol. In order to produce lignocellulosic bioethanol economically attractive, use of cheaper substrates, inexpensive pretreatment techniques, use of overproducing and recombinant strains for maximized ethanol tolerance and yields, better recovery processes, effective bioprocess integration, exploitation of side products, and reduction of energy and waste may be encouraged. A cohesive and dedicated approach can help in realizing large-scale commercial production of lignocellulosic bioethanol. This chapter is an exhaustive overview of this aspect of lignocellulosic biofuel production. Special case studies have been included to highlight the probable strategies that may be adopted in future.

7.1 INTRODUCTION

Current focus of research is in the development of clean technologies that use a sustainably produced feedstock [1]. Contextually Biofuels, represent a significant potential for sustainability and growth of industrialized countries because they can be generated from locally available renewable material. Serious technical barriers like, ineffective biomass to fuel conversion efficiency, limited supply of key enzymes employed, use of high energy demanding process operation, limit the near-term commercial application of advanced biofuels technologies. Despite an immense future potential, large-scale expansion of advanced biofuels technologies seems unlikely unless further research and development lead to lowering these barriers. The hidden promise of biofuels is locked up in lignocellulosic biomass. Lignocellulose is a sturdy, microstructure of cellulose fibers covered by the heterogeneous hemicellulose and pectin and associated with lignin. The macromolecular structure of lignocellulose has been depicted in Fig. 7.1.

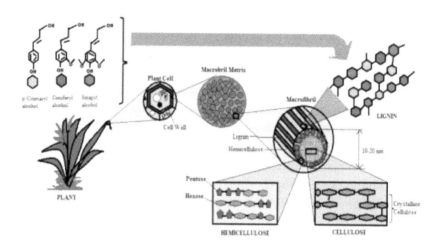

FIGURE 7.1   Molecular structure of lignocellulosic material [126].

Globally the availability of lignocellulosic raw materials is fairly consistent, unlike the fossil fuel reserves. Therefore, production of biofuels and biochemicals using nonedible lignocellulosic feedstock may prove advantageous to the society [2]. The significant benefits include production of energy from renewable and sustainable feedstock, aiding in the carbon dioxide fixation and pacifying the pertinent global warming issue, mobilizing local economic development minimizing air pollution from conventional mode of use of such feedstock by burning, decreasing the dependency on imported oils, and creating employment. Lignocellulose biomass by virtue of their high carbohydrate content holds tremendous potential for large-scale biofuel production, and need to be exploited for commercial production [3, 4]. The recalcitrance of lignocellulosic feedstock to chemicals or enzymes may be attributed to numerous natural factors. The degree of lignifications, the structural heterogeneity and complexity of cell-wall constituents, affect the conversion process. Enzymatic action is impeded while acting on an insoluble substrate. Additionally, the crystallinity and restricted solvent accessibility affect liquid penetration and/or enzyme accessibility and activity.

To realize the true potential of the lignocellulosic feedstock the various aspects of the biofuel production have to be dealt into. Low ethanol yield at the lab scale compounded with the high production costs have till date prevented the commercial scale production of this fuel [3, 4]. However, it is anticipated that with the maturity of the technology the cost would be reduced. For biofuel production from lignocellulosic biomass, it is therefore imperative to identify the technology bottlenecks and cost-contributing factors. Critical analysis of the most recent translational research strategies can make the overall process practical and profitable at an industrial level.

7.2 LIGNOCELLULOSIC SOURCES

Several groups of lignocellulosic are available throughout the world. These groups of raw materials are differentiated by their origin, composition and structure into:

7.2.1 FORESTLAND MATERIALS

Forestland materials include mainly woody biomass. These feedstocks are classified into broad categories of either softwoods or hardwoods. Softwoods originate from conifers and gymnosperm trees [5] and possess lower densities and grow faster than the hardwoods. Gymnosperm trees include mostly evergreen species such as pine, cedar, spruce, cypress, fir, hemlock and redwood [6]. Hardwoods are angiosperm trees and are mostly deciduous [7] and include trees such as poplar, willow, oak, cottonwood and aspen [8]. Unlike agricultural biomass, woody raw materials offer flexible harvesting times, avoid long latency periods of storage [9], possess more lignin and less ash content (close to zero), high density and minimal ash content than agricultural residues. These unique characteristics of woody biomass in conjunction to lower pentose content make them very attractive to cost-effective transportation over agricultural biomass and hence more favorable for greater bioethanol conversion if recalcitrance is surmounted[9]. Forestry wastes such as sawdust from sawmill, slashes, wood chips and branches from dead trees have also been used as biofuel feedstocks [10].

7.2.2   AGRICULTURAL RESIDUES

Crop residues are mostly comprised of agricultural wastes such as corn stover, corn stalks, rice and wheat straws as well as sugarcane bagasse [11]. Pasture includes food crops and grassland that encompass primarily agricultural residues, that is, nonfood crops and switch grass and alfalfa [12]. These substrates avoid "food versus fuel" controversy and are known to possess several advantages [13, 14]. Increased growth of these plants occurs in marginal and erosive soils and with nitrogen fertilization [15]. C4 biofuel feedstock improvement is only at its inception stage [16]. These materials contain more hemicellulosic material than woody biomass (approximately 25–35%) [17]. These residues help to avoid reliance on forest woody biomass and thus reduce deforestation (nonsustainable cutting plants). Unlike trees, crop residues are characterized by a short-harvest rotation that renders them more consistently available to bioethanol production [18, 19].

7.2.3   MUNICIPAL AND INDUSTRIAL WASTES

Municipal and industrial wastes are also potential recyclable cellulosic materials that can originate either from residential or nonresidential sources such as food wastes and paper mill sludge [12, 20]. Environmental problems associated with the disposal of garbage household, processing papers, food-processing by-products, black liquors and pulps [21] are controlled by their effective utilization.

7.2.4   MARINE ALGAE

Interest in algae as a potential biofuel feedstock has received an increasing support [11] in the recent times but its development as a viable and scalable source commercial enterprise remained limited. Marine algae biomass is regaining interest as a biofuel feedstock due to the rapid biorefineries expansion, which has lead to a shortage on current energy crops. Algae would also be a feedstock for other possible applications involving bio-crude oils, bio-plastics and recovered livestock coproducts [22].

Furthermore, algae feedstock with its thin cellulose layer has a higher carbohydrate composition [23] and does not compete directly with foods. It does not require agricultural land or use of freshwater to be cultivated. It consumes a high level of CO_2 during its growth, which makes it environmentally attractive as a CO_2 sink [24].

7.3   LIGNOCELLULOSIC BIOMASS COMPOSITION

Lignocellulosic material can generally be divided into three main components: cellulose (30–50%), hemicellulose (15–35%) and lignin (10–20%) [25–28] (Fig. 7.2). Cellulose and hemicelluloses make up approximately 70% of the entire biomass and are tightly linked to the lignin component through covalent and hydrogenic bonds that make the structure highly robust and resistant to any treatment [18, 27, 29].

7.3.1   HEMICELLULOSE

Hemicellulose is an amorphous and variable structure formed of heteropolymers including hexoses (D-glucose, D-galactose and D-mannose) as well

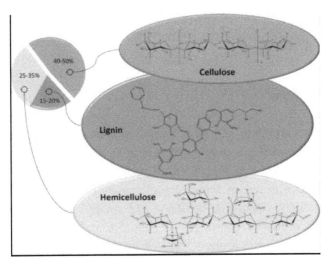

FIGURE 7.2   Lignocellulosic composition [125].

as pentose (D-xylose and L-arabinose) and may contain sugar acids (uronic acids) namely, D-glucuronic, D-galacturonic and methylgalacturonic acids [30, 31]. Its backbone chain is primarily composed of xylan b(1/4)-linkages that include D-xylose (nearly 90%) and L-arabinose (approximately 10%) [28]. Branch frequencies vary depending on the nature and the source of feedstocks. Softwood has typically glucomannans while hardwood is more frequently composed of xylans [30]. Xylan is the most abundant hemicellulose component but its composition varies in each feedstock [32]. The diversity of hemicellulose sugars, advocates the use of a wide range of enzymes for its complete hydrolysis into free monomers.

7.3.2   CELLULOSE

Cellulose is the most prevalent organic polymer and is approximately 30% of the plant composition [11]. Cotton, flax and chemical pulp represent the purest sources of cellulose while soft and hardwoods contain approximately 45% cellulose [17, 25]. Cellulose is a structural linear component of a plant's cell wall. It consists of a long-chain of glucose monomers linked b(1/4)-glycosidic bonds. The extensive hydrogen linkages among molecules lead to a crystalline and strong matrix structure [33]. Numerous Cross-linking hydroxyl groups constitute the microfibrils, which give the molecule more strength and compactness. To be converted from crystalline to amorphous texture in water, cellulose requires 320 °C as well as a pressure of 25 MPa [34].

7.3.3   LIGNIN

Lignin is an aromatic and rigid biopolymer with a molecular weight of 10,000 Da bonded via covalent bonds to xylans (hemicellulose portion) conferring rigidity and high level of compactness to the plant cell wall [27]. It is composed of three phenolic monomers of phenyl propionic alcohol namely, coumaryl, coniferyl and sinapyl alcohol. Forest woody biomass is primarily composed of cellulose and lignin polymers. Softwood barks have the highest level of lignin while grasses and agricultural residues contain the lowest level of lignin [17, 25]. Conversely, crop residues such as corn stover, rice and wheat straws are comprised mostly of a hemicellulosic heteropolymer that includes a large number of 5-carbon pentose sugars of primarily xylose [35]. Lignin components are gaining importance because of their dilution effect on the hydrolysis process once solids are added to a fed batch hydrolytic or fermentation bioreactor in addition to their structure and concentration effects that would affect potential hydrolysis [36].

7.4 BIOFUELS OBTAINED FROM LIGNOCELLULOSICS

7.4.1 BIODIESEL

Biodiesel is the second most abundant renewable liquid fuel [37]. Its preparation is outlined schematically in Fig. 7.3 given below. First generation biodiesel [38, 39] is produced by esterification of fatty acids or transesterification of oils (triglycerides) with alcohols (normally methanol and ethanol) using a basic [40–42] or acidic catalyst [43, 44]. The fatty esters are separated from glycerol by decantation and purified for direct use as fuels. Another alternative to process oil into biofuels is hydro-treating [45] which can be carried out synergistically in the existing petroleum refinery infrastructure through mixing and coprocessing of vegetable oils with petroleum derived feedstocks [46, 47]. The main drawback of oil-based processes is the availability of inexpensive feedstocks. Challenges in biodiesel research are focused on the development of new low cost catalytic processes that allow the use of low quality or waste oils (e.g., used fryer oil). Other sources include different nonedible oils (e.g., cynara [48], Jatropha, Karanja [49]) and triglycerides derived from algae [50]. Lignocellulosic feedstocks despite their complexity are currently being used to produce biofuels from biomass.

7.4.2 OTHER LIQUID FUELS OBTAINED FROM LIGNOCELLULOSICS

Lignocellulosic biomass is converted to hydrocarbon fuels by the removal of oxygen, combined with the formation of C–C bonds to control the molecular weight of the final hydrocarbons. The various strategies adopted have been depicted in the Fig. 7.4.

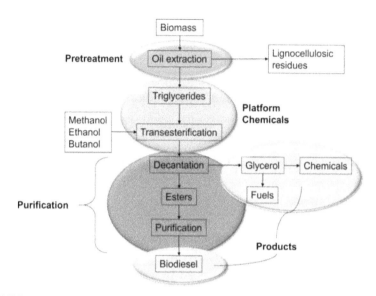

FIGURE 7.3   Process schematic: biodiesel production [125].

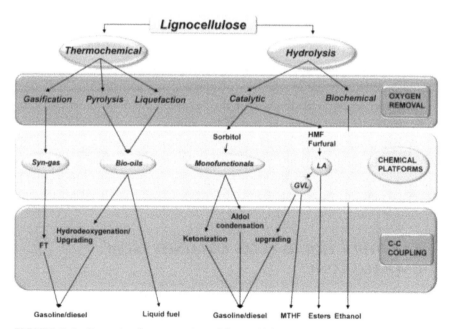

FIGURE 7.4   Strategies for conversion of lignocellulosic biomass to liquid biofuels by thermochemical and hydrolysis routes [125].

Two definite approaches include: (i) Gasification combined with Fischer–Tropsch synthesis, and (ii) Other pyrolysis and liquefaction. Other approaches include: (iii) Aqueous-phase reforming, (iv) An approach in which aqueous solutions of sugars formed by treatment of lignocellulosic biomass undergo catalytic dehydration to produce furan compounds, such as furfural and hydroxymethylfurfural (HMF). These furanic aldehydes can then be used as feedstocks for aldol-condensation reactions over basic catalysts to produce hydrocarbons suitable for Diesel fuel applications. (v) Finally, the levulinic acid platform, in which lignocellulosic biomass first undergoes treatment in acid solutions to produce levulinic acid. The aqueous solution of levulinic acid (in the presence of formic acid) then undergoes catalytic reduction to g-valerolactone (GVL), which serves as an intermediate for the production of nonane for Diesel fuel or the production of branched alkanes with molecular weights appropriate for jet fuel.

7.4.3 BIOETHANOL

Bioethanol is the most abundantly produced biofuel. It accounts for more than 94% of total biofuel production. The production of bioethanol, has been outlined in Fig. 7.5.

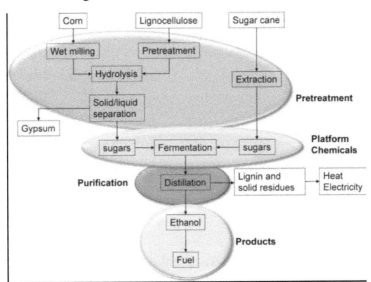

FIGURE 7.5   Process schematic: bioethanol production [125].

Lignocellulosic biomass has been considered an appropriate feedstock to produce sustainable, second generation bioethanol. An active area of research is to improve the suitability of this feedstock and the recently included algal feedstock. However, these potential alternatives are still at an early stage of development, and more research on feedstock availability and process optimization is necessary.

7.5 BIOCHEMICAL ROUTE OF BIOETHANOL PRODUCTION FROM LIGNOCELLULOSIC FEEDSTOCKS

Lignocellulosic biomass can be transformed into bioethanol via two different approaches (i.e., biochemical or thermochemical conversion) as shown in the Fig. 7.6 [51].

Both routes involve degradation of the recalcitrant cell wall structure of lignocellulose into fragments of lignin, hemicellulose and cellulose, hydrolysis of polysaccharides into sugars and finally conversion and purification of bioethanols [52, 53]. However, these conversion routes do not fundamentally follow similar techniques or pathways. Unlike the thermochemical

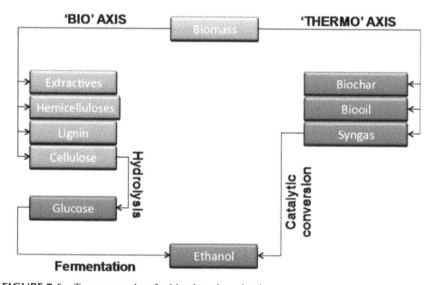

FIGURE 7.6   Two approaches for bioethanol production.

route, biochemical conversion involves physical (i.e., size reduction) or/and thermo-chemical with possible biological pretreatment [54]. Biochemical pretreatment is mainly used to overcome recalcitrant material and increase surface area to optimize cellulose accessibility to cellulases [54–56]. The upstream operation is followed by enzymatic or acidic hydrolysis of cellulosic materials (cellulolysis) and conversion of hemicellulose into monomeric free sugars (saccharification) subsequent to biological fermentation where sugars are fermented into ethanol and then purified via distillation [52, 57]. Concurrently, lignin, the most recalcitrant material of cell walls is combusted and converted into electricity and heat [53].

Overall, biochemical approaches include four unit-operations (as shown in Fig. 7.7) that can be summarized as:

i.   pretreatment to degrade the recalcitrant structures of lignocelluloses to open it up for further action,

ii. enzymatic hydrolysis of the polymers into monomeric sugars, and

iii. fermentation of the sugars into alcohol,

iv. separation and recovery of bioethanol.

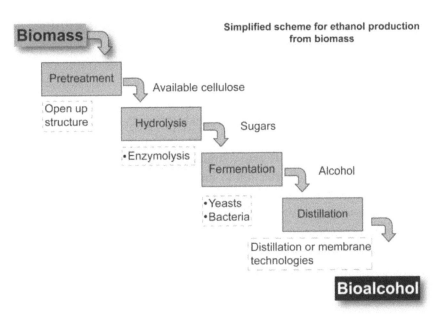

FIGURE 7.7   Simplified representation of biochemical conversion of Lignocellulosic to bioethanol.

Currently the biochemical route is the most commonly used process [58]. Following is the detailed discussion on the salient steps of the biochemical route of conversion of lignocellulosic biomass to bioethanol.

7.5.1   PRE-TREATMENT OF LIGNOCELLULOSIC FEED-STOCK

Pretreatment of recalcitrant lignocellulosic biomass is essential to degrade lignin and facilitation of cellulose and hemicelluloses digestion. Pretreatment or "prehydrolysis" is the process of exposing the cellulose component and making it more susceptible to enzymatic hydrolysis. Various pretreatment techniques involve a synergism between heat action, medium pH and duration of exposure under process conditions. All of these decrease cellulose crystallinity, and consequently, make it more susceptible to action of cellulases [59]. Prima focus of pretreatment research is on developing methods, which are mild, effective, cost-effective and environment-friendly. An effective pretreatment method should:

- degrade the crystalline matrix structure, separate the lignin and facilitate enzymatic action,
- minimize production of toxic by-products or inhibitors detrimental to the enzymes or fermentation microbes,
- reduce loss of the polymers like cellulose and hemicelluloses,
- make the process economical,
- be ecofriendly.

Several pretreatment methods include: physical, physicochemical, chemical or biological. Variation in the structural and compositional of the lignocellulosic matter of different sources, envisages the choice of different pretreatment methods or selection of more than one pretreatment method. In conclusion the selection of pretreatment method depends a lot on the source of the matter, its composition, the subsequent method of hydrolysis the pretreated matter will be subjected to, process parameters, etc.

7.5.2   HYDROLYSIS

During this reaction, the released polymer sugars, cellulose and hemicellulose are hydrolyzed into free monomer molecules readily available for fermentation conversion to bioethanol [60]. The success of the hydroly-sis step is essential to the effectiveness of a pretreatment operation [61]. There are two different types of hydrolysis processes that involve either acidic (sulfuric acid) or enzymatic reactions [62].

7.5.2.1 Acid Hydrolysis

The acidic reaction can be divided into dilute or concentrated acid hydro-lysis. Concentrated acid hydrolysis, the more prevalent method, has been considered to be the most practical approach [63]. Unlike dilute acid hydro-lysis, this process is not followed by high concentrations of inhibitors nor production of a high yield of free sugars (90%); however, it requires large quantities of acid as well as costly acid recycling, which makes it com-mercially less attractive [64].

7.5.2.2 Enzymatic Hydrolysis

While acid pretreatment results in a formation of reactive substrates when acid is used as a catalyst, acid hydrolysis causes significant chemi-cal dehydration of the monosaccharide's formed such that aldehydes and other types of degradation products are generated [65]. This particular issue has driven development of research to improve cellulolytic-enzymes and enzymatic hydrolysis. Effective pretreatment is fundamental to a successful enzymatic hydrolysis [66]. During the pretreatment process, the lignocellulosic substrate enzymatic digestibility is improved with the increased porosity of the substrate and cellulose accessibility to cellu-lases. *Trichoderma reesei* is one of the most efficient and productive fungi used to produce industrial grade cellulolytic enzymes. The most common cellulase groups produced by *T. reesei* that cleave the b(1,4) glycosidic bonds are b-glucosidase, endoglucanases and exoglucanases [67]. How-ever, cellulase enzymes exposed to lignin and phenolic-derived lignin are subjected to adverse effects [68–70] and have demonstrated that phenolic derived lignin have the most inhibitory effects on cellulases. In addition to phenolic components effect on cellulases, lignin has also an adverse effect on cellulases. As mentioned previously, the lignin adverse effect has two aspects including nonproductive adsorption and the limitation of the accessibility of cellulose to cellulase.

Although considerable genetic modifications (GMs) have been deployed to transform lignin effects, lignin has been shown to be a potential source of self sustaining-energy and added value components. Consequently, several research studies have determined practical approaches in eliminating inhibition of cellulases without involving GM approaches. Lui et al. [71] have demonstrated that the application of metal components namely, Ca (II) and Mg (II) via lignin-metal complexation substantially enhanced enzymatic hydrolysis. Additionally, Erickson et al. [72] have reported the importance of additives namely, surfactants and bovine serum albumin (BSA) in blocking lignin interaction with cellulases. Sewalt et al. [70] have reported that the adverse effect of lignin on cellulases can be surmounted by ammonization and various N compounds. Moreover, the enzymatic treatment can be accomplished simultaneously with the engineered cofermentation microbial process known as simultaneous saccharification and fermentation (SSF) [71, 73]. This process has been of interest since the late 1970 s for its effectiveness to minimize cellulolytic product inhibition and subsequently increase alcohol production [74]. Typically, separate hydrolysis and fermentation (SHF) processes involve the inhibition of the hydrolytic enzymes (cellulases) by saccharide products such as glucose and cellobiose. Unlike SHF, the SSF process combines hydrolysis and fermentation activities simultaneously and hence keeps the concentration of saccharides too low to cause any considerable cellulase inhibition [74].

7.5.2.3   Fermentation

Pretreatment and hydrolysis processes are designed to optimize the fermentation process [61]. This natural, biological pathway depending on the conditions and raw material used requires the presence of microorganisms to ferment sugar into alcohol, lactic acid or other end products [74, 60]. Typically, during batch fermentation *S. cerevisiae* ferments hexose sugars, mainly glucose, into ethanol in a large tank via the Embdene Meyerhof pathway under anaerobic conditions and controlled temperature. In yeast fermentation this is always accompanied by formation of CO_2 by-products and supplemented by nitrogen to enhance the reaction. This conventional strain is optimal at a temperature of approximately 30°C and resists a high osmotic pressure. It demonstrates tolerance to low pH levels of 4.0 as well as inhibitory products. *S. cerevisiae* can generate a high yield of ethanol (12.0–17.0% w/v; 90% of the theoretical) from hexose sugars.

Traditionally, separate hydrolysis and fermentation (SHF) sequential steps are used in bioethanol production. However, there is particular interest in targeting bioethanol production that can be derived from lignocellulosic biomass materials where both hexose and pentose sugars are available from the hemicellulose fraction.

Despite its broad tolerance to stressful bioethanol process conditions, *S. cerevisiae* is not able to ferment sugars other than hexose. Unfortunately, the biomass material includes a large proportion of hemicellulosic biomass that contains mainly pentose sugars such as D-xylose [78]. Moreover, an optimal fermentative microorganism should be tolerant to a high ethanol concentration and to chemical inhibitors formed during pretreatment and hydrolysis process. Inability of *S. cerevisiae* to ferment pentose sugars, has been attempted to be pacified by conducting extensive efforts to develop genetically engineered microorganisms that are capable of fermenting pentose and hexose sugars simultaneously. An optimal fermentative microorganism should be able to use both hexose and pentose simultaneously with minimal toxic end-products formation.

Different techniques including SSF and consolidated bioprocessing (CBP) have been developed to ensure the combination of hydrolysis (step 3) and fermentation (step 4) in one single reactor and thus, reduce product inhibition and operation costs. In addition to continuing downstream steps, CBP processing integrates both fermentation and cellulase formation in one fermentative/cellulolytic microorganism [79]. However, despite the extensive range of prokaryotic and eukaryotic microorganisms that have been shown to be able to produce ethanol from sugars, most of them remain limited in terms of sugars cofermentation, ethanol yield and tolerance to chemical inhibitors, high temperature and ethanol. Hydrolysis of Lignocellulosic materials not only produces fermentable sugars but also toxic compounds, such as weak acids, furan aldehydes and phenolics [80]. These substances and the presence of other stress factors often lead to reduced capacity or complete failure of the yeast for ethanol production.

For economic production of low value bulk chemical like ethanol, suitable fermentation conditions are essential to improve the ability of yeast to utilize the lignocellulosic substrate. Rapid ethanol fermentation methods to take place even at inhibiting conditions, that is, with the presence of fermenting toxic media or lignocellulosic hydrolysates are being developed. High cell concentrations inside the reactor were achieved by two methods – yeast encapsulation and fermentation in membrane bioreactors (MBR). The limitations of rapid yeast ethanol fermentation were attempted to be overcome by using these techniques at increased stress conditions like elevated cultivation temperatures, high inhibitor levels, such as high furfural concentration, high acetic acid concentrations and high hydrolysate dilution rates.

7.5.2.4   Separation and Purification of Bioethanol

Bioethanol obtained from a fermentation conversion requires further separation and purification of ethanol from water through a distillation process. Fractional distillation is a process implemented to separate ethanol from water based on their different volatilities. This process consists simply of boiling the ethanol water mixture. Because the boiling point of water is higher than the ethanol-boiling point, ethanol gets converted to steam before water. Thus, water can be separated via a condensation procedure and ethanol distillate recaptured at a concentration of 95% [81]. Typically, most large-scale industries and biorefineries use a continuous distillation column system with multiple effects [82]. Liquid mixtures are heated and allowed to flow continuously all along the column. At the top of the column, volatiles are separated as a distillate and residue is recovered at the bottom of the column.

7.6   MAIN CHALLENGES DURING LIGNOCELLULOSIC FEEDSTOCKS BIOCONVERSION TO BIOETHANOL

To bring down the production cost, several challenges in converting lignocellulosic biomass to biofuels need to be addressed. These challenges are in the areas of:

7.6.1   FEEDSTOCK PRODUCTION

Compositional changes in plant cell wall and differences in ultrastructure greatly influence the pretreatment and the resultant pretreated biomass sugar conversion. Also, the same type of biomass harvested from the same field in different years display changes in biomass composition (due to environmental conditions). This variance poses a challenge in adjusting the processing conditions and directly influences the biofuels yield. The cost of feedstocks significantly influences the cost of biofuel production. It's a challenge to convince farmers who cultivate grains for living to switch to bioenergy feedstocks.

Also most of the lignocellulosic materials are seasonal and need to be available in large quantities to be considered as feedstocks for biofuel industries which are itself yet another challenge. Biomass harvesting is an energy intensive process that requires large machinery and demands large amounts of fuel for transportation [84, 84]. Soil contamination of biomass is considered as one of the biggest challenges in biomass harvesting. Other key challenges are the moisture content of biomass and the amount of biomass that can be harvested from the field [85].

About one-third of biofuel production cost is associated with biomass cost. The cost of biomass is directly proportional to the yield (ton per ha) [86], which is influenced by soil fertility, location, and genetics. The existing challenge is to maintain a continuous supply of the feedstocks to the industries for the production process at a reduced cost.

Solution to these challenges includes plant breeding and genetic engineering techniques that have been applied to increase the yield of several potential energy feedstocks [87, 88]. The following are gene alterations by which growth can be promoted in different energy crops: (i) photosynthetic genes, (ii) transcription factors, (iii) cell cycle machinery, (iv) hormone metabolism, (v) lignin modification [89], and (vi) microRNA. In addition to these genetic manipulations, biomass yield can be improved by manipulating pathways in both abiotic and biotic stress [90, 91].

7.6.2   FEEDSTOCK LOGISTICS

The biomass supply chain include several key processing steps, which are: (i) collection, (ii) storage, (iii) preprocessing (densification by compaction, pelleting, and briquetting), (iv) transportation (from field to biorefinery), and (v) postprocessing at the biorefinery [92, 93]. These supply chain steps directly impact the cost of feedstock delivery. Texture variance, seasonal availability, low bulk density, and distribution over a large area are major challenges in transporting lignocellulosic biomass to biorefineries [94]. Transportation cost is also influenced by the moisture content, distance from the field to biorefinery, available infrastructure, available on-site technology, and the mode of transportation (rail or road) [95, 96].

7.6.3   CHALLENGES IN BIOCHEMICAL PROCESS OF LIGNOCELLULOSIC BIOMASS CONVERSION

Although lignocellulosic biomass is a potential feedstock for biorefineries, its recalcitrant structure and complexity remains a major economic and technical obstacle to lignocellulosic-based biofuel production [97]. The resilience of lignocellulosic materials is due to their composition and physicochemical matrix. The organization of vascular, epicuticular waxes as well as the amount of sclerenchymatous and the complexity of matrix molecules, contribute to the compactness and strength of the cellulosic material [98].

Furthermore, lignocellulosic materials as discussed previously are composed principally of three components namely, cellulose, hemicellulose and lignin. Together the polysaccharides, cellulose and hemicelluloses serve as initial substrates for subsequent saccharification and fermentation. However, these components are encapsulated via a tight covalent and hydrogen link to the lignin seal [99]. These tight bonds not only give the cell wall its compact structure but limit enzyme access to the surface area. Moreover, cellulose, a polymer of glucose molecules linked via b(1/4) -glycosidic bonds confers to cellulose a crystalline and compact structure [100].

Hemicellulose, the amorphous part of the cell wall, is composed of different hexoses and pentose sugars including xylose and arabinose bonded through xylans b(1/4)-linkages. These varieties of sugars polymers and linkages between molecules impose more complexities to the cell wall and therefore the hydrolysis process necessitates numerous cost-prohibitive enzymes to cleave polysaccharides entirely into fermentable sugar fragments.

Additionally, components including primarily xylo-oligosaccharides produced from hemicelluloses hydrolysis have been shown to be inhibitory to cellulase enzymes [101]. Although xylose causes a higher level of inhibition to cellulase enzymes than xylan, soluble xylo-oligomers are considered the inhibitoriest to cellulase and substantially influence enzymatic hydrolysis [102, 103]. Hence, the removal of these components in addition to organic acids and phenolics is desired in an attempt to achieve an efficient cellulose conversion via enzymatic hydrolysis [104]. Thus, a successful and low-cost ethanol bioconversion is closely related to the efficiency of the pretreatment step.

An effective pretreatment also requires a reduction of energy consumption with minimum toxic inhibitory products formation [105, 106]. However, in addition to these complexities and differences between components within the lignocellulosic material, lignocellulose composition from each type of biomass varies depending on the origin and geographical location. Not all types of lignocellulosic feedstocks require the same pretreatment strategy. These heterogeneities have an important impact on the choice of pretreatments and the downstream processes [107]. Currently, the SPORL treatment is of interest for its broad-spectrum ability on acting in both softwood and strong hardwood materials [108, 109]. This pretreatment degrades high lignin forest material with a limited formation of hydrolysis inhibitors [110].

Other issues include, establishment of biofuel and biochemical standards, biofuel distribution, societal acceptance, and environmental impact minimization. All of these challenging areas require expertise in agronomy, biomass logistics, biomass conversion, process engineering, chemistry, conversion technology, genetic engineering, microbial fermentation, economics, and environmental science.

7.7   RECENT INNOVATIVE APPROACHES TO COMBAT THE BOTTLENECKS IN LIGNOCELLULOSIC BIOCONVERSION TO BIOETHANOL

Several attempts have been tried to improve the quality of enzyme hydrolysis and fermentation. These include: multienzyme action, noncatalytic additives, high solids operation, multimicrobial systems, strain improvement, simultaneous pretreatment and saccharification, and efficient design of bio-reactors. Some of these are discussed in the following sub-sections.

7.7.1   ENZYME CONCOCTION APPROACH

Major costs are incurred by the enzymes, which are used to release and scarify the polysaccharides from the complex lignocellulosic structure during the bioethanol production. This is a great challenge for cost effective bioethanol production. Available cellulases suffer from inhibition due to different end products. Hence, more potent and efficient enzyme preparations are needed for the enzymatic saccharification process at reduced costs.

A large number of cellulases and other complementary enzymes will enable tailoring such cocktails. Pretreatment and the residual lignocellulosic properties is a potent determining factor in tailoring of cellulase mixtures for improvements in overall performance. Rapid screening of hydrolase and transferase functions will be required to screen for activity (e.g., endo-, exo-, betaglucosidase). With the advent of liquid handling automation, miniaturization of enzyme assays, and high throughput screening methodologies, large enzyme libraries from diverse organisms can be evaluated for their enzymic action. High throughput methods are being developed to implement protein engineering for enhancement of enzyme performance on biomass. Despite extensive research on cellulases, there are major gaps in our understanding of the actual mechanism involved in the hydrolysis of crystalline cellulose, mediated by the synergistic action of various constituents of the cellulase enzyme complex including the accessory enzymes. Approaches like enzyme engineering, reconstitution of enzyme mixtures and bioprospecting for superior enzymes are gaining importance [111]. The current scenario, however, also warrants the need for research and development of integrated biomass production and conversion systems.

7.7.2   DIRECT GLUCOSE PRODUCTION USING *CLOSTRIDIUM THERMOCELLUM* CULTURES SUPPLEMENTED WITH A THERMOSTABLE β-*GLUCOSIDASE*

Cost of Cellulases continues to be one of the major determining factors associated with the lignocellulose hydrolysis process. *Clostridium thermocellum* is an anaerobic, thermophilic, cellulolytic bacterium that produces cellulosomes that are effective plant cell walls degraders. Designer cellulosomes [112–115] and reconstruction of recombinant cellulosomes [116, 117] have been studied for enhancing the hydrolytic action of the strain. The strains ability to grow is higher than its aerobic counterpart while its cellulose yield is lower [118]. Consolidated bioprocessing is one effective proven method capable of resolving these issues to resolve these problems. It relies on *C.thermocellum* to ferment substrate to desired products (for example, ethanol) in one step without adding externally produced enzymes. This approach is widely recognized as for the final configuration for low-cost hydrolysis and fermentation of cellulosic biomass tis method is gaining popularity. Weak tolerance to ethanol and low ethanol yield obtained with this strain limit its immediate and direct application to industrial ethanol fermentation from cellulose [119]. It has been observed that significant enhancement of its cellulolytic activity from the *C. thermocellum* S14 strain [120] could be obtained when in combination with a thermostable β-glucosidase from *Thermoanaerobacter brockii* (CglT) [121].

Moreover the end-product cellobiose, inhibits degradation for which reason the end product action has to eliminated for the successful action of this strain. In biological simultaneous enzyme production and saccharification (BSES) saccharification cultures of *C. thermocellum* cultures supplemented with thermostable β-glucosidases are employed with no additional incorporation of cellulolytic enzymes. Direct production of glucose from cellulosic materials is obtainable due to supplementation of cellulose degrading cultures with β-glucosidase. Exclusive glucose accumulation of glucose occurred when *C. thermocellum* was cultured with a thermostable β-glucosidase under a high cellulose load. This approach can successfully resolve a significant barrier to economical production of bio-based chemicals and fuels from lignocellulosic biomass.

7.7.3 SINGLE STEP BIOETHANOL PRODUCTION BY EMPLOYING NOVEL STRAINS

A process strategy that aims to encompass the critical cost-increasing item is the consolidated bioprocessing approach [122, 123]. In CBP an organism or a mixed culture of organisms produces the requisite enzymes for hydrolysis of the polymers of lignocellulosic biomass and subsequently ferments the C5 and C6 sugars into ethanol or other valuable products without the extraneous addition of cellulolytic or hemicellulolytic enzymes. Several mesophilic and thermophilic cellulolytic as well as noncellulolytic microorganisms with engineered cellulase activity are under development for the application in CBP [122, 124]. Co-cultivation of the thermophilic cellulolytic bacterium *Clostridium thermocellum* with noncellulolytic thermophilic anaerobes at temperatures of 55–60 °C are being used in CBP. *Caldicellulosiruptor sp.* strain DIB 004C a recently discovered strain is found to be capable of producing impressive yields of ethanol from lignocellulose in fermentors. Thus the cocultures of new *Caldicellulosiruptor* strains with new Thermoanaerobacter strains establish the importance of using specific strain combinations for high ethanol yields. These thus prove effective constituents of CBP pathway for ethanol production and represent a crucial turning point for development of a highly integrated commercial ethanol production process.

7.7.4 USE OF SUPER COMPUTER IN BIOETHANOL PRODUCTION

Computer simulation analysis seem the ultimate means to make detailed structural, dynamic and functional descriptions of otherwise difficult complex biological molecular systems [125]. The simulation procedure involves entering the exhaustive information from experiment and high-level calculations (such as quantum chemistry) to generate an atomic-detail model of the system and its interactions. Subsequently high-performance computer clusters and supercomputers may be employed to follow the system evolution when it is subjected to realistic forces. The technique would aid in revealing the mechanisms of plant cell-wall deconstruction in exquisite detail. This in turn would facilitate a thorough understanding of the structures, dynamics, and degradation pathways of lignocellulosic materials. Because biomass is a multicomponent and complex multiscale material, clarifying its structure is particularly challenging. Advanced and experimental imaging characterization techniques, such as atomic force microscopy and neutron scattering, which span a wide range of length scales (from angstroms to micrometers) may be employed to elucidate the multicomponent and multiscale structure. Differentiation of individual components from the embedded phases within the biomaterial, such as crystalline and amorphous cellulose, lignin, and hemi-cellulose may also be achieved thereby facilitating the overall bioethanol production.

7.8 SUMMARY

With the abundantly available biomass waste, the development of new technologies that will make use of biomass for biofuels and materials production represents an important opportunity to fully use our resources. Rigorous research in Ligno-cellulosic technology must be conducted to overcome significant challenges in various aspects. Development of efficient techniques to process lignocellulosic biomass into its core components will facilitate research on the production of specific biomass derived sugars, building block chemicals and ultimately value-added commodity chemicals by promoting effective utilization of all feedstock fractions. Research activities undertaken to achieve this include, improvement of the perennial biomass, manipulation of lignin structure and plant cell walls so as to achieve optimal sugar release, trials with new enzymes and improvement of the yeast strains to ferment sugars. An in-depth understanding of the mechanism of interlocking of the sugars in plant structures would further facilitate in the design and development of enzymatic digestion processes for release of sugars by enzymes. This would lead to an effective biofuel production. Currently we lack effective enzymes to digest all kinds of woody materials. Intensive research studies on the digestive process of a range of wood boring microbes, and exploitation of these microbes for extensive industrial application is advocated.

Pretreatment, the most costly step is of particular concern due to the high recalcitrance of lignocellulosic raw materials. It has been deemed imperative to design a general pretreatment combination that would be effective against a wide range of cellulosic material and hence deal with feedstock variability.

Emerging technologies including CBP represent potential improvements as they reduce operation steps as well as chemical inhibitors and can be enhanced by lignin, energy-self-sustaining coproducts. These processes are typically associated with thermophilic and cellulolytic microorganisms with some of them possessing fermentative abilities in addition to their hydrolytic properties. Genetically engineered conventional strains, *S.cerevisiae* and ethanologenic *Z. mobilis* may be preferred for their higher alcohol tolerance and yield. In conjunction to rapid molecular biology techniques, mathematical modeling including biotechnology risk assessment (BRA) can be used to ensure greater predictability for limiting antibiotic resistant microflora and GMO dissemination during operation. While technological accomplishments and multiple research coalition efforts are still progressing, an efficient combination of the most advanced systems analysis and economical techniques designed to cope with lignocellulosic feedstock versatility should emerge as the option of choice in an attempt to achieve optimal biofuel performance with minimal negative environmental, social or economic impacts needs to be adopted and enhanced.

KEYWORDS

- **Biofuels**
- **Bottlenecks**
- **Hydrolysis**
- **Lignocellulosics**
- **Pretreatment**
- **Renewable**

REFERENCES

1. Tilman, D., Socolow, R., Foley, J. A., Hill, J., Larson, E., Lynd, L., Pacala, S., Reilly, J., Searchinger, T., Somerville, C., Williams, R. (2009). Energy. Beneficial biofuels–the food, energy, and environment trilemma. *Science, 325,* 270–271.
2. Greenwell, H. C., Loyd-Evans, M., Wenner, C. (2012). Biofuels, science and society. *Inter. Foc. 3,* 1–4.
3. Lynd, L. R., Cushman, J. H., Nichols, R. J., Wyman, C. E. (1991). Fuel ethanol from cellulosic biomass. *Science. 251,* 1318–1323.

4. Linoj Kumar, N. V., Sameer, M. Alternative feedstock for bioethanol production in India, in Biofuels: toward a greener and secure energy future, ed by Bhojvaid PP. The Energy and Resources Institute, New Delhi, 89–103 (2006).

5. Hoadley, R. B. Understanding wood: a craftsman's guide to wood technology.2nd ed. Newtown, CT: Taunton Press; 2000.

6. Boone, R. S., Kozlik, C. J., Bois, C. J. P. J., Wengert, P. P. J. E. M. Dry kiln schedules for commercial woods temperate and tropical. Gen. Tech. Rep. FPL_GTR_57.Madison, WI: U. S. Department of Agriculture, Forest Service, Forest Products Laboratory; 1988.

7. Markwardt, L. J., Wilson, T. R. C. Strength and related properties of woods grown in the United States. Tech. Bull. 479. Washington, DC: U.S. Department of Agriculture, Forest Service. U. S. Government Printing Office; 1935.

8. Kennedy, J. H. E. Cottonwood, an American wood. Washington, DC: U. S.Department of Agriculture, Forest Service; 1985.

9. Zhu, J. Y., Pan, H. J. (2010). Woody biomass treatment for cellulosic ethanol production:technology and energy consumption evaluation. *Bioresour. Technol. 101,* 4992–5002.

10. Perlack, R. D., Wright, L., Turhollow, L. A., Graham, R. L., Stokes, B., Erbach, D. C. Biomass as feedstock for a bioenergy and bioproducts industry: the technical feasibility of a billion-ton annual supply. Oak Ridge National Laboratory Report ORNL/TM-2005/66. Oak Ridge, TN: US Dept. of Energy; 2005.

11. U. S. Department of Energy Biomass Program. http://www1.eere.energy.gov/biomass/pdfs/biomass_deep_dive_pir.pdf. 2009.

12. Hu, G., Heitmann, J. A., Rojas, O. J. (2008). Feedstock pretreatments strategies for producing ethanol from wood, bark, and forest residues. *BioRes. 3,* 270–294.

13. Hill, J., Nelson, E., Tilman, D., Polasky, S., Tiffany, D. Environmental, economic, and energetic costs and benefits of biodiesel and ethanol biofuels. *Proc. Nat. Acad. Sci. USA. 103,* 11206–11210.

14. Chandel, A. K., Singh, V. (2011). Weedy lignocellulosic feedstock and microbial metabolic engineering: advancing the generation of 'Biofuel'. *App. Microbiol. Biotechnol. 89,* 1289–1303.

15. Muir, J. P., Sanderson, M. A., Ocumpaugh, W. R., Jones, R. M., Reed, R. L. (2001). Biomass production of 'Alamo' switchgrass in response to nitrogen, phosphorus, and row spacing. *Agron. J. 93,* 896–901.

16. Byrt, C. S., Grof, C. P. L., Furbank, R. T. (2011). C_4 Plants as biofuel feedstocks: optimizing biomass production and feedstock quality from a lignocellulosic perspective. *J. Integr. Plant. Biol. 53,* 120–135.

17. Demirbas, A. (2005). Bioethanol from cellulosic materials: a renewable motor fuel from biomass. *Energy. Source. 27,* 327–337.

18. Knauf, M., Moniruzzaman, M. Lignocellulosic biomass processing. *Persp. Int. Sugar. J.* (2004). *106,* 147–150.

19. Kim, S., Dale, B. E. (2005). Global potential bioethanol production from wasted crops and crops residues. *Biomass. Bioenerg. 29,* 361–375.

20. Cardona, C. A., Quintero, J. A., Paz, I. C. (2009). Production of bioethanol from sugarcane bagasse: status and perspectives. *Bioresour. Technol. 101,* 4754–4766.

21. Khanna, M. A billion tons of biomass a viable goal but at high price. *Am J Agric Econ*, http://news.illinois.edu/news/11/0216biomass_MadhuKhanna.html; 2011.

22. Emerging markets, ALGAE (2020). http://www.emerging-markets.com/algae/; 2011.

23. Rodolfi, L., Zitelli, G. C., Bassi, N., Padovani, G., Biondi, N., Bionini, G., et al. (2009). Microalgae for oil: strain selection, induction of lipid synthesis and outdoor mass cultivation in a low-cost photo-bioreactor. *Biotechnol. Bioeng. 102*, 100–112.

24. Harel, A. Noritech seaweed biotechnology Inc. Algae World Conference.Rotterdam, NL. 2009.

25. Pettersen, R. C. (1984). The chemical composition of wood. In: Rowell RM, editor. The chemistry of solid wood. Advances in chemistry series, vol. 207. Washington, DC: American Chemical Society; p. 115e6.

26. Badger, P. C. In: Jannick J, Whipsekey A, editors. (2000). Trends in new crops and new uses. Alexandria, VA: ASHS Press; p. 17e21.

27. Mielenz, J. R. (2001). Ethanol production from biomass: technology and commercialization status. *Curr. Opin. Microbiol. 4,* 324–325.

28. Girio, F. M., Fonseca, C., Carvalheiro, F., Duarte, L. C., Marques, S., Bogel-Lukasic, R. (2010). Hemicellulose. *Bioresour. Technol. 101,* 4775–4800.

29. Edye, L. A., Doherty, W. O. S. Fractionation of a lignocellulosic material. PCT Int Appl 2008; 25.

30. McMillan, J. D. (1993). Pretreatment of lignocellulosic biomass. In: Himmel ME, Baker JO, Overend RP, editors. Enzymatic conversion of biomass for fuel production. Washington, D.C: American Chemical Society; p. 292e323.

31. Saha, B. D. (2003). Hemicellulose bioconversion. *J. Ind. Microbiol. Biotechnol. 30,* 279–291.

32. Aspinall, G. O. (1980).Chemistry of cell wall polysaccharides. In: Preiss J, editor. The biochemistry of plants (a comprehensive treatise). Carbohydrates structure and function, vol. 3. NY: Academic Press; p. 473e500.

33. Ebringerova, A., Hromadkova, Z., Heinze, T. (2005). Hemicellulose. *Adv. Polym. Sci. 186,* 1–67.

34. Deguchi, S., Mukai, S. A., Tsudome, M., Horikoshi, K. (2006). Facile generation of fullerene nanoparticles by hand-grinding. *Adv. Mater. 18,* 729–732.

35. Foody, B. E., Foody, K. J. (2006). Development of an integrated system for producing ethanol from biomass. In: Klass DL, editor. Energy from biomass and waste. Chicago: Institute of Gas Technology; 1991. p. 1225e43.

36. Ladisch, M. R., Mosier, N. S., Kim, Y., Ximenes, E., Hogsett, D. (2010). Converting cellulose to biofuels. SBE special supplement biofuels. CEP *106 (3),* 56e63

37. ENERS Energy Concept. Production of biofuels in the world.Lausanne, 2009. http://www.plateforme-biocarburants.ch/en/infos/production.php?id = bioethanol.

38. Fukuda, H., Kondo, A., Noda, H. (2001). Biodiesel Fuel Production by Transesterification of Oils.*J. Biosci. Bioeng. 92,* 405–416.

39. Ma, F. R., Hanna, M. A. (1999). Biodiesel production: a review. *Bioresour. Technol. 70,* 1–15.

40. Kim, H. J., Kang, B. S., Kim, M. J., Park, Y. M., Kim, D. K., Lee, D. S., Lee, K. Y. (2004). Transesterification of vegetable oil to biodiesel using heterogeneous base catalyst. *Catal. Today. 93–95,* 315–320.

41. Di Serio, M., Ledda, M., Cozzolino, M., Minutillo, G., Tesser, R., Santacesaria, E. (2006). Transesterification of soybean oil to biodiesel by using heterogeneous basic catalysts. *Ind. Eng. Chem. Res. 45,* 3009–3014.

42. Granados, M. L., Poves, M. D. Z., Alonso, D. M., Mariscal, R., Galisteo, F. C., Moreno-Tost, R., Santamaria, J., Fierro, J. L. G. (2007). Biodiesel from surflower oil by using activated calcium oxide. *Appl.Catal., B. 73,* 317–326.

43. Lotero, E., Liu, Y. J., Lopez, D. E., Suwannakarn, K., Bruce, D. A., Goodwin, J. G. (2005). Synthesis of biodiesel via acid catalysis. *Ind. Eng. Chem. Res. 44,* 5353–5363.

44. Melero, J. A., Iglesias, J., Morales, G. (2009). Heterogenous acid catalysts for Biodiesel production:current status and future challenges. *Green. Chem. 11,* 1285–1308

45. Huber, G. W., O'Connor, P., Corma, A. (2007). Processing biomass in conventional oil refineries: Production of high quality diesel by hydrotreating vegetable oils in heavy vacuum oil mixtures.*Appl. Catal. A. 329,* 120–129.

46. Wheals, A. E., Basso, L. C., Alves, D. M. G., Amorim, H. V. (1999). Fuel ethanol after 25 years. *Trends. Biotechnol. 17,* 482–487.

47. Li, L. X., Coppola, E., Rine, J., Miller, J. L., Walker, D. (2010). Catalytic Hydrothermal Conversion of Triglycerides. Non-ester Biofuels. *Energy. Fuel. 24,* 1305–1315.

48. Fern'andez, J., Curt, M. D., Aguado, P. L. (2006). Industrial applications of *Cynara cardunculus L.* for energy and other uses. *Ind. Crops Prod. 24,* 222–229.

49. Patil, P. D., Deng, S. (2009). Optimization of biodiesel production from edible and nonedible vegetable oils. *Fuel. 88,* 1302–1306.

50. Chisti, Y. (2007). Biodiesel from microalgae. *Biotechnol. Adv. 25,* 294–306.

51. Demirbas, A. (2007). Progress and recent trends in biofuels. *Prog. Energy. Combust. Sci. 33,* 1–18.

52. Chandel, A. K., Chan, E., Rudravaram, R., Narasu, M. L., Rao, L. V., Ravindra, P. (2007). Economics and environmental impact of bioethanol production technologies: an appraisal. *Biotechnol. Mol. Biol. Rev. 2,* 14–32.

53. Gamage, J., Howard, L., Zisheng, Z. (2010). Bioethanol production from lignocellulosic biomass. *J. Biobased. Mater. Bioenerg. 4,* 3–11.

54. Yang, B., Wyman, C. E. (2008). The key to unlocking low-cost cellulosic ethanol. Biofuels. Bioprod. Bioref. 2, 26–40.

55. Zhu, J. Y., Pan, H. J. (2010). Woody biomass treatment for cellulosic ethanol production:technology and energy consumption evaluation. *Bioresour. Technol. 101,* 4992–5002.

56. Yang, B., Wyman, C. E. (2008). The key to unlocking low-cost cellulosic ethanol. *Biofuels. Bioprod. Bioref. 2,* 26–40.

57. Zhu, J. Y., Wang, G. S., Pan, X. J., Gleisner, R. (2009). Specific surface to evaluate the efficiencies of milling and pretreatment of wood for enzymatic saccharification. *Chem. Eng. Sci. 64,* 474–485.

58. Fehrenbacher, K. Logen suspends U. S. cellulosic ethanol plant plans [accessed October 22], http:earth2tech.com/2008/06/04/iogen-suspends-uscellulosic-ethanol-plant-plans-/; 2009.

59. Lynd, L. R., Weimer, P. J., Van Zyl, W. H., Pretorius, I. S. Microbial cellulose utilization: fundamentals and biotechnology. *Microbiol. Molecul. Biol.* Rev. (2002). *66,* 506–577.

60. Chandel, A. K., Chan, E., Rudravaram, R., Narasu, M. L., Rao, L. V., Ravindra, P. (2007). Economics and environmental impact of bioethanol production technologies: an appraisal. *Biotechnol. Mol. Biol. Rev. 2*, 14–32.

61. Gamage, J., Howard, L., Zisheng, Z. (2010). Bioethanol production from lignocellulosic biomass. *J. Biobased. Mater. Bioenerg. 4*, 3–11.

62. Nazhad, M. M., Ramos, L. P., Paszner, L., Saddler, J. N. (1995). Structural constraints affecting the initial enzymatic hydrolysis of recycled paper. *Enzyme. Microbiol. Technol. 17*, 66–74.

63. Torget, R., Walter, P., Himmel, M., Grohmann, K. (1991). Dilute acid pretreatment of short rotation woody and herbaceous crops. *Appl. Biochem. Biotechnol. 4–25*, 115–126.

64. Hamelinck, C. N., Hooijdonk, G., Faaij, A. P. C. (2005). Ethanol from lignocellulosic biomass: techno-economic performance in short-, middle- and long-term. *Biomass. Bioenerg. 28*, 384–410.

65. Sun, Y., Cheng, J. (2002). Hydrolysis of lignocellulosic materials for ethanol production: a review. *Bioresour. Technol. 83*, 1–11.

66. Hendriks, A. T. W. M., Zeeman, G. (2009). Pretreatments to enhance the digestibility of lignocellulosic biomass. *Bioresour. Technol. 100*, 10–18.

67. Eggeman, T., Elander, R. T. (2005). Process and economic analysis of pretreatments technologies. *Bioresour. Technol. 96*, 2019–2025.

68. Larsson, S., Palmqvist, E., Hahn-Hagerdal, B., Tengborg, C., Stenberg, K., Zacchi G., et al. (1999). The generation of fermentation inhibitors during dilutes acid hydrolysis of softwood. *Enzyme. Microbiol. Technol. 24*, 151–159.

69. Ximenes, E., Kim, Y., Mosier, N., Dien, B., Ladisch, M. (2010). Inhibition of cellulases by phenols. *Enzyme. Microbiol. Technol. 46*, 170–176.

70. Sewalt, V. J. H., Glasser, W. G., Beauchemin, K. A. (1997). Lignin impact on fiber degradation. 3. Reversal of inhibition of enzymatic hydrolysis by chemical modification of lignin and by additives. *J. Agric. Food. Chem. 45*, 1823–1828.

71. Cao, N. J., Krishnan, M. S., Du, J. X., Gong, C. S., Ho, N. W. Y. (1996). Ethanol production from corn cob pretreated by the ammonia steeping process using genetically engineered yeast. *Biotechnol. Lett. 118*, 1013–1018.

72. Erickson, T., Borjesson, J., Tjerneld, F. (2002). Mechanism of surfactant effect in enzymatic hydrolysis of lignocelluloses. *Enzyme. Microbiol. Technol. 31*, 353–364.

73. Bisaria, V. S., Ghose, T. K. (1981). Biodegradation of cellulosic materials: substrate, microorganisms, enzymes and products. *Enzyme. Microbiol. Technol. 3*, 90–104.

74. Kumar, S., Singh, S. P., Mishra, I. M., Adhikari, D. K. (2009). Recent advances in production of bioethanol from lignocellulosic biomass. *Chem. Eng. Technol. 32*, 517–526.

75. Hahn-Hägerdal, B., Karhumaa, H. B. K., Fonseca, C., Spencer-Martins, I., Gorwa-Grauslund, M. F. (2007). Toward industrial pentose-fermenting yeast strains. *Appl. Microbiol. Biotechnol. 74*, 937–953.

76. Bayrock, D., Ingledew, W. M. (2001). Changes in steady state on introduction of a Lactobacillus contaminant to a continuous culture ethanol fermentation. *J. Ind. Microbiol. Biotechnol. 27*, 39–45.

77. Claassen, P. A. M., Lopez, C., Sijtsma, A. M., Weusthuis, L., Van Lier, R. A., Van Niel, J.B, et al. (1999). Utilization of biomass for the supply of energy carriers. Appl. Microbiol. Biotechnol. *52,* 741–755.

78. Martin, C., Galbe Wahlbom, M., Hagerdal, B. H., Jonsson, J. L. (2002). Ethanol productionfrom enzymatic hydrolysates of sugarcane bagasses using recombinantxylose using Saccharomyces cerevisisae. *Enzyme. Microbiol. Technol. 31 (2),* 274–282.

79. Ladisch, M. R., Mosier, N. S., Kim, Y., Ximenes, E., Hogsett, D. (2010). Converting cellulose to biofuels. *SBE special supplement biofuels. CEP, 6 (3),* 56–63.

80. Demain, A. L., Newcomb, M., Wu, J. H. D. (2005). Cellulase, clostridia and ethanol. *Microbiol. Mol. Biol. Rev. 69,* 124–154.

81. Cardona, C. A., Sanchez, O. J. (2007). Fuel ethanol production: process design trends and integration opportunities. *Bioresour. Technol. 98,* 2415–2457.

82. Kent, N. L., Evers, A. D. (1994). Malting, brewing and distiling. In: Kent's technology of cereals. 4th ed., vol. 4. Cambridge: Woodhead Publishing; p. 218–232.

83. McKendry, P. (2002). Energy production from biomass (part 1): overview of biomass. *Bioresour. Technol. 83,* 37–46.

84. Thorsell, S., Epplin, F. M., Huhnke, R. L., Taliaferro, C. M. (2004). Economics of a coordinated biorefinery feedstock harvest system: lignocellulosic biomass harvest cost. *Biomass. Bioenerg. 27,* 327–337.

85. Huggins, D. R., Karow, R. S., Collins, H. P., Ransom, J. K. (2011). Introduction: evaluating long- termimpacts of harvesting crop residues on soil quality. *Agron. J. 103,* 230–233.

86. Duffy, M. D., Nanhou, V. Y. Switchgrass Production in Iowa: Economic Analysis, Special publication for Cahriton Valley Resource Conservation District, Iowa State University Extension Publication, Iowa State University, 2002.

87. Rojas, C. A., Hemerly, A. S., Ferreira, P. C. G. (2010). Genetically modified crops for biomass increase. Genes and strategies. *GM. Crop. 1,* 137–142.

88. Dubouzet, J.G, ; Strabala, T. J., Wagner, A. (2013). Potential transgenic routes to increase tree biomass. Plant. Sci. *212,* 72–101.

89. Vanholme, R., Morreel, K., Darrah, C. et al. (2012). Metabolic engineering of novel lignin in biomass crops. *New. Phytol. 196,* 978–1000.

90. Sticklen, M. (2006). Plant genetic engineering to improve biomass characteristics for biofuels. *Curr. Opin. Biotechnol. 17,* 315–319.

91. Bhatnagar-Mathur, P., Vadez, V., Sharma, K. K. (2008). Transgenic approaches for abiotic stress tolerance in plants: retrospect and prospects. *Plant. Cell. Report. 27,* 411–424.

92. Sokhansanj, S., Hess, J. R. "Biomass supply logistics and infrastructure," in Biofuels:Methods and Protocols, J. R.Mielenz, Ed., vol. 581 of Methods in Molecular Biology, chapter 1, pp. 1–25, Springer, Berlin, Germany, 2009.

93. Sultana, A., Kumar, A., Harfield, D. (2010). Development of agripellet production cost and optimum size. *Bioresour. Technol. 101,* 5609–5621.

94. Caputo, A. C., Palumbo, M., Pelagagge, P. M., Scacchia, F. (2005). Economics of biomass energy utilization in combustion and gasification plants: effects of logistic variables. *Biomass. Bioenerg. 28,* 35–51.

95. Kumar, A., Sokhansanj, S., Flynn, P. C. (2006). Development of a multicriteria assessment model for ranking biomass feedstock collection and transportation systems. Appl. Biochem. Biotechnol. *129*, 71–87.

96. Badger, P. C. Biomass transport systems, in Encyclopedia of Agricultural, Food and Biological Engineering, vol. 1, pp. 94–98, 2003.

97. Yu, Z., Zhang, B., Yu, F., Xu, G., Song, A. (2012). A real explosion: the requirement of steam explosion pretreatment. *Bioresour. Technol. 121*, 335–341.

98. Hideno, A., Inoue, H., Tsukahara, K. et al. (2009). Wet disk milling pretreatment without sulfuric acid for enzymatic hydrolysis of rice straw. *Bioresour. Technol. 100*, 2706–2711.

99. Dadi, A. P., Schall, C. A., Varanasi, S. (2007). Mitigation of cellulose recalcitrance to enzymatic hydrolysis by ionic liquid pretreatment. *Appl. Biochem. Biotechnol.* 137–140, 407–421.

100. Kaliyan, N., Morey, R, V. (2010). Natural binders and solid bridgetype binding mechanisms in briquettes and pellets made from corn stover and switchgrass. *Bioresour. Technol. 101*, 1082–1090.

101. Zhang Y., Fu, X., Chen, H. (2012). Pretreatment based on twostep steam explosion combined with an intermediate separation of fiber cells-optimization of fermentation of corn straw hydrolysates. *Bioresour. Technol. 121*, 100–104.

102. Zheng, Y., Lin, H. M., Tsao, G. T. (1998). Pretreatment for cellulose hydrolysis by carbon dioxide explosion. *Biotechnol. Progr. 14*, 890–896.

103. S'anchez, O. J., Cardona, C. A. (2008). Trends in biotechnological production of fuel ethanol from different feedstocks. Bioresour. Technol. *99*, 5270–5295.

104. Sharma, B., Ingalls, R. G., Jones, C. L., Khanchi, A. (2013). Biomass supply chain design and analysis: basis, overview, modeling, challenges, and future. *Renew. Sust. Energy Rev. 24*, 608–627.

105. Dubouzet, J. G., Strabala, T. J., Wagner, A. (2013). Potential transgenic routes to increase tree biomass. *Plant Sci. 212*, 72–101.

106. Zhu, X., Li, X., Yao, Q., Chen, Y. (2011). Challenges and models in supporting logistics system design for dedicated-biomass-based bioenergy industry. *Bioresour. Technol. 102*, 1344–1351.

107. Balan, V., da Costa Sousa, L., Chundawat, S. P. S., Vismeh, R., Jones, A. D., Dale, B. E. (2008). Mushroom spent straw: a potential substrate for an ethanol-based biorefinery. J. Indust. Microbiol. Biotechnol. *35*, 293–301.

108. Dale, B. E., Ong, R. G. (2012). Energy, wealth, and human development: why and how biomass pretreatment research must improve. *Biotechnol. Progr. 28*, 893–898.

109. Balan, V., Bals, B., Chundawat, S. P. S., Marshall, D., Dale, B. E. Lignocellulosic biomass pretreatment using AFEX," in Biofuels: Methods and Protocols, Mielenz, J. R. Ed., vol. 581ISRN Biotechnology 27 of Methods in Molecular Biology, pp. 61–77, Humana Press, Springer Science + Business Media, LLC, 2009.

110. Chundawat, S. P. S., Bals, B., Campbell, T. et al.,"Primer on ammonia fiber expansion pretreatment," in Aqueous Pretreatment of Plant Biomass for Biological and Chemical Conversion to fuels and Chemicals, C.Wyman, Ed., Wiley Series in Renewable Resources, chapter 9, pp. 169–195, JohnWiley&Sons, NewYork, NY, USA, 1st edition, 2013.

111. Mohanram, S., Amat, D., Choudhary, J., Arora, A., Nain, L., et al. (2013). Novel perspectives for evolving enzyme cocktails for lignocellulose hydrolysis in biorefineries. *Sust. Chem. Process. 1,* 15.

112. Bayer, E., Belaich, J., Shoham, Y., Lamed, R. The cellulosomes: multienzyme machines for degradation of plant cell wall polysaccharides. *Annu. Rev. Microbiol.* (2004). *58,* 521–554.

113. Moraïs, S., Barak, Y., Caspi, J., Hadar, Y., Lamed, R., Shoham, Y., Wilson, D. B., Bayer, E. A. (2010). Cellulase-xylanase synergy in designer cellulosomes for enhanced degradation of a complex cellulosic substrate. *M. Bio.* 1.

114. Bayer, E. A., Morag, E., Lamed, R. (1994). The cellulosome–a treasure-trove for biotechnology. *Trends. Biotechnol. 12,* 379–386.

115. Doi, R. H., Kosugi, A. (2004). Cellulosomes: plant-cell-wall-degrading enzyme complexes. *Nat. Rev. Micro. 2,* 541–551.

116. Krauss, J., Zverlov, V. V., Schwarz, W. H. In vitro reconstitution of the complete Clostridium thermocellum cellulosome and synergistic activity on crystalline cellulose. *Appl. Environ. Microbiol.* (2012). *78,* 4301–4307.

117. Demain, A., Newcomb, M., Wu, J. (2005). Cellulase, clostridia, and ethanol. *Microbiol. Mol. Biol. Rev. 69,* 124–154.

118. Tachaapaikoon, C., Kosugi, A., Pason, P., Weonukul, R., Ratanakhanokchai, K., Kyu, K., Arai, T., Murata, Y., Mori, Y. (2012). Isolation and characterization of a new cellulosome-producing *Clostridium thermocellum* strain. *Biodegradation. 23,* 57–68.

119. Lynd, L., Weimer, P., Van Zyl, W., Pretorius I: Microbial cellulose utilization: fundamentals and biotechnology. *Microbiol. Mol. Biol. Rev.* (2002). *66,* 506–577.

120. Weonukul, R., Kosugi, A., Tachaapaikoon, C., Pason, P., Ratanakhanokchai, K., Prawitwong, P., Deng, L., Saito, M., Mori, Y. (2012). Efficient saccharification of ammonia soaked rice straw by combination of *Clostridium thermocellum* cellulosome and *Thermoanaerobacter brockii* β-glucosidase. *Bioresour. Technol. 107,* 352–357.

121. Svetlitchnyi, V. A., Kensch, O., Falkenhan, D. A., Korseska, S. G., Lippert, N., Prinz, M., Sassi, J., Schickor, A., Svetlitchnyi, S. C. et al. (2013). Single-step ethanol production from lignocellulose using novel extremely thermophilic bacteria. *Biotechnol. Biofuel. 6,* 31

122. Olson, D. G., McBride, J. E., Joe Shaw, A., Lynd, L. R. Recent progress in consolidated bioprocessing. *Curr. Opin. Biotechnol.* (2011). *23,* 1–10.

123. Lynd, L. R., Weimer, P. J., van Zyl, W. H. Pretorius I. S. (2002). Microbial cellulose utilization: fundamentals and biotechnology. *Microbiol. Mol. Biol. Rev. 66,* 506–577.

124. La Grange, D. C., den Haan, R., van Zyl, W. H. (2010). Engineering cellulolytic ability into bioprocessing organisms. *Appl. Microbiol. Biotechnol. 87,* 1195–1208.

125. Alonso, D. M., Bond, J. Q., Dumesic, J. A. Catalytic conversion of biomass to biofuels. *Green. Chem.* (2010). *12,* 1493–1513.

126. Roy, A., Kumar, A. (2013). Pretreatment Methods of Lignocellulosic Materials for Biofuel Production: A Review. *J. Emerg. Trend. Eng. Appl. Sci. 4,* 181–193.

PART 4

BIODIESEL PRODUCTION

CHAPTER 8

BIODIESEL: PRODUCTION, OPPORTUNITIES AND CHALLENGES

DEEKSHA SACHDEVA

Department of Biochemistry, Punjab Agricultural University, Ludhiana, Punjab, India–141004; E-mail: sachdeeksha@gmail.com

CONTENTS

8.1   INTRODUCTION

Bioenergy is an energy source derived from organic matter or biomass. It can be as low-tech as burning wood on an open fire or as high-tech as developing advanced production methods for biofuels. The unifying factor with all bioenergy sources, however, is that the carbon they contain (and emit when used) comes from plants, and those plants obtain their carbon from the atmosphere. Hence, bioenergy is considered to be renewable.

In the light of the risks posed by climate change and diminishing oil reserves, energy transition from fossil fuels to low-carbon sources of energy seems imminent. Solar, wind, geothermal and biofuels all play an important role in this transition.

The main drivers for biofuels and bioenergy development are [1]:
• need to reduce greenhouse gas (GHG) emissions;
• energy security concerns;
• rising oil prices; and
• the opportunity to support the countries' rural economy.

Biodiesel, a methyl ester, is an alternative to diesel that is made from a triglyceride (like vegetable oil) and either ethanol or methanol. Biodiesel production is based on transesterification of vegetable oils and fats through the addition of alcohol and a catalyst. Feedstock includes rapeseeds, sunflower seeds, soy seeds and palm oil seeds from which the oil is extracted chemically or mechanically. Advanced processes include the replacement of methanol of fossil origin, by bioethanol to produce fatty acid ethyl ester instead of fatty acid methyl ester (the latter being the traditional bio-diesel).

8.2   GENERATIONS OF BIOFUELS

8.2.1   FIRST GENERATION BIOFUELS

First Generation Biofuels are the ones, which are produced directly from food crops by extracting the oils from them and is used for biodiesel synthesis through fermentation. Crops such as oilseed rape, sunflower seeds, etc., are used for biodiesel production. However, first generation biofuels have a number of associated problems. One major issue with first generation biofuels is "fuel v/s food". As the majority of biofuels are produced directly from food crops, the rise in demand for biofuels has led to global increase in food prices.

8.2.2   SECOND GENERATION BIOFUELS

Second Generation Biofuels have been developed to overcome the limitations of first generation biofuels. They are produced from nonfood crops such as wood, organic waste, food crop waste and specific biomass (ligno-cellulosic biomass) crops. Low-cost crops and forest residues, wood process wastes and the organic fraction of municipal solid wastes can be used as ligno-cellulosic feedstocks. This eliminates the main issue with the first generation biofuels.

8.2.3   THIRD GENERATION BIOFUELS

The third generation of biofuels is based on improvements in the production of biomass. It takes advantage of specially engineered energy crops such as algae as its energy source. The algae are cultured to act as a low-cost, high energy and renewable feedstock. Algae can also be grown using land and water unsuitable for food production therefore reducing the strain on already depleted water sources. Another benefit of algae based biofuels is that the fuel can be manufactured into a wide range of fuels such as diesel, petrol and jet fuel.

Production of biofuels from algae depends on the lipid content of the microorganisms. Algae species such as *Chlorella* are targeted because of their high lipid content and high productivity [2, 3]. Algae produces 1 to 7 g/L/d of biomass in ideal growth conditions [3]. This requires large volumes of water, which is a problem for cold countries such as Canada where temperature is below 0°C during a significant part of year. High water content is again a problem while extracting lipids from the algal biomass, which needs dewatering via either centrifugation or filtration before extracting lipids. Lipids extracted from algae can be processed via transesterification [4].

8.2.4 FOURTH GENERATION BIOFUELS

Fourth Generation Biofuels involves the use of genetically modified organisms or metabolic engineering of microorganisms to as to achieve sustainable production of biofuels. Since gene expression and metabolic pathways are well understood in *E. coli* as compared to other microorganisms therefore, it is relatively easy to apply genetic engineering and metabolic engineering to improve the oil accumulation in bacteria [5]. It was reported that a metabolically engineered *E. coli* could produce biodiesel directly and the concentration of fatty acid esters obtained was 1.28 g/1 [6].

8.3 SOURCES OF BIODIESEL OR FEEDSTOCK

Biodiesel can be produced by animal fats, cooking waste and vegetable oils. Various edible and non edible vegetable oils like rice bran oil, coconut oil, *Jatrophacurcas*, castor oil, cottonseed oil, mahua, karanja which are either surplus and are nonedible type, can be used for the preparation of biodiesel [7] (Fig. 8.1, Table 8.1).

8.3.1 BIODIESEL FROM WASTE COOKING OIL (WCO)

The high cost of biodiesel is mainly due to the cost of virgin vegetable oil. Therefore, it is not surprising that the biodiesel produced from vegetable oil (e.g., pure soybean oil) costs much more than petroleum based

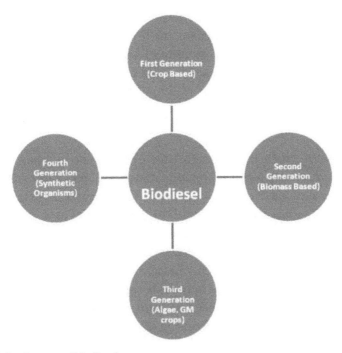

FIGURE 8.1   Sources of biodiesel.

TABLE 8.1   Sources of Biodiesel

Vegetable oils	Non-edible oils	Animal fats	Other sources
• Soybean	• Jatropha	• Fish oil	• Algae
• Sunflower	• Palm	• Poultry fat	• Microalgae
• Rapeseed	• Almond		• Cooking oil (yellow grease)
• Canola	• *Brassica napus*		
• Coconut	• *Brassica carinata*		• Bacteria
• Cotton seed	• Camelina		• Fungi

diesel. Therefore, it is necessary to explore ways to reduce production costs of biodiesel. The use of waste frying oil, instead of virgin oil, to produce biodiesel is an effective way to reduce the raw material cost because waste-frying oil is estimated to be about half the price of virgin oil [8]. Uddin et al. [9] reported the synthesis of biodiesel from waste cook oil by a three-step method. Transesterification gives lower yield than three-step method. In the three-step method, the first step is saponification of the oil followed by acidification to produce FFA and finally esterification of FFA to produce biodiesel. The reaction yield was 79%.

8.4   CHEMICAL NATURE OF BIODIESEL

Biodiesel is produced from vegetable oils and fats by transesterification process. Fats and oils are primarily water-insoluble and hydrophobic in nature, composed of one mole of glycerol and three moles of fatty acids and commonly called as triglycerides [10]. Triglycerides have good amount of oxygen in their structure. When three fatty acids, in triglycerides, are identical it is called simple triglycerides. When they are dissimilar, they are called mixed triglycerides. Oils found in different sources have different fatty acid compositions. The fatty acids vary not only in their carbon chain length but also in the number of unsaturated bonds in the carbon chain. The fatty acids commonly found in vegetable oils are stearic, palmitic, oleic, linoleic and linolenic. The other fatty acids, which are also present in oils and fats, are myristic (tetradecanoic), palmitoleic, arachidic, linolenic and octadecatetranoic [11]. Chemically, biodiesel is mono-alkyl esters of long chain fatty acids from renewable lipid sources. In other words, biodiesel is ester based oxygenated fuel from renewable biological sources. It can be used in compression ignition engines with little or no modifications [11, 12]. On the contrary, petroleum derived diesel, called as petrodiesel, is composed of 75% saturated hyrdocarbon or 25% aromatic hydrocarbon, with average chemical formula $C_{12}H_{23}$ [13].

Depending on the fatty acid composition, the quality of fatty acid methyl esters of oil (biodiesel) was predicted. Fatty acid methyl ester of oils of 26 species including *Azadirachta indica*, *Calophyllum inophyllum*, *Jatropha curcas* and *Pongamia pinnata* were found most suitable for use as biodiesel and they meet the major specification of Biodiesel standards of USA, Germany and European Standard Organization [14]. The fatty acid methyl esters of another 11 species meet the specification of Biodiesel standard of USA only. These selected plants have great potential for biodiesel [15].

Thus, oil, ester and diesel have different number of carbon and hydrogen compound. Diesel has no oxygen atom while biodiesel is oxygenated fuel. The large size of vegetable oil molecules (typically three or more times larger than hydrocarbon fuel) and the presence of oxygen in the molecules suggests that some fuel properties of vegetable oil derived biodiesel would differ from those of hydrocarbon fuels [16].

8.5 SYNTHESIS OF BIODIESEL BY TRANSESTERIFICATION

Biodiesel is produced by a simple process called transesterification reaction. Transesterification is the process in which the methanol is reacted with triglycerides of vegetable oil to form glycerol and methyl ester. Biodiesel is chemically fatty acid esters. A catalyst is often used to improve the reaction rate and yield. The reaction is represented below:

$$
\begin{array}{cccccc}
CH_2OCOR & & & CH_2OH & & RCOOR \\
| & & & | & & | \\
CH_2OCOR & +\,3ROH & \xrightarrow{\ \ CATALYST\ \ } & CH_2OH & + & RCOOR \\
| & & & | & & | \\
CH_2OCOR & & & CH_2OH & & RCOOR \\
\end{array}
$$

FAT/OIL   ALCOHOL   FATTY ACID ESTERS   GLYCEROL

8.6 FACTORS AFFECTING RATE OF REACTION

A number of factors affect the conversion, product yield and quality [17]. Three parameters have effect on the transesterification namely [18],

- temperature (T);
- time (t); and
- ratio of oil to alcohol.

8.7   CATALYSIS OF TRANSESTERIFICATION

8.7.1   BASE OR ACID CATALYSIS

Transesterification can be catalyzed by either base or acid. In base-catalyzed transesterification, NaOH or KOH is used as catalyst, while in acid-catalyzed transesterification, acid is employed as catalyst. Most studies have focused on base-catalyzed transesterification of triglycerides. For acid-catalyzed systems, sulfuric acid has been the most investigated catalyst but other acids such as HCl, BF_3, H_3PO_4 and organic sulfonic acids have also been used [19].

8.7.2   ENZYMATIC CATALYSIS

Enzymatic synthesis of biodiesel employs lipases as catalyst. Various types of alcohols -primary, secondary and straight- or branched-chain can be employed in transesterification using lipases as catalysts. Lipases are known to act on long chain fatty alcohols better than short chain ones. Thus the efficiency of transesterification of triglycerides with methanol is likely to be very low compared to that with ethanol [20].

8.8   MICROBIAL SYNTHESIS OF BIODIESEL

There is a strong need to explore oil sources especially nonedible oils so as to achieve the rapid production of biodiesel. Microbial oils hold great potential in this regard. Microbial oils or single cell oils are produced by some oleaginous microorganisms such as yeast, fungi, bacteria and microalgae [21]. The content of fat in these oleaginous microorganisms can reach over 80% of cellular dry weight [22]. Hence these microbial oils can be used as feedstocks for biodiesel production. There are number of advantages associated with using microbial oils as compared to other vegetable oils and animal fats: short life cycle, less labor required, not dependent on season, climate and easier to scale up [23].

Autotrophic microalgae use carbon dioxide as the carbon source and sunlight as the energy for oil accumulation. Examples of autotrophic microalgae include *Chlorella vulgaris, Botryococcus braunii, Navicula pelliculosa, Scenedsmus acutus, Crypthecodinium cohnii, Dunaliella primolecta, Monallanthus salina, Neochloris oleoabundans, Phaeodactylum tricornutum* and *Tetraselmis sueica* can accumulate oils [24, 25].

There are several yeast species, such as *Cryptococcus albidus, Lipomyces lipofera, Lipomyces starkeyi, Rhodosporidium toruloides, Rhodotorula glutinis, Trichosporon pullulan,* and *Yarrowia lipolytica,* which are also reported to accumulate oils under certain cultivation conditions [25, 26]. Cultivation conditions, such as C/N ratio, nitrogen resources, temperature, pH, oxygen and concentration of trace elements and inorganic salts, has been reported to influence the oil accumulation. The more the nitrogen compounds in medium, the lesser would be oil content in cells. Mainul et al. [27] reported that upon increasing C/N ratio from 25 to 70, oil content increased from 18% to 46%.

Bacteria belonging to the actinomycetes group such as species of *Mycobacterium, Rhodococcus* and *Nocardia* are able to synthesize intracellular triacylglycerides [5, 28]. Genetic engineering is also applied to improve fatty acid or biodiesel synthesis in *E.coli* by improving the acetyl-CoA biosynthetic pathway (i.e., by cloning of acetyl- CoA carboxylase genes, malonyl-CoA synthetase, a malonate transporter from *Rhizobium,* and thioesterase enzyme from plant) [6, 29].

8.9   BLENDS

Biodiesel can be blended in any ratio with petroleum diesel fuel. A blend of 20% biodiesel 80% diesel fuel is called B20. B20 is becoming a popular alternative fuel for U.S. fleets. A fleet is a group of vehicles operated from a central location such as buses or delivery trucks. B20 is appealing because it reduces emissions effectively, is reasonably priced and requires no engine modifications [30].

8.10   BENEFITS OF BIODIESEL

There are several benefits of biodiesel as a blended fuel in diesel engines [31, 32]:

- Biodiesel burns cleaner than petroleum diesel. Biodiesel's exhaust is free of lead, sulfur dioxide, halogens and thus has reduced particulate matter, unburnt hydrocarbons, carbon monoxide and carbon dioxide [33].
- Biodiesel is biodegradable and nontoxic. Biodiesel is as biodegradable as sugar and less toxic than table salt [34].
- There are only limited or no needed modifications to current engines to be able to use biodiesel, as it allows longer engine life and produces higher cetane number. Cetane number (CN) is a measurement of the combustion quality of diesel fuel during compression ignition. It is an important factor in determining the quality of diesel fuel. The higher the cetane number, the more easily the fuel will ignite.
- Biodiesel gives almost equal engine performance and mileage as petroleum based diesel, dissipating engine heat faster, with more hours between oil changes.
- Biodiesel has a lower flash point than petroleum diesel and thus helps prevent damaging fires.

8.11   OTHER USES OF BIODIESEL

Biodiesel has number of applications other than engine fuel [30]. For example:
- Biodiesel can be used in diesel-fueled heaters, lanterns and stoves. It can be substituted for kerosene in lanterns and stoves.
- Biodiesel can be used as substitute for model aircraft fuel in model aircraft engines with glow plugs.
- Biodiesel can be used as an all-purpose machinery lubricant.
- Biodiesel can be used to clean up petroleum oil spills on land or in water [35].
- Biodiesel can be used as a solvent for nonautomotive paint, spray paint and other adhesive chemicals [36].
- Biodiesel can be used as a nontoxic crop adjuvant [36].

8.12 ENERGY BALANCE RATIO

Biodiesel has a positive energy balance ratio. An energy balance ratio is a comparison of the energy stored in a fuel to the energy required to grow, process and distribute the fuel. The energy balance ratio of biodiesel is at least 2.5 to 1 [37]. For every one unit of energy put into the fertilizer, pesticides, fuel, feedstock, extraction, refining, processing and transporting of biodiesel, there are at least 2.5 units of energy contained in the biodiesel [30].

8.13 CHALLENGES

One of the biggest challenges associated with biodiesel production is its cost and limited availability of fat and oil resources. There are two aspects of the cost of biodiesel, the costs of raw material (fats and oils) and the cost of processing. The costs of raw materials accounts for 60–75% of the total cost of biodiesel fuel [10]. The use of waste cooking oil can lower the cost significantly. However, the quality of used cooking oil can be bad [38].

Biofuel production has pushed up prices of some food crops, an expected outcome when they are also used as feedstock. To avoid the food-or-fuel dilemma, biofuels should be produced from food crops only after additional feedstocks are made available for this purpose. This can be achieved by either attaining higher yields in currently cropped areas, or expanding cultivation. Non-edible oils can also be used for making biodiesel fuel.

8.14 CONCLUSION

Biodiesel is an ecofriendly diesel fuel prepared from domestic renewable resources, for example, vegetable oils and animal fats. Biodiesel is becoming popular worldwide as an alternative energy source because it is nontoxic, biodegradable and nonflammable. Increasing population and hence increased consumption of fossil fuels is the main driving force behind developing this technology. In order to displace large quantities of petroleum with biomass oils there should be well-coordinated research projects at national and international level.

KEYWORDS

- **Biodiesel**
- **Cetane number**
- **Generations of biofuels**
- **Greenhouse gas**
- **Waste cooking oil**

REFERENCES

1. Enrique, R., Chalico, T. A. (2009). *Opportunities and Challenges For Biofuel Production in Latin America.* CIFOR Environment Briefs.
2. Liang, Y., Sarkanyet, N., Cui, Y. (2009). Biomass and lipid productivities of *Chlorella Vulgaris* under autotrophic, heterotrophic, and mixotrophic growth conditions. *Biotechnol. Lett. 31*, 1043–1049.
3. Chen, C. Y., Yeh, K. L., Aisyah, R., Lee, D. J., Chang, J. S. (2011). Cultivation, photobioreactor design, and harvesting of microalgae for biodiesel production: A critical review. *Bioresource. Technol. 102*, 71–81.
4. Tran, N., Bartlett, J., Kannangara, G., Milev, A., Volk, H., Wilson, M. (2010). Catalytic upgrading of biorefinery oil from microalgae. *Fuel. 89*, 265–274.
5. Alvarez, H. M., Steinbuchel, A. (2002). Triacylglycerols in prokaryotic microorganisms. *Appl. Microbiol. Biotechnol. 60*, 367–376.
6. Kalscheuer, R., Stolting, T., Steinbuchel, A. (2006). Microdiesel: *Escherichia coli* engineered for fuel production. *Microbiology. 152*, 2529–2536.
7. Malhotra, R. K., Das, L. M. (1999). Biofuels as blending components for motor gasoline and diesel fuels. *J. Sci. Ind. Res. 61,* 99.
8. Zheng, S., Kates, M., Dube, M. A., Mcleon, D. D. (2006). Acid-catalyzed production of biodiesel from waste frying oil. *Biomass. Bioenergy. 30*, 267–272.
9. Uddin, M. R., Ferdous, K., Uddin, M. R., Khan, M. R., Islam, M. A. (2013). Synthesis of biodiesel from waste cooking oil. *Chem. Eng. Sci. 1,* 22–26.
10. Sonntag, N. O. V. (1979). Structure and composition of fats and oils. In: Bailey's industrial oil and fat products, Swern D., Ed., John Wiley and Sons, New York; Vol. 1, pp. 1.
11. Singh, S. P., Singh, D. (2010). Biodiesel production through the use of different sources and characterization of oils and their esters as the substitute of diesel: a review. *Renew. Sus. Energy. Rev. 14*, 200–216.
12. Demirbas, A. (2002). Diesel fuel from vegetable oil via transesterification and soap pyrolysis. *Energy. Sources. 24*, 835–41.
13. ATSDR: *Toxicological Profile For Fuel Oils 1995*; U. S. Department of Health and Human Services, Public Health Service, Agency for Toxic Substances and Disease Registry (ATSDR): Atlanta, GA, 1995.

14. Mohibbe, A. M., Waris, A., Nahar, N. M. (2005). Prospects and potential of fatty acid methyl esters of some nontraditional seed oils for use as biodiesel in India. *Biomass. Bioenergy. 29,* 293–302.

15. Anonymous. The wealth of India: raw materials, vol. IX. Publication & Information Directorate, Council of Scientific & Industrial Research; New Delhi, 1972.

16. Goering, C. E., Schwab, A. W., Daugherty, M. J., Pryde, E. H., Heakin, A. J. (1982). Fuel properties of 11oils. *Trans. ASAE. 25,* 1472–83.

17. Divya, B., Tyagi, V. K. (2006). Biodiesel: Source, Production, Composition, Properties and Its Benefits. *J. Oleo. Sci. 55,* 487–502.

18. Kapilakarn, K., Peugtong, A. (2007). A Comparison of Costs of Biodiesel Production From Transesterication. *Int. Energy. J. 8,* 1–6.

19. Liu, K. S. (1994). Preparation of Fatty-Acid Methyl Esters for Gas- Chromatographic Analysis of Lipids inBiological-Materials. *J. Am.Oil Chem. Soc. 71,* 1179–1187.

20. Shimada, Y., Sugihara, A., Minamigawa, Y., Higasbiyama, K., Akimoto, K., Fujikawa, S., Komemushi, S., Tominaga, Y. (1998). Enzymatic enrichment of arachidonic acid from Mortierella single cell oil. *J. Am. Oil. Chem. Soc. 75,* 1213–17.

21. Ma, Y. L. (2006). Microbial oils and its research advance. *Chin. J. Bioprocess. Eng. 4,* 7–11.

22. Meng, X., Yang, J., Xu, X., Zhang, L., Nie Q., Xian, M. (2009). Biodiesel production from oleaginous microorganisms. *Renew. Energy. 34,* 1–5.

23. Li, Q., Wang, M. Y. (1997). Use food industry waste to produce microbial oil. Science and Technology of Food Industry. *6,* 65–69.

24. Chisti, Y. (2007). Biodiesel from microalgae. *Biotechnol. Adv. 25,* 294–306.

25. Liang, X. A., Dong, W.B; Miao, X. J., Dai, C. J. (2006). Production technology and influencing factors of microorganism grease. *Food. Res. Dev. 27,* 46–47.

26. Liu, S. J., Yang, W. B., Shi; A. H. (2000). Screening of the high lipid production strains and studies on its flask culture conditions. *Microbiology. 27,* 93–97.

27. Mainul, H., Philippe, J. B., Louis, M. G., Alain, P. (1996). Influence of Nitrogen and Iron Limitations on Lipid Production by Cryptococcus curvatus Grown in Batch and Fed-batch Culture. *Process. Biochem. 31,* 355–361.

28. Li, Q., Du, W., Liu, D. (2008). Perspectives of microbial oils for biodiesel production. *Appl. Microbiol. Biotechnol. 80,* 749–756.

29. Adamczak, M., Bornscheuer, U. T., Bednarski, W. (2009). The application of biotechnological methods for the synthesis of biodiesel. *Eur. J. Lipid. Sci. Technol. 111,* 808–813.

30. Tickell, J. *From the fryer to the Fuel Tank: The Complete guide to using Vegetable Oil as an alternative Fuel.* 3rd edition, Tickell Energy Consulting, Tallahassee, Florida, 2000; pp. 1–159.

31. Zhang, Y., Dub, M. A., McLean, D. D., Kates, M. (2003). Biodiesel production from waste cooking oil: Process design and technological assessment. *Elsevier. J. Bioresour. Technol. 89,* 1–16.

32. Idusuyi, N., Ajide, O. O., Abu, R. (2012). Biodiesel as an alternative energy resource in South-west Nigeria. *Int. J. Sci. Technol. 5,* 323–327.

33. Korbitz, Werner. The technical, energy and environmental properties of Biodiesel. Korbitz Consulting Vienna, Austria, 1993.

34. Walker, Kerr. (1994). Biodiesel from rapeseed. In *Journal of the Royal Agricultural Society of England*. Vol 155, 43–44.

35. Randall, V. W. *Cytosol Process: Shoreline oil spill remediation with a vegetable based biosolvent*. Presented at the fourth international marine biotechnology conference, September 1997.

36. Ahmed, I., Clements, L. D., Van Dyne, D. L. Non-fuel industrial uses of soybean oil-based esters: Final Report. Development systems/Applications International, January 31, 1997.

37. Ahmed, I., Decker, J., Morris, D. *How much energy does it take to make a gallon of soydiesel;* Institute for Local Self-Reliance, January 1994.

38. Murayama, T. (1994). Evaluating vegetable oils as a diesel fuel. *Inform. 5,* 1138–1145.

CHAPTER 9

AN OVERVIEW OF REACTOR DESIGNS FOR BIODIESEL PRODUCTION

VIJAY KUMAR GARLAPATI, LAKSHMI SHRI ROY, and RINTU BANERJEE

Microbial Biotechnology and DSP Laboratory, Agricultural and Food Engineering Department, IIT Kharagpur, Kharagpur, West Bengal – 721302

CONTENTS

ABSTRACT

Nowadays, the idea of biofuels through utilization of biomass has been triggering in a exponential phase due to social, moral and environmental reasons for using hydrocarbon based crops for bioenergy. The concept of second- and third-generation biofuels from nonedible crops and wastes gains absolute social acceptance, still remains a big question. Usually, the biodiesel produced through transesterification of oils will produce in a batch mode with longer reaction times. The process economics of the biodiesel can be enhanced by implementing the continuous processes which results in a pure product (reduces down stream steps) through improved mixing with lower residence time (reduction in reactor volume). Ample scope is available for the use of process intensification techniques with an objective of making the synthesis economically viable and one such intensification approach is based on the use of different bioreactor configurations. This chapter is a comprehensive overview of current technologies and appropriate options for scale-up development, providing the basis for a proposal for the exploitation of various reactor configurations to optimize biodiesel production.

9.1   INTRODUCTION

Current energy policies emphasize cleaner, efficient and environmentally friendly technologies. Their target is to increase the supply and usage of energy [1–4]. Thus, developments in alternative renewable energy sources have become indispensable for sustainable environmental and economic growth. Among the explored alternative energy sources, considerable attention has been focused on biodiesel because it is widely available from inexhaustible feedstocks that can effectively reduce its production cost.

Biodiesel, which is also known as fatty acid methyl ester (FAME), is a mixture of monoalkyl esters of long-chain fatty acids derived from renewable lipid feedstocks, such as vegetable oil and animal fats. Because biodiesel has physical properties similar to diesel fuels it has established its commercial value in the automobile markets of Europe, US, Japan, Brazil and India [5]. Moreover, the implementation of the "directive on the promotion of the use of biofuels" for transport in the EU (Directives 2003/30/EC) mandated the increased use of biofuels to power transportation from 2% to 5.75% between 2005 and 2010, triggering a huge demand for biodiesel [6]. Unlike conventional diesel fuel, biodiesel offers several advantages, including renewability, higher combustion efficiency [7], cleaner emission [5], higher cetane number, higher flash point, better lubrication [8] and biodegradability [9]. Depending on the climate, local soil conditions and availability, various biolipids have been used in different countries as feedstocks to produce biodiesel.

Biolipid feedstocks can be divided into four categories: virgin vegetable oils, waste vegetable oils, animal fats and nonedible oils. Virgin vegetable oil feedstock refers to rapeseed, soybean, sunflower and palm oil [10], while waste vegetable oil refers to these oils that have been used in cooking and are no longer suitable for human consumption [11, 12]. Animal fats include tallow, lard and yellow grease [13] while the nonedible oils include Jatropha [14, 15], neem oil, castor oil, tall oil [10] and microalgae [16].

9.2 BIODIESEL PRODUCTION METHODS

Four primary production methodologies for producing biodiesel have been studied extensively. The high viscosity, acid number, free fatty acid content, oxidation and polymerization problems of vegetable oils makes them unsuitable for direct usage in diesel engines. Several modification techniques, such as dilution, microemulsion, pyrolysis and transesterification have been used to reduce the viscosity of oil [17]. Of these processes, transesterification is the most widely used; this method involves the alcoholysis of vegetable oil to produce alkyl ester. Generally, the mechanism consists of three consecutive reversible reaction steps. The first step involves the conversion of triglycerides (TG) to diglycerides (DG) and later to monoglycerides (MG). Subsequently, the monoglycerides are converted to glycerol. Each reaction step produces an alkyl ester. Thus, a total of three alkyl esters are produced in the transesterification process [18]. The typical tranesterification reactions were depicted in Fig. 9.1.

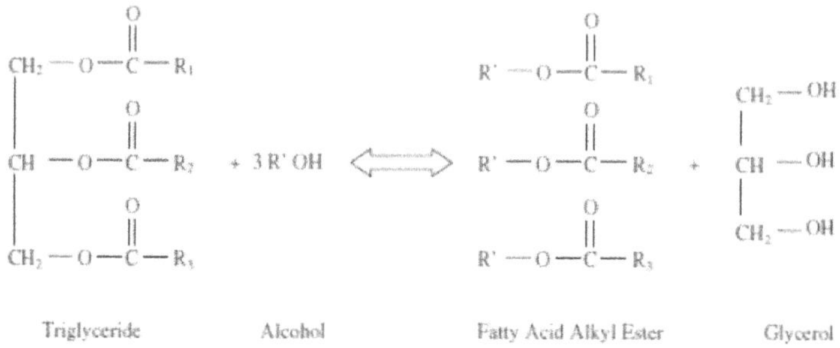

FIGURE 9.1   Typical Transesterification reaction [19].

Common transesterification reactions that are used to produce bio-diesel include [20]:
- homogeneous catalyzed transesterification;
- heterogeneous catalyzed transesterification;
- enzymatic catalyzed transesterification; and
- supercritical technology.

Each of these methods has its own disadvantages that eventually limit the economic feasibility and low environmental impact of the entire bio-diesel production process.

9.3   LIMITATIONS OF CONVENTIONAL BIODIESEL PRODUCTION TECHNOLOGY

In the conversion of vegetable oil by the transesterification process, the reversible reaction between the reactant and product indicates that the formation of biodiesel is highly dependent on the proportion of the reac-tant and the conditions of the transesterification process. According to Le Chatelier's principle, large quantities of alcohol are needed to shift the equilibrium of the reaction to the product side and increase the yield of biodiesel [21]. Unfortunately, high consumption of alcohol is associated with higher production cost. The consumption of alcohol could be reduced by using acid or alkaline catalysts, which could improve the reaction rate and biodiesel yield. However, homogeneous acid solutions that catalyze transesterification processes, such as sulfuric [22], hydrochloric [23], or sulphonic acids [24] have been largely ignored because they increase the time consumption of the process, require a higher reaction temperature and are corrosive by nature. Although the use of homogeneous alkaline catalysts, such as sodium and potassium hydroxide could overcome these limitations, it has been reported that the alkaline catalyzed reaction is sensitive to the purity of the reactant. The presence of water and free fatty acids in the raw feedstock could induce a saponification process in which the free fatty acid produced by the hydrolysis of triglycerides reacts with the alkaline catalyst to form soap. The dissolved soap in the glycerol phase would increase the solubility of methyl ester in the glycerol and complicate the subsequent separation process [25]. Also, the removal of either the homogeneous acidic or alkaline catalyst using hot distilled water would eventually result in the need to dispose of wastewater [26].

Heterogeneous catalyst has been viewed as an alternative solution to replace the homogeneous catalyst because it is noncorrosive and environmentally benign. The heterogeneous catalytic reaction usually faces a mass transfer resistance problem due to the presence of three-phase system (triglycerides, alcohol and solid catalyst) in the reaction mixture which limits the pore diffusion process and reduces the active site availability for the catalytic reaction which in turn decreases the reaction rate [27, 28]. The lipase-mediated transesterification of oils to biodiesel is an attractive option due to usage of environment-friendly catalyst (lipase) and facilitates easier separation of products [29]. Due to requirement of longer reaction times and unfavorable product yields this process not yet commercialized compared to the alkaline catalyst. It has been reported that the enzymatic transesterification process requires 24 hr to achieve a biodiesel yield of 90% [30]. Most importantly, the major obstacle to this process is the high cost of the enzyme. The enzyme also requires very specific reaction conditions because the denaturation of the enzyme and its deactivation as a result of feed impurity could decrease its efficiency [31].

Supercritical alcohol transesterification provides a new path for the production of biodiesel without the aid of a catalyst. The supercritical condition could overcome the mass transfer limitation by enabling the mixture of triglyceride and alcohol to become a homogeneous phase [32]. However, the major drawbacks of this noncatalytic process are its large energy requirement and its infeasibility for large-scale industrial application because of the increased production cost imposed by the high reaction temperature and pressure [33]. Moreover, the supercritical process is potentially hazardous and requires attention to personal risk and safety.

Usually the catalytic and noncatalytic transesterification downstream processes will receive unreacted reactant and catalyst along with the biodiesel and glycerol. Ineffective downstream processing of biodiesel may cause severe diesel engine problems, such as plugging of filters, coking on injectors, carbon deposits, excessive engine wear, oil ring sticking, engine knocking, and thickening and gelling of lubricant oil [34]. The downstream processing of biodiesel technology (glycerol separation, catalyst neutralization and biodiesel purification) alone constituted over 60–80% of the total cost of a transesterification process plant [35]. In addition, the multiple separation and purification stages could cause loss of the biodiesel, resulting in a decrease in the pure biodiesel yield.

9.4   PROCESS INTENSIFICATION TECHNOLOGIES FOR BIODIESEL PRODUCTION

The process intensification technologies for biodiesel production can alleviate the limitations associated with biodiesel production technology such as mass transfer, thermodynamic equilibrium, wastewater and multiple downstream processing steps [26, 27, 33, 35, 36]. These technologies involve the use of novel reactors or coupled reaction/separation processes to enhance the reaction rate and to reduce the residence time [37].

Recently, novel reactors have been developed and applied to improve the mixing and mass/heat transfer between the oil and methanol in biodiesel production [37]. Those novel reactors are:

- micro-channel reactor;
- oscillatory flow reactor;
- rotating/spinning tube reactor;
- cavitational reactor;
- sonochemical reactor;
- microwave reactor; and
- membrane reactor.

9.4.1 MICRO-CHANNEL REACTOR

The microchannel reactor can achieve a rapid reaction rate because it has a high volume/surface ratio, short diffusion distance and fast and efficient heat dissipation and mass transfer [37, 38]. The microchannel reactors used in biodiesel production include microchannel reactors with T- or Y-flow structures, zigzag microchannel reactors [38] and slit channel reactor [39]. The typical flow diagram of the microstructured reactor used for biodiesel production was depicted in Fig. 9.2. Because of the high heat transfer rate, it has been reported that the microchannel reactor consumed less energy than the conventional stirrer reactor [37]. However, the microchannel reactor suffers from the drawback of low production throughput, which is attributed to the limitations of the microfabrication technology that is used to produce the microchannel. Furthermore, the high investment cost of the microchannel reactor prohibits the addition of more reactors in parallel to amplify the production of biodiesel [39].

9.4.2 OSCILLATORY FLOW REACTOR

The oscillatory flow reactor is a type of continuous plug flow reactor (PFR) [41] in which the orifice plate baffles is equally spaced, and a piston is used to produce oscillatory flow [37]. The flow diagram of

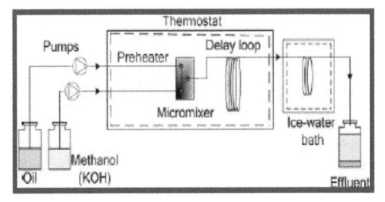

FIGURE 9.2 Flow diagram of the microstructured reactor used for the biodiesel production [40].

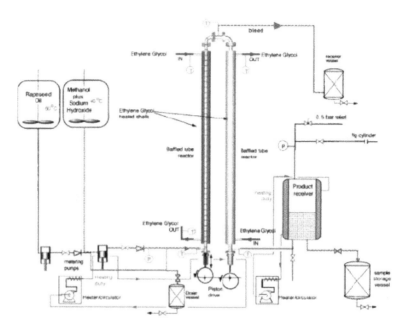

FIGURE 9.3   Flow diagram of Biodiesel production in OFR [42].

biodiesel production in oscillatory flow reactor consists of two vertically positioned jacketed QVF tubes connected at the top by an inverted QVF U-tube, which was fitted with a manual purge value. The feeds were pumped from the feed vessels using a metering pump and temperatures were maintained by Eurotherm temperature controllers. The oscillation of fluid facilitates a 'nutating cam' arrangement. OFR is well suited for laboratory use (Fig. 9.3) [42]. The combination of baffles and oscillatory motion intensifies the radial mixing by the formation of periodic vortices in the bulk fluid, causing an increase in mass and heat transfer while maintaining plug flow [37, 41]. In addition, the oscillatory flow reactor can also improve the residence time distribution (RTD) and multiphase suspension [43]. Because the oscillatory flow reactor can achieve long residence times, it can be designed with a smaller length to diameter ratio, which eventually helps to improve the economy of biodiesel production because of the smaller "footprint," lower capital, reduced pumping cost and ease of control [37].

9.4.3 ROTATING/SPINNING TUBE REACTOR

The rotating/spinning tube reactor is a shear reactor containing two tubes. The inner tube rotates rapidly within the concentric stationary outer tube. Both tubes are separated by a narrow annular gap, which produces Couette flow when the reactants are introduced (Fig. 9.4) [44]. Because of the high shear rate, the reactants are mixed and move through the gap as a coherent thin film. This thin film provides a large interfacial contact area to enhance the reaction rate between the oil and the methanol. As a result, less mixing power and reaction time are required to produce biodiesel using a rotating/spinning reactor compared to a conventional reactor. This type of reactor is suitable to handle feedstocks with high FFA because the residence time is short [37].

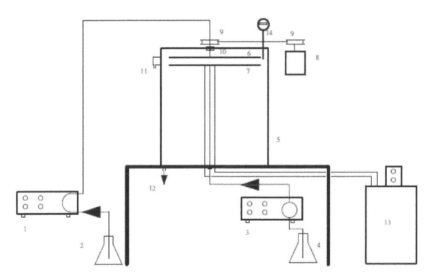

FIGURE 9.4 Experimental setup of the intensive spinning disk reactor for biodiesel synthesis. (1) Peristaltic pump; (2) canola oil vessel; (3) digital piston pump; (4) sodium methoxide vessel; (5) cylinder; (6) rotating disk; (7) stationary disk; (8) variable-speed DC motor; (9) pulley; (10) bearing; (11) sampling point; (12) products drainage; (13) heating circulator; (14) thermometer [44].

9.4.4 CAVITATIONAL REACTOR

The cavitational reactor is another type of novel reactor that has been used successfully in biodiesel production [37, 45–47]. Cavitation is defined as the generation of cavities followed by their growth and violent collapse, causing high local energy densities, temperatures and pressures [37, 45]. Cavitation enhances the mass transfer rate of the reaction by creating conditions of local intense turbulence and liquid microcirculation currents in the reactor [37, 45, 46]. Cavitational reactors can be classified into two types: hydrodynamic cavitation and acoustic cavitation [37, 46]. Hydrodynamic cavitation can be generated by using a restriction component, such as an orifice plate, a throttling valve or a venture, placed in a liquid flow [45, 46]. At the constriction area, the kinetic energy or velocity of the liquid increases, but the local pressure decreases [45]. The typical hydrodynamic cavitation reactor used for biodiesel production was depicted in Fig. 9.5 [46, 48]. It consists of a reservoir or a collecting tank, which

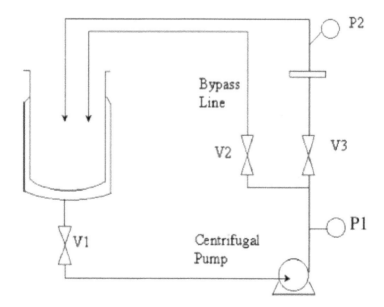

FIGURE 9.5 Schematic representation of the experimental setup for hydrodynamic cavitation reactor [46, 48].

was connected to the multistage centrifugal pump. The discharge from the pump branches into two lines which help in control of the inlet pressure and the inlet flow rate into the main line housing the orifice with the help of valves). The operating temperature of the reactor was maintained to near ambient conditions by circulating water within the jacket surrounding the tank.

A hydrodynamic cavitation reactor is more effective for mixing of immiscible liquids [47]. The mixing efficiency of a hydrodynamic cavitation reactor has been reported to be 160–400 times higher than that of the conventional mixing method [37]. Therefore, the hydrodynamic cavitation reactor consumes half of the energy required by conventional mechanical stirring [47].

9.4.5 SONOCHEMICAL REACTOR

A reactor that generates cavitation by ultrasound is known as a sono-chemical reactor [45] or an acoustic cavitation reactor [37, 49]. Ultra-sound causes a series of compression and rarefaction cycles by alternately compressing and stretching the molecular spacing of the medium [50]. The sonochemical reactor used for biodiesel production consists of an ultrasonic bath equipped with transducers (three) at the bottom of the tank arranged in a triangular pitch. It operates at an irradiating frequency of 20 kHz and power dissipation of 120 W and temperatures were maintained around 28°C (Fig. 9.6) [46]. Low-frequency ultrasound irradiation is useful for the emulsification of immiscible liquids, such as methanol and oil. Emulsification is a result of the induced collapse of cavitation bubbles that disrupt the phase boundary of methanol and oil [51]. Emulsions with large interfacial areas provide more reaction sites for trans-esterification and eventually increase the reaction rate [52]. It has been reported that the operating parameters, such as temperature, pressure, reaction time and catalyst concentration, are significantly reduced in ultrasound-assisted transesterification [53, 54]. However, sonochemical reactors suffer from erosion and particle shedding at the delivery tip surface because of the high surface energy intensity [45]. Also, the scale-up of a sonochemical reactor is relatively more difficult than it is for a

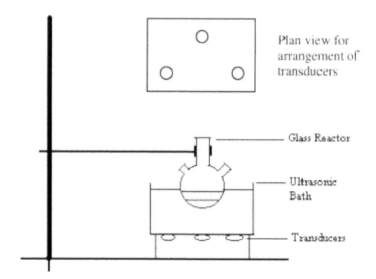

FIGURE 9.6   Schematic representation of sonochemical reactor generating acoustic cavitation [46].

hydrodynamic cavitation reactor because the former relies on a source of vibration [37].

9.4.6   MICROWAVE REACTOR

A new reactor design, to operate efficiently, is crucial. This is an area where microchem reactors particularly shine. Microchem reactors are based on the use of microwave process technology. It is a developing field of chemical processing that exploits rapid reaction rates with higher energy efficiency. This approach can result in poor control system performance for more typical conditions. The microwave reactor is another intensification technology for biodiesel production. The main function of a microwave reactor is not to improve the mixing of oil and methanol but to use its irradiation to transfer energy directly into the reactants and thus accelerate the transesterification. Because both polar and ionic components are available in the mixture of oil and methanol/alcohol, a microwave reactor plays an important role in the more efficient heating of reactants to the desired temperature because of the energy interactions at the molecular level [55].

Compared to a conventional thermal heating reactor, a microwave reactor is able to achieve similar biodiesel conversion with a shorter reaction time and in a more energy-efficient manner [37].

Modified microwave reactors from domestic microwave oven are not scalable from laboratory small-scale synthesis to industrial multi kilogram production, as they do not allow studying the process at a higher scale [56]. An efficient model for biodiesel reactor is not easy, because the microwave-assisted chemistry is poorly understood [57]. And biofuels systems are continuously subject to disturbances covering a wide range of conditions [58]. Instantaneous condition is a function of time and of the robustness of the system with respect to imminent disturbances. Transesterification reaction is highly nonlinear processes [59] and there are many effects parameters on temperature control, like heat transfer on microwave cavity, oil viscosity changes that lead to changes in flow rate, uniformity of reactant concentration, reactant temperature, power and reflected power in microwave cavity, a coolant temperature and ambient temperature. Therefore, it is a challenge to design a robust controller for nonlinear chemical process subject to model uncertainties [60]. If system modeling based on conventional mathematical tools is not well suited for dealing with ill-defined and uncertain systems, the presence of uncertainties can make a mismatch between the formulated mathematical model and the true process, which may degrade the control performance and would lead to serious stability problems consequently.

The entire above-mentioned novel reactors intensify the transesterification by either enhancing the mixing of oil and methanol or improving the heat transfer between the two liquid phases. However, none of these novel reactors, except the membrane reactor, is able to overcome the limitation caused by chemical equilibrium in transesterification. Therefore, the membrane reactor offers another interesting process intensification technology for biodiesel production

9.4.7 MEMBRANE REACTOR

A membrane reactor is also known as a membrane-based reactive separator [61]. According to IUPAC, a membrane reactor is defined as a device that combines reaction and separation in a single unit [62]. Usually, the membrane reactors were classified based on the reactor design (extractor, distributor or contactor), the membrane used in the reaction (organic, inorganic, porous or dense membrane), whether it is an inert or catalytic membrane reactor and the reaction that occurs in membrane reactor (such as dehydrogenation [62], esterification [62, 63], water dissociation [62] or wastewater treatment [64, 65]. In addition to providing the separation, a membrane reactor also enhances the selectivity and yield of the reaction [61].The two basic configurations of membrane reactors were depicted in Fig. 9.7 [66].

Figure 9.7a shows the membrane reactor system in which the membrane reactor appears as an external process unit whereas in Fig. 7b type, the reactor combines with membrane separator as a single unit [67]. Compared with conventional biodiesel production process, the main advantage of membrane reactor system is capital and operating cost reduction, which is due to elimination of intermediate processing steps [61, 67]. Having this process advantages, the membrane reactors occupied a prominent place in biodiesel production technologies [36, 68, 69]. Undeniably, membrane reactors can be considered to be an emerging technology for biodiesel production. In order to successfully develop and commercialize membrane reactors in the biodiesel industry, knowledge is required in three major

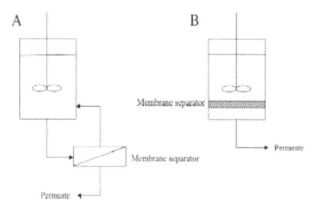

FIGURE 9.7 Membrane reactor basic layout. (a) A conventional membrane reactor system; (b) An integrated membrane reactor system [66].

fields: catalysis, membrane technology and reactor engineering. However, the desired properties, such as the mechanical properties and surface morphology of the membrane (especially in polymeric membranes),
used in biodiesel production have not been fully studied. Also, most of the
reported transesterification reactions via catalytic polymeric membranes
have been performed in situations in which the membrane was cut into
small squares and loaded together with the reactants [70–72]. Therefore,
the ability of the synthesized polymeric membrane to separate glycerol
from the product stream remains unstudied. Additionally, the engineering
aspects of the membrane reactor have been minimally studied because
most publications have only offered proofs of concepts. Biodiesel production using membrane reactors is still running under nonoptimal conditions. Therefore, it is a challenge to choose the best possible combination
between catalyst and membrane. Optimization studies and modeling
will be needed to advance the membrane reactor into commercial operation. Although high biodiesel yield can be obtained via catalytically inert
membrane (mainly the microporous ceramic and carbon membranes), a
water-washing step is still needed to purify the produced biodiesel. The
purification problem can be reduced by using catalytically active membranes (constructed from polymeric membranes) in the flow configuration
studied by Guerreiro et al. [24, 70]; Sarkar et al. [73] and Zhu et al. [72].
However, the polymeric membranes face the problem of low mechanical
strength. It has been reported that the tested polymer membranes break
before a high conversion of biodiesel could be achieved [24]. Therefore,
more attention is needed to the selection of membrane and operating conditions to avoid membrane failure.

9.5   CONCLUSIONS

The obstacles faced by conventional biodiesel production processes have
hindered the production of biodiesel in a way that is fully economically
feasible and environmentally friendly. Very fast biodiesel production at
high flow rates with low pressure drops can be achieved in microstructured reactors assembled with a micromixer and designed delay loops.
With the advantages of a high throughput and a small footprint of the

microstructured reactor, this biodiesel production technique makes distributed energy production and refining on a truck quite possible. The twin-disk, spinning disk reactor demonstrated greatly increased rates of reaction for biodiesel synthesis, at a range of temperatures when compared with a stirred tank reactor. On other hand cavitational reactors appears to be very effective for intensification of chemical processing operations and harnessing the spectacular effects of cavitation, chemical as well as mechanical, for physical and chemical processing applications would lead to considerable economic savings. The future of sonochemical reactors lies in the design of multiple frequencies multiple transducer-based reactors whereas for hydrodynamic cavitation reactors, orifice plate type configuration appears to be most suitable. The emergence of membrane technology offers a solution for future biodiesel production that has the potential to be developed into large-scale commercial processes. In addition to the typical transeserification conditions (reaction temperature, methanol-to-oil ratio and catalyst concentration, etc.), the process parameters, such as reactant flow rate, transmembrane pressure, membrane thickness and pore size, are reported to have great impacts on the yield of biodiesel as well, and they cannot be ignored.

KEYWORDS

- **Biodiesel**
- **Bioreactor**
- **Cavitational**
- **Membrane**
- **Sonochemical**
- **Transesterification**

REFERENCES

1. Hammond, G. P., Kallu, S., McManus, M. C. (2008). Development of biofuels for the UK automotive market. *Appl. Energy. 85,* 506–515.
2. Hoekman, S. K. (2008). Biofuels in the U.S. Challenges and opportunities. *Renew. Energy. 34,* 14–22.

3. Monni, S., Raes, F. (2008). Multilevel climate policy: the case of the European Union, Finland and Helsinki. *Environ. Sci. Policy. 11,* 743–755.
4. Sawyer, R. F. (2009). Science based policy for addressing energy and environmental problems. *Proc. Combust. Inst. 32,* 45–56.
5. Janaun, J., Ellis, N. (2010). Perspectives on biodiesel as a sustainable fuel. *Renew. Sust. Energ. Rev. 14,* 1312–1320.
6. Mabee, W. E. (2007). Policy options to support biofuel production. *Adv. Biochem. Eng. Biotechnol. 108,* 329–357.
7. Fazal, M. A., Haseeb, A. S. M. A., Masjuki, H. H. (2011). Biodiesel feasibility study: an evaluation of material compatibility; performance; emission and engine durability. *Renew. Sust. Energ. Rev. 15,* 1314–1324.
8. Lin, L., Cunshan, Z., Vittayapadung, S., Xiangqian, S., Mingdong, D. (2011). Opportunities and challenges for biodiesel fuel. Appl. Energy. 88, 1020–1031.
9. Wardle, D. A. (2003). Global sale of green air travel supported using biodiesel. *Renew. Sust. Energ. Rev. 7,* 1–64.
10. Demirbas, A. (2008). New liquid biofuels from vegetable oils via catalytic pyrolysis. *Energy. Educ. Sci. Technol. 21,* 1–59.
11. Conservation ADoE. (2011). How can I dispose of used cooking oils and fats?. Retrieved 13th March, from. http://www.dec.state.ak.us/eh/docs/fss/cooking.pdf.
12. Lam, M. K., Lee, K. T., Mohamed, A. R. (2010). Homogeneous, heterogeneous and enzymatic catalysis for transesterification of high free fatty acid oil (waste cooking oil) to biodiesel: a review. *Biotechnol. Adv. 28,* 500–518.
13. Atadashi, I. M., Aroua, M. K., Aziz, A. A. (2010). High quality biodiesel and its diesel engine application: a review. *Renew. Sust. Energ. Rev. 14,* 1999–2008.
14. Shuit, S. H., Lee, K. T., Kamaruddin, A. H., Yusup, S. (2010). Reactive extraction and in situ esterification of *Jatropha curcas* L. seeds for the production of biodiesel. *Fuel. 89,* 527–530.
15. Yee, K. F., Tan, K. T., Abdullah, A. Z., Lee, K. T. (2009). Life cycle assessment of palm biodiesel: revealing facts and benefits for sustainability. *Appl. Energy. 86,* S189–8196.
16. Ahmad, A. L., Yasin, N. H. M., Derek, C. J. C., Lim, J. K. (2011). Microalgae as a sustainable energy source for biodiesel production: a review. *Renew. Sust. Energ. Rev. 15,* 584–593.
17. Andrade, J. E., Pérez, A., Sebastian, P. J., Eapen, D. (2011). A review of bio-diesel production processes. *Biomass. Bioenerg. 35,* 1008–1020.
18. Sharma, Y. C., Singh, B. Development of biodiesel from karanja, a tree found in rural India. *Fuel* (2008). *67,* 1740–1742.
19. Lim, S., Teong, L. K. (2010). Recent trends, opportunities and challenges of biodiesel in Malaysia: an overview. *Renew. Sus. Energ. Rev. 14,* 938–954.
20. Marchetti, J. M., Miguel, V. U., Errazu, A. F. (2007). Possible methods for biodiesel production. *Renew. Sust. Energ. Rev. 11,* 1300–1311.
21. Othman, R., Mohammad, A. W., Ismail, M., Salimon, J. (2010). Application of polymeric solvent resistant nanofiltration membranes for biodiesel production. *J. Membr. Sci. 348,* 287–297.

22. Sahoo, P. K., Das, L. M., Babu, M. K. G., Naik, S. N. (2007). Biodiesel development from high acid value polanga seed oil and performance evaluation in a CI engine. *Fuel. 86,* 448–454.

23. Boucher, M. B., Unker, S. A., Hawley, K. R., Wilhite, B. A., Stuart, J. D., Parnas, R. S. (2008). Variables affecting homogeneous acid catalyst recoverability and reuse after esterification of concentrated omega-9 polyunsaturated fatty acids in vegetable oil triglycerides. *Green. Chem. 10,* 1331–1336.

24. Guerreiro, L., Castanheiro, J. E., Fonseca, I. M., Martin-Aranda, R. M., Ramos, A. M., Vital, J. (2006). Transesterification of soybean oil over sulfonic acid function-alised polymeric membranes. *Catal. Today. 118,* 166–171.

25. Vicente, G., Martínez, M., Aracil, J. (2004). Integrated biodiesel production: a comparison of different homogeneous catalysts systems. *Bioresour. Technol. 92,* 297–305.

26. Xie, W., Li, H. (2006). Alumina-supported potassium iodide as a heterogeneous catalyst for biodiesel production from soybean oil. *J. Mol. Catal. A: Chem. 255,* 1–9.

27. Mbaraka, I. K., Shanks, B. H. (2006). Conversion of oils and fats using advanced mesoporous heterogeneous catalysts. *J. Am. Oil. Chem. Soc. 83,* 79–91.

28. Liu, X., He, H., Wang, Y., Zhu, S., Piao, X. (2008). Transesterification of soybean oil to biodiesel using CaO as a solid base catalyst. *Fuel. 87,* 216–221.

29. Jegannathan, K. R., Abang, S., Poncelet, D., Chan, E. S., Ravindra, P. (2008). Production of biodiesel using immobilized lipase—a critical review. *Crit. Rev. Biotechnol. 28,* 253–264.

30. Oda, M., Kaieda, M., Hama, S., Yamaji, H., Kondo, A., Izumoto, E, et al. (2005). Facilitatory effect of immobilized lipase-producing *Rhizopus oryzae* cells on acyl migration in biodiesel-fuel production. *Biochem. Eng. J. 23,* 45–51.

31. Dizge, N., Aydiner, C., Imer, D. Y., Bayramoglu, M., Tanriseven, A., Keskinler, B. (2009). Biodiesel production from sunflower, soybean, and waste cooking oils by transesterification using lipase immobilized onto a novel microporous polymer. *Bioresour. Technol. 100,* 1983–1991.

32. Pinnarat, T., Savage, P. E. (2008). Assessment of noncatalytic biodiesel synthesis using supercritical reaction conditions. *Ind. Eng. Chem. Res. 47,* 6801–6808.

33. Yin, J. Z., Xiao, M., Song, J. B. (2008). Biodiesel from soybean oil in supercritical methanol with cosolvent. *Energy. Convers. Manage. 49,* 908–912.

34. Demirbas, A. (2007). Progress and recent trends in biofuels. Prog. Energy. Combust. Sci. 33:1–18.

35. Tai-Shung, N. C. (2007). Development and purification of biodiesel. *Sep. Purif. Technol. 20,* 377–381.

36. Cao, P., Dubé, M. A., Tremblay, A. Y. (2008). Methanol recycling in the production of biodiesel in a membrane reactor. *Fuel. 87,* 825–833.

37. Qiu, Z., Zhao, L., Weatherley, L. Process intensification technologies in continuous biodiesel production. *Chem. Eng. Process.* (2010). 49, 323–330.

38. Wen, Z., Yu, X., Tu, S. T., Yan, J., Dahlquist, E. (2009). Intensification of biodiesel synthesis using zigzag microchannel reactors. *Bioresour. Technol. 100,* 3054–3060.

39. Kalu, E. E., Chen, K. S., Gedris, T. (2011). Continuous-flow biodiesel production using slit-channel reactors. *Bioresour. Technol. 102,* 4456–61

40. Sun, P., Wang, B., Yao, J., Zhang, L., Xu, N. (2010). Fast synthesis of biodiesel at high throughput in microstructured Reactors. *Ind. Eng. Chem. Res. 49*, 1259–1264.

41. Phan, A. N., Harvey, A. P., Rawcliffe, M. (2011). Continuous screening of base-catalyzed biodiesel production using new designs of mesoscale oscillatory baffled reactors. *Fuel. Process. Technol. 92*, 1560–1567.

42. Harvey, A. P., Mackley, M. R., Seliger, T. (2003). Process intensification of biodiesel production using a continuous oscillatory flow reactor. *J. Chem. Technol. Biotechnol. 78*, 338–341.

43. Zheng, M., Skelton, R. L., Mackley, M. R. (2007). Biodiesel reaction screening using oscillatory flow meso reactors. *Process. Saf. Environ. Prot. 85*, 365–371.

44. Qiu, Z., Petera, J., Weatherley, L. R. (2012). Biodiesel synthesis in an intensified spinning disk reactor. *Chem. Eng. J. 210*, 597–609.

45. Gogate, P. R., Kabadi, A. M. (2009). A review of applications of cavitation in biochemical engineering/biotechnology. *Biochem. Eng. J. 44*, 60–72.

46. Kelkar, M. A., Gogate, P. R., Pandit, A. B. (2008). Intensification of esterification of acids for synthesis of biodiesel using acoustic and hydrodynamic cavitation. *Ultrason. Sonochem. 15*, 188–194.

47. Pal, A., Verma, A., Kachhwaha, S. S., Maji, S. (2010). Biodiesel production through hydrodynamic cavitation and performance testing. *Renew. Energy. 35*, 619–624.

48. Gogate, P. R. (2008). Cavitational reactors for process intensification of chemical processing applications: A critical review. *Chem. Eng. Process. 47*, 515–527.

49. Wu, Z., Ondruschka, B., Zhang, Y., Bremner, D. H., Shen, H., Franke, M. (2009). Chemistry driven by suction. *Green. Chem. 11*, 1026–1030.

50. Colucci, J., Borrero, E., Alape, F. (2005). Biodiesel from an alkaline transesterification reaction of soybean oil using ultrasonic mixing. *J. Am. Oil. Chem. Soc. 82*, 525–530.

51. Rokhina, E. V., Lens, P., Virkutyte, J. (2009). Low-frequency ultrasound in biotechnology: state-of-the-art. *Trends. Biotechnol. 27*, 298–306.

52. Chand, P., Chintareddy, V. R., Verkade, J. G., Grewell, D. (2010). Enhancing biodiesel production from soybean oil using ultrasonics. *Energy. Fuel. 24*, 2010–2015.

53. Deshmane, V. G., Gogate, P. R., Pandit, A. B. (2008). Ultrasound-assisted synthesis of biodiesel from palm fatty acid distilate. *Ind. Eng. Chem. Res. 48*, 7923–7927.

54. Kalva, A., Sivasankar, T., Moholkar, V. S. (2008). Physical mechanism of ultrasound-assisted synthesis of biodiesel. *Ind. Eng. Chem. Res. 48*, 534–544.

55. Barnard, T. M., Leadbeater, N. E., Boucher, M. B., Stencel, L. M., Wilhite, B. A. (2007). Continuous-flow preparation of biodiesel using microwave heating. *Energy. Fuel. 21*, 1777–1781.

56. Vyas, A., Verma, J., Subrahmanyam, N. (2010). A review on fame production processes. *Fuel. 89*, 1–9.

57. Zhu J., Palchik, O., Chen, S., Gedanken, A. (2000). Microwave assisted preparation of CdSe, PbSe, and Cu_2-Xse nanoparticles. *J. Phys. Chem. 104*, 7344–7347.

58. Mu, D. T., Seager, P., Rao, J., Park, F., Zhao, A. (2011). A resilience perspective on biofuel production. *Integr. Enviro. Assess. Manage. 7*, 348–359.

59. Sohpal, V., Singh, A., Dey, A. (2011). Fuzzy modeling to evaluate the effect of temperature onbatch transesterification of *jatropha curcas* for biodiesel production, *Bull. Chem. React. Eng. Catal. 6*, 31–38.

60. Nithya, S. A., Gour, N., Sivakumaran, T. K., Radhakrishnan, T., Balasubramanian, N., Anantharaman. (2008). Design of intelligent controller for nonlinear processes. *Asian. J. Appl. Sci. 1,* 33–45.

61. Sanchez Marcano, J. G., Tsotsis, T. T. (2002). The coupling of the membrane separation process with a catalytic reaction. Catalytic Membranes and Membrane Reactor. Weinheim: Wiley-VCH; p. 5–14.

62. Caro, J. (2008). Catalysis in Micro-structured membrane reactors with nano-designed membranes. *Chin. J. Catal. 29,* 1169–1177.

63. Buonomenna, M. G., Choi, S. H., Drioli, E. (2010). Catalysis in polymeric membrane reactors: the membrane role. *Asia. Pac. J. Chem. Eng. 5,* 26–34.

64. Drioli, E., Fontananova, E., Bonchio, M., Carraro, M., Gardan, M., Scorrano, G. (2008). Catalytic membranes and membrane reactors: an integrated approach to catalytic process with a high efficiency and a low environmental impact. *Chin. J. Catal. 29,* 1152–1158.

65. Ertl, H., Knozinger, H., Schuth, F., Weitkamp, J. (2008). Catalytic membrane reactors. Handbook of heterogeneous catalysis. Weinheim: Wiley-VCH; 2198–241.

66. Lipnizki, F., Field, R. W., Ten, P. K. (1999). Pervaporation-based hybrid process: a review of process design, applications and economics. *J. Membr. Sci. 153,* 183–210.

67. Shuit, S. H., Ong, Y. T., Lee, K. T., Subhash, B., Tan, S. H. (2012). Membrane technology as a promizing alternative in biodiesel production: A review. *Biotechnol. Adv. 30,* 1364–1380.

68. Baroutian, S., Aroua, M. K., Raman, A. A. A., Sulaiman, N. M. N. (2011). A packed bed membrane reactor for production of biodiesel using activated carbon supported catalyst. *Bioresour. Technol. 102,* 1095–1102.

69. Dubé, M. A., Tremblay, A. Y., Liu, J. (2007). Biodiesel production using a membrane reactor. *Bioresour. Technol. 98,* 639–647.

70. Guerreiro, L., Pereira, P. M., Fonseca, I. M., Martin-Aranda, R. M., Ramos, A. M., Dias, J. M. L, et al. (2010). PVA embedded hydrotalcite membranes as basic catalysts for biodiesel synthesis by soybean oil methanolysis. *Catal. Today. 156,* 191–197.

71. Shi, W., He, B., Ding, J., Li, J., Yan, F., Liang, X. (2010). Preparation and characterization of the organic–inorganic hybrid membrane for biodiesel production. *Bioresour. Technol. 101,* 1501–1505.

72. Zhu, M., He, B., Shi, W., Feng, Y., Ding, J., Li. J, et al. (2010). Preparation and characterization of PSSA/PVA catalytic membrane for biodiesel production. *Fuel. 89,* 2299–2304.

73. Sarkar, B., Sridhar, S., Saravanan, K., Kale, V. (2010). Preparation of fatty acid methyl ester through temperature gradient driven pervaporation process. *Chem. Eng. J. 62,* 609–615.

CHAPTER 10

STUDIES ON THE EFFECT OF ANTIOXIDANTS ON THE LONG-TERM STORAGE STABILITY AND OXIDATION STABILITY OF *PONGAMIA PINNATA* AND *JATROPHA CURCUS* BIODIESEL

A. OBADIAH, and S. VASANTH KUMAR

Department of Chemistry, Karunya University, Coimbatore – 641114, Tamilnadu, India

CONTENTS

ABSTRACT

There is increasing interest in developing alternative energy resources. An immediately applicable option is replacement of diesel fuel by bio-diesel, which consists of the simple alkyl esters of fatty acids. With little modification, diesel engine vehicles can use biodiesel fuels. Biodiesel has been defined as the fatty acid methyl esters (FAME) or FA ethyl esters derived from vegetable oils or animal fats (Triglycerides, TG) by transesterification with methanol or ethanol. Its main advantages over fossil fuel are that it is renewable, biodegradable, and nontoxic. Its contribution to greenhouse gases is minimal, since the emitted CO_2 is equal to the CO_2 absorbed by the plants to create the TG. They can also be used as heating oil. Conversely, they do present other techni-cal challenges, such as low cloud points and elevated NO_x emissions. The use of biodiesel is encouraged by governments across the world to improve energy supply security, reduce greenhouse gas emissions, and boost rural incomes and employments. One of the major drawbacks for the quality of biodiesel and its widespread commercialization is its oxidation stability. Unlike petroleum diesel fuel, the nature of bio-diesel makes it more susceptible to oxidation or autoxidation during long-term storage. Storage conditions, exposure to water, and exposure to oxygen, which is naturally present in the ambient air, influence the rate of oxidation.

This study investigates the impact of various synthetic pheno-lic antioxidants on the oxidation stability and storage stability of Pon-gamia (karanja) biodiesel (PBD) and Jatropha curcus biodiesel (JBD). The results of Rancimat experiments show that the induction point (IP) increased substantially on adding certain antioxidants to the Pongamia biodiesel. The study reveals pyrogallol (PY) to be the best antioxidant, which showed the best improvement in the oxidative stability of PBD and JBD. The storage conditions employed were: (i) ordinary glass bot-tle with open space (OGOS), (ii) ordinary glass bottle with closed space

(OGCS), (iii) ordinary glass bottle with closed space containing nitrogen (OGCSN), (iv) amber glass bottle with open space (AGOS), (v) amber glass bottle with closed space (AGCS), and (vi) amber glass bottle with closed space containing nitrogen (AGCSN). Ambient humidity was between 41% and 72%. The samples were analyzed weekly. The storage stability studies were carried out according to the ASTM standard procedure ASTM 4625 @30 °C/50 weeks by adding different antioxidants like BHT, BHA, PY, GA and TBHQ. The induction period of PBD and JBD were determined with and without the addition of antioxidants like BHA, BHT, TBHQ, PY and GA, used at different concentrations (100, 200, 500, 1000, 2000, 3000 and 5000 ppm). Biodiesel samples stored in amber glass bottles with closed space containing nitrogen gave the best results as compared to the other five conditions tested. Pyrogallol was found to be the best commercial antioxidant for both *Pongamia* and *Jatropha* biodiesels as it increased the induction time from 0.05 hr to 4 hr, at 1000 ppm. Kinematic viscosity and acid values are good indicators of storage stability of biodiesels. BHT, TBHQ, PY, GA, and BHA, at concentrations of 1000 ppm, improved the storage stability of biodiesel.

10.1 INTRODUCTION

The word "antioxidant" has become increasingly popular in modern society as it gains publicity through mass media coverage of its health benefits. The dictionary definition of antioxidant is rather straightforward but a traditional annotation would define antioxidant as "a substance that opposes oxidation or inhibits reactions promoted by oxygen or peroxides, many of these substances (as the tocopherols) being used as preservatives in various products (as in fats, oils, food products, and soaps for retarding the development of rancidity, in gasoline and other petroleum products for retarding gum formation and other undesirable changes, and in rubber for retarding aging)." A more biologically relevant definition of antioxidants is "synthetic or natural substances added to products to prevent or delay their deterioration by action of oxygen in air [1].

Biodiesel, defined as fatty acid mono-alkyl esters made from vegetable oil or animal fat, is an alternative fuel for combustion in compression–ignition (diesel) engines. Several recent reviews have reported on the technical characteristics of biodiesel. In short, biodiesel is made from domestically renewable feedstock, is environmentally innocuous, is relatively safe to handle (high flash points), and has an energy content, specific gravity, kinematic viscosity (KV), and cetane number (CN) comparable to petroleum middle distillate fuels (petro diesel) [2, 3].

With a production of almost 1 million tons in Europe, FAME more generally called biodiesel, has become a fast growing renewable liquid biofuel within the European Community. Just like vegetable oils or fats, fatty acid methyl esters undergo degradation over time, mainly influenced by temperature and oxygen. Degradation products of biodiesel, such as insoluble gums and sediments, or the formation of organic acids and aldehyde may cause engine and injection problems [3–5].

The bis-allylic configurations, where the central methylene group is activated by the two double bonds (i.e., -CH=CH-CH$_2$-CH=CH-), react with oxygen via the autoxidation mechanism, with the radical chain reaction steps of initiation, propagation, chain branching, and termination. During these reaction steps, several products can be formed, such as peroxides and hydro peroxides, low molecular weight organic acids, aldehydes and keto compounds, alcohols, as well as high molecular weight species (dimers, trimers, and cyclic acids) *via* polymerization mechanisms. The use of anti-oxidant additives can help slow the degradation process and improve fuel stability up to a point [6–9]. Fuel properties degrade during long-term storage as follows: (i) oxidation or autoxidation from contact with ambient air; (ii) thermal or thermal-oxidative decomposition from excess heat; (iii) hydrolysis from contact with water or moisture in tanks and fuel lines; or (iv) microbial contamination from migration of dust particles or water droplets containing bacteria or fungi into the fuel [10]. Monitoring the effects of autoxidation on biodiesel fuel quality during long-term storage presents a significant concern for biodiesel producers, suppliers, and consumers [11–13].

10.1.1 BIODIESEL STORAGE AND OXIDATION STABILITY STUDIES

The parameter of oxidation stability has been fixed at a minimum limit of a 6-hour induction period at 110°C [14, 15]. The method adopted for determination of the oxidation stability is the so-called Rancimat method, which is commonly used in the vegetable oil sector. Especially high contents of unsaturated fatty acids, which are very sensitive to oxidative degradation, lead to very low values for the induction period. Thus, even the conditions of fuel storage directly affect the quality of the product. Several studies showed that the quality of biodiesel over a longer period of storage strongly depends on the tank material as well as on contact to air or light. Increase in viscosities and acid values and decreases in induction periods have been observed [16] during such storage. Although there are numerous publications on the effect of natural and synthetic antioxidants on the stability of oils and fats used as food and feed, little is available on the effect of antioxidants on the behavior of FAME used as biodiesel. To retard oxidative degradation and to guarantee a specific stability, it becomes necessary to find appropriate additives for biodiesel. Simkovsky [17] studied the effect of different antioxidants on the induction period of rapeseed oil methyl esters at different temperatures but did not find significant improvements. Schober [18] tested the influence of the antioxidant TBHQ on the peroxide value of soybean oil methyl esters during storage and found good improvement of stability. Canakci [19] described the effect of the antioxidants TBHQ and α-tocopherol on fuel properties of methyl soyate and found beneficial effects on retarding oxidative degradation of the sample. Das [20] described effect of commercial antioxidants used in kharanja biodiesel for storage stability. Recently Karavalakis [21] described the effect of synthetic phenolic antioxidants used for storage stability and oxidative stability. The storage stability of different biodiesel blends with automotive diesel treated with various phenolic antioxidants has also been investigated over a storage time of 10 weeks.

In the previous studies, numerous methods for assessing the oxidation status of biodiesel have been investigated, including acid value, density, and kinematic viscosity. The peroxide value may not be suitable because, after an initial increase, it decreases due to secondary oxidation reactions, although the decrease likely affects only samples oxidized beyond what may normally be expected. Thus there is the possibility of the fuel having undergone relatively extensive oxidation but displaying an acceptable peroxide value. The peroxide value is also not included in biodiesel standards. Acid value and kinematic viscosity, however, are two facile indicators for rapid assessment of biodiesel fuel quality as they continuously increase with deteriorating fuel quality [22]. In this chapter, oxidative and storage stability of biodiesel was investigated using commercially available antioxidants.

The aim of the present study was to investigate the oxidative and storage stabilities of *Pongamia pinnata* (PBD) and *Jatropha curcus* (JBD) biodiesels with five commercially available antioxidants such as BHA, BHT, GA, TBHQ and PY. The Rancimat procedure and the ASTM procedure were employed to investigate oxidation stability and storage stability, respectively, of the biodiesels. Fuel properties such as acid value (AN) and KV of PBD and JBD were determined at regular periods of time, using the selected antioxidants at different concentrations.

10.2 EXPERIMENTAL

10.2.1 BIODIESEL STORAGE CONDITION

Normally, 200 mL of PBD and JBD biodiesel samples were stored in open Borosil glass bottles of 250 mL capacity and kept indoors, at a temperature of either 30°C or 42°C. The samples were exposed to air under daylight condition. The storage studies on PBD and JBD using commercially available antioxidants are presented in this section. Experiments were carried out at different storage conditions. The storage conditions employed were: (i) ordinary glass bottle with open space (OGOS); (ii) ordinary glass bottle with closed space (OGCS); (iii) ordinary glass bottle with closed space containing nitrogen (OGCSN); (iv) amber glass bottle with open space (AGOS); (v) amber glass bottle with closed space (AGCS); and (vi) amber glass bottle with closed space containing nitrogen (AGCSN). Ambient humidity was between 41% and 72%. The samples were analyzed weekly.

10.2.2 DETERMINATION OF OXIDATIVE STABILITY

Oxidative stability (OS) of biodiesel sample was studied with a Rancimat 873 instrument (Metrohm, Switzerland), as followed by Tang [9] Das [20]

Karavalakis [21] and Knothe [22]. The sample was heated at a constant temperature of 110°C with excess airflow which passed through a conductivity cell filled with distilled water. During this oxidation process volatile acids were formed which were carried along with the airflow resulting in a gradual increase in the conductivity of distilled water present in the conductivity cell. After a period of time, called 'induction period,' the gradual increase leads to a sudden increase in the conductivity of the distilled water. The induction period of PBD and JBD were determined with and without the addition of antioxidants like BHA, BHT, TBHQ, PY and GA, used at different concentrations (100, 200, 500, 1000, 2000, 3000 and 5000 ppm).

10.2.3 EVALUATION OF STORAGE STABILITY

To evaluate the storage stability, the ASTM procedures D-4625 – 25 weeks and D4625 – 12 weeks were employed. In the ASTM D4625 – 25 weeks procedure, the KV, PV and AN values were determined at 30°C every week over a period of 25 weeks. In the ASTM D4625 – 12 procedure, the KV, PV and AN values were monitored at 43°C every week over a period of 12 weeks. The storage container used was 250 mL bottle containing 200 mL of biodiesel and 50 mL of free space. The ambient humidity in the storage room was 41% to 72%.

10.3 RESULTS AND DISCUSSION

10.3.1 BIODIESEL STORAGE STUDIES

The study on karanja oil methyl ester was carried out by Das et al. by using three antioxidants namely butylated hydroxy anisole, butylated hydroxy toluene and propyl gallate [20]. They varied the load level of the three antioxidants from 100 to 1000 ppm and noticed decrease in peroxide values as the concentration of antioxidants increased. Peroxide value is not listed as a parameter in the biodiesel fuel specification, though viscosity and acid value were among the specifications listed within PS121 (provisional fuel standard guideline for biodiesel) and are known to be affected by the autoxidation of biodiesel. Changes in viscosity and acid value were therefore monitored in this study, to understand the oxidative stability of *Pongamia* biodiesel. Dunn examined the effects of oxidation under controlled accelerated conditions on fuel properties of methylsoyate, where the author employed only TBHQ and α-tocopherol as antioxidants, which were found to have beneficial effects on retarding oxidative degradation of methylsoyate biodiesel [18]. Karavalakis [21] in their study evaluated the impact of biodiesel concentration in diesel fuel on the stability of the final blend. They have also discussed the effect of sulfur content in the base diesel on the oxidation stability of the blend (Fig. 10.1).

Results of storage studies of *Pongamia* and *Jatropha* biodiesel using commercially available antioxidants are presented in this chapter. The chemical structure of the selected antioxidants are given below.

Experiments were carried out at different storage conditions. The storage conditions employed were: (i) ordinary glass bottle with open space (OGOS), (ii) ordinary glass bottle with closed space (OGCS), (iii) ordinary

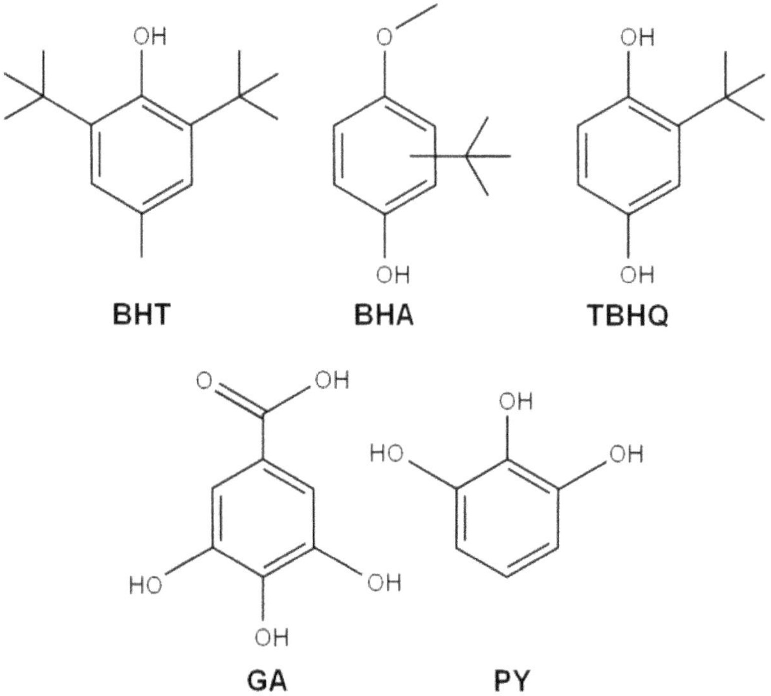

FIGURE 10.1   Chemical structure of commercially available antioxidants.

glass bottle with closed space containing nitrogen (OGCSN), (iv) amber glass bottle with open space (AGOS), (v) amber glass bottle with closed space (AGCS), and (vi) amber glass bottle with closed space containing nitrogen (AGCSN). The effect of five different antioxidants on the oxidative stability and storage stability of *Pongamia* and *Jatropha* biodiesel were evaluated using standard Rancimat procedure (EN 14214) and recommended ASTM procedure D-4625 – 25 weeks. Storage in amber glass bottle with nitrogen atmosphere was found to be the best storage condition exhibiting minimum oxidation. Figure 10.2 shows the variance of kinematic viscosity, acid value and peroxide value with respect to time in the presence of nitrogen atmosphere.

Studies using various concentrations reveal that 1000 ppm is sufficient for a storage time of 25 weeks. Figures 10.2 and 10.3 shows the variance of kinematic viscosity, acid value and peroxide value with respect to time in the presence of nitrogen atmosphere. The effects of commercial antioxidants on *Pongamia* and *Jatropha* biodiesel at different atmospheric conditions and various concentrations of antioxidants were studied and are discussed below.

10.3.1.1   *Jatropha* Biodiesel With BHA as Additive

In the absence of antioxidants (neat) JBD exhibits a KV value of 14.76 mm^2/s, PV of 9.3 mg/kg and AV of 2.47 mg KOH/g. The storage studies also reveal that in the absence of antioxidants PBD shows a KV value of 14.7 mm^2/s, PV of 6.38 mg/kg and AV of 1.68 mg KOH/g. The results presented below are from the storage studies carried out under AGCSN conditions. All other storage conditions showed higher values of KV, PV and AV)

10.3.1.2   *Pongamia* Biodiesel With BHA as Additive

Results of storage studies of *Pongamia* and *Jatropha* biodiesel using BHA as the antioxidant are shown in Figures 10.4 and 10.5. Various storage conditions such as: (i) OGOS, (ii) OGCS, (iii) OGCSN, (iv) AGOS, (v) AGCS, and (vi) AGCSN were tested. Storage in amber glass bottle with nitrogen atmosphere was found to be the best storage condition exhibiting minimum oxidation.

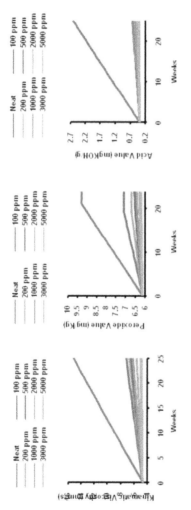

FIGURE 10.2 Storage studies using pyrogallol as additive in *Pongamia* biodiesel.

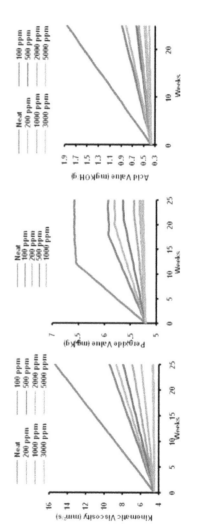

FIGURE 10.3 Storage studies using pyrogallol as additive in *Jatropha* biodiesel.

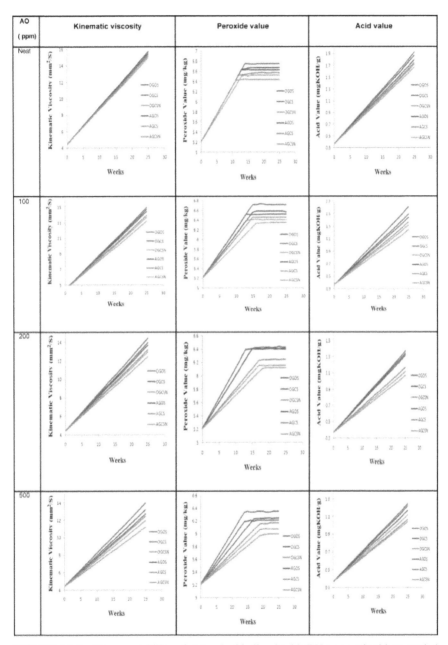

FIGURE 10.4   Storage stability of *Jatropha* biodiesel with BHA as antioxidant carried out for 25 weeks.

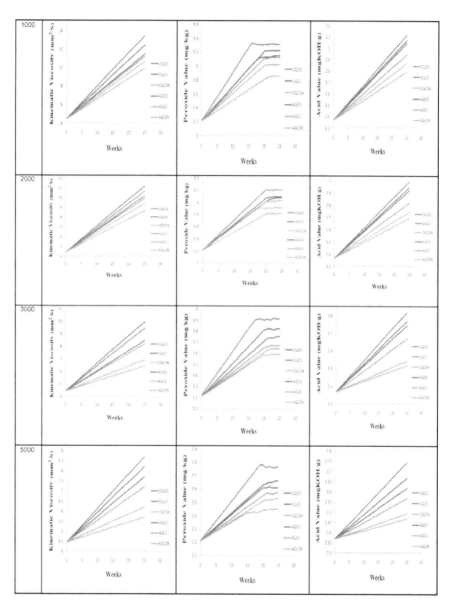

FIGURE 10.4 (*Continued*)

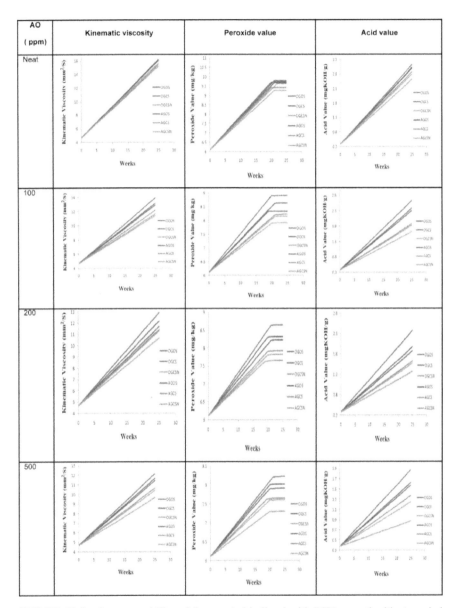

FIGURE 10.5   Storage stability of *Pongamia* biodiesel with BHA as antioxidant carried out for 25 weeks.

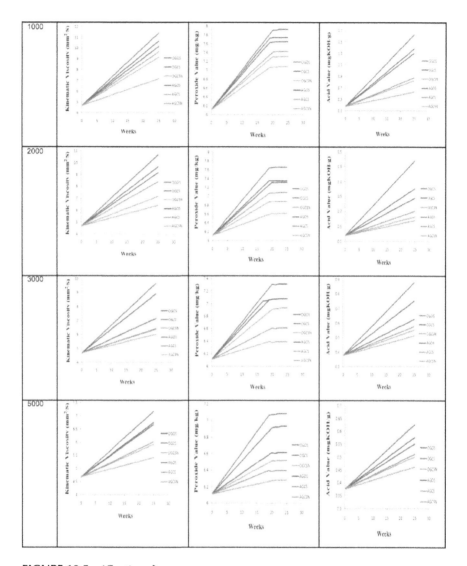

FIGURE 10.5   *(Continued)*

Results reveals that JBD with 5000 ppm BHA, under nitrogen atmosphere stored in a amber glass bottle exhibited a KV of 5.21 mm²/s, PV of 5.42 mg/kg and AV of 0.44 mg KOH/g. The storage studies also reveal that PBD with 5000 ppm of BHA showed KV of 5.27 mm²/s, PV of 6.2 mg/kg and AV of 0.44 mg KOH/g. The storage stability increased when the concentration of antioxidant was increased.

10.3.1.3   *Jatropha* Biodiesel With BHT as Additive

Storage stability of *Jatropha* biodiesel with BHT as antioxidant carried out for 25 weeks is shown in Fig. 10.6.

10.3.1.4   *Pongamia* Biodiesel With BHT as Additive

Results of storage studies of *Pongamia* and *Jatropha* biodiesel using BHT as the antioxidant under various storage conditions such as: (i) OGOS, (ii) OGCS, (iii) OGCSN, (iv) AGOS, (v) AGCS, and (vi) AGCSN are presented above (Figs. 10.7 and 10.8). Storage in amber glass bottle with nitrogen atmosphere was found to be the best storage condition exhibiting minimum oxidation. Figure 10.6 and 10.7 show the variance of kinematic viscosity, acid value and peroxide value with respect to time in the presence of nitrogen atmosphere.

JBD with 5000 ppm BHT under nitrogen atmosphere stored in a amber glass bottle exhibited a KV of 4.7 mm²/s, PV of 5.28 mg/kg and AV of 0.39 mg KOH/g. The study also reveals that PBD with 5000 ppm BHT exhibits KV of 5.1 mm²/s, PV of 6.22 mg/kg and AV of 0.41 mg KOH/g. The storage stability increased when the concentration of antioxidants was increased.

10.3.1.5   *Jatropha* Biodiesel With GA as Additive

Storage stability of *Jatropha* biodiesel with GA as antioxidant carried out for 25 weeks is shown in Fig. 10.8.

10.3.1.6   *Pongamia* Biodiesel With GA as Additive

Results of storage studies of *Pongamia* and *Jatropha* biodiesel using GA as the antioxidant under various storage conditions such as, (i) OGOS, (ii) OGCS, (iii) OGCSN, (iv) AGOS, (v) AGCS, and (vi) AGCSN are presented above. Storage in amber glass bottle with nitrogen atmosphere was found to be the best storage condition exhibiting minimum oxidation.

Figures 10.8 and 10.9 show the variance of kinematic viscosity, acid value and peroxide value with respect to time in the presence of nitrogen

256 — Bioenergy: Opportunities and Challenges

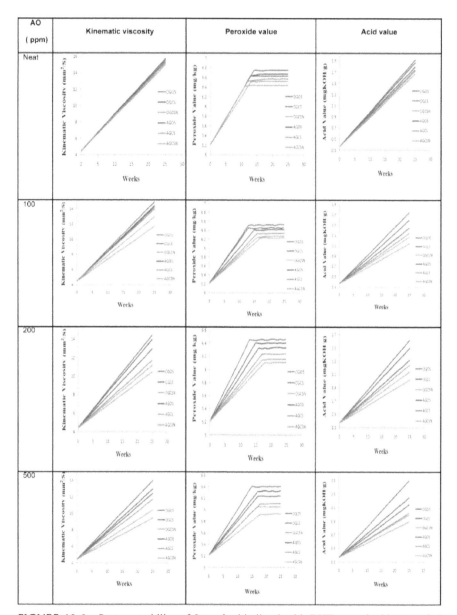

FIGURE 10.6 Storage stability of Jatropha biodiesel with BHT as antioxidant carried out for 25 weeks.

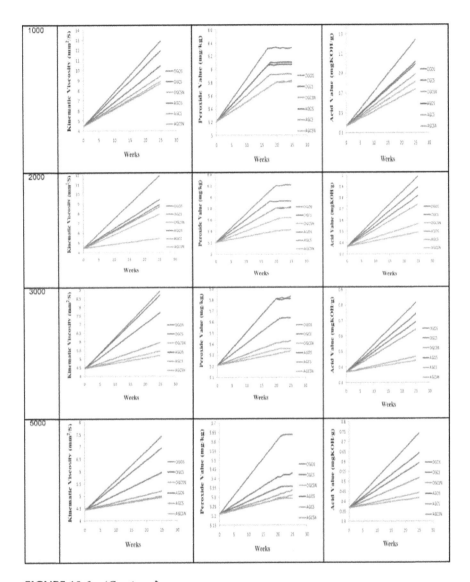

FIGURE 10.6 (*Continued*)

atmosphere. JBD with 5000 ppm GA, under nitrogen atmosphere stored in a amber glass bottle, exhibited a KV of 5.4 mm²/s, PV of 5.33 mg/kg and AV of 0.45 mg KOH/g. The storage studies also reveals that 5000 ppm of GA in PBD shows KV of 5.1 mm²/s, PV of 6.22 mg/kg and AV of 0.41 mg KOH/g. The storage stability increased with an increase in the concentration of GA.

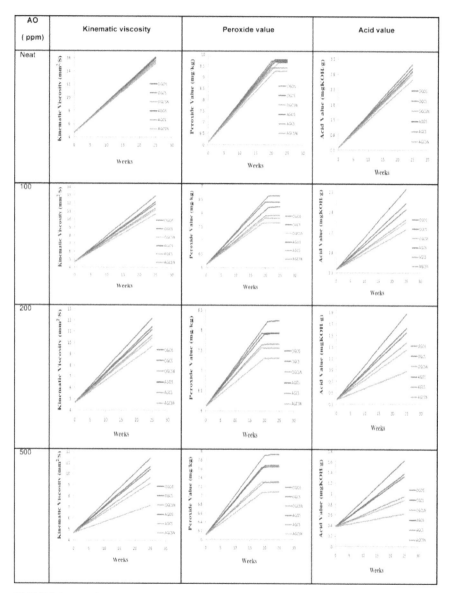

FIGURE 10.7   Storage stability of Pongamia biodiesel with BHT as antioxidant carried out for 25 weeks.

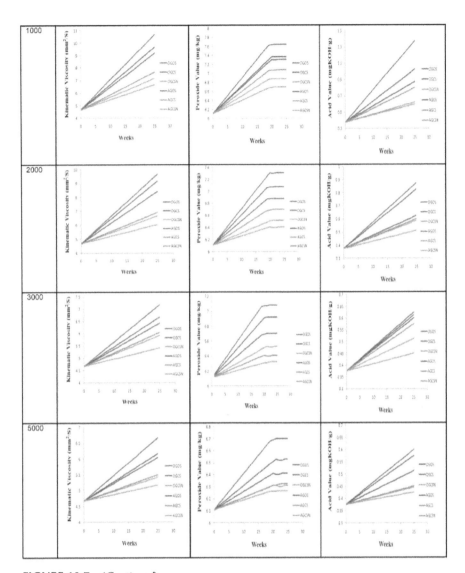

FIGURE 10.7 (*Continued*)

10.3.1.7 *Jatropha* Biodiesel With TBHQ as Additive

Storage stability of *Jatropha* biodiesel with TBHQ as antioxidant carried out for 25 weeks is shown in Fig. 10.10.

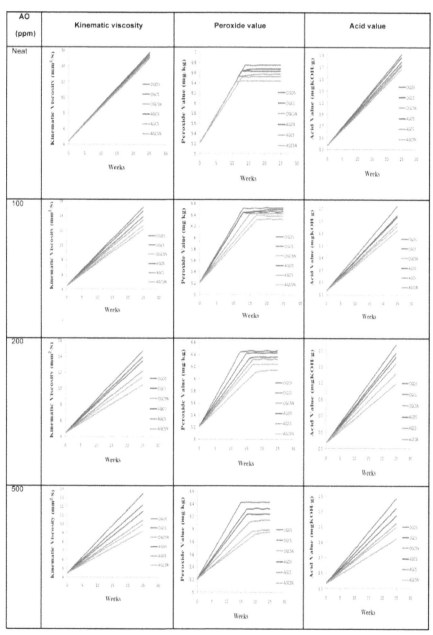

FIGURE 10.8   Storage stability of *Jatropha* biodiesel with GA as antioxidant carried out for 25 weeks.

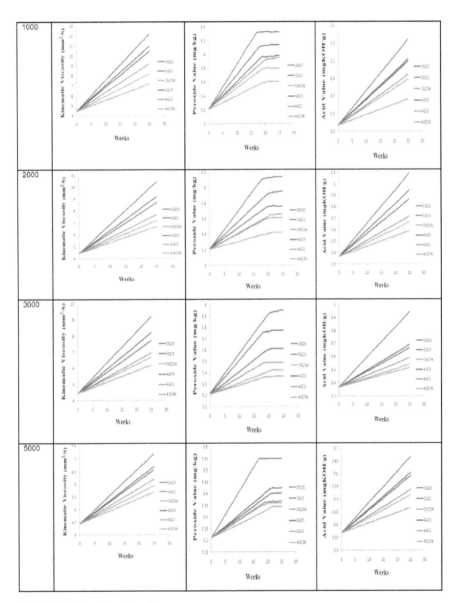

FIGURE 10.8   (*Continued*)

10.3.1.8   *Pongamia* Biodiesel With TBHQ as Additive

Results of storage studies of *Pongamia* and *Jatropha* biodiesel using TBHQ as the antioxidant were investigated under different storage condi-

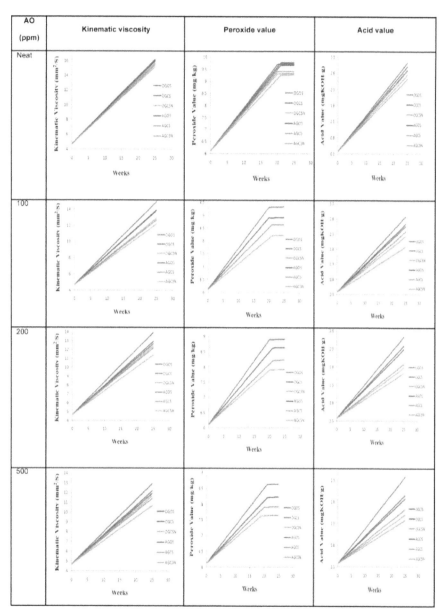

FIGURE 10.9   Storage stability of *Pongamia* biodiesel with GA as antioxidant carried out for 25 weeks.

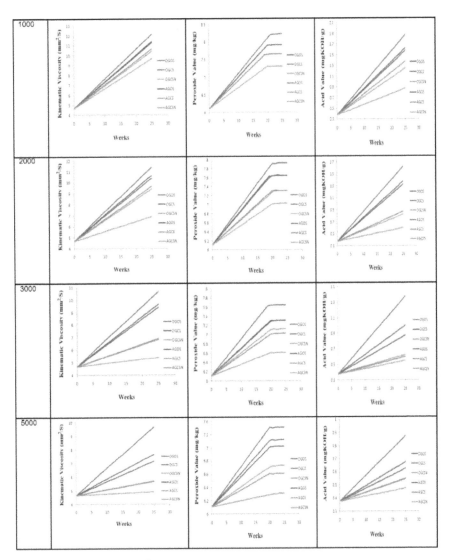

FIGURE 10.9   (*Continued*)

tions such as, (i) OGOS, (ii) OGCS, (iii) OGCSN, (iv) AGOS, (v) AGCS, and (vi) AGCSN. Storage in amber glass bottle with nitrogen atmosphere was found to be the best storage condition exhibiting minimum oxidation.

Figures 10.10 and 10.11 shows that JBD with 5000 ppm TBHQ, under nitrogen atmosphere stored in a amber glass bottle, exhibited a KV of 4.48

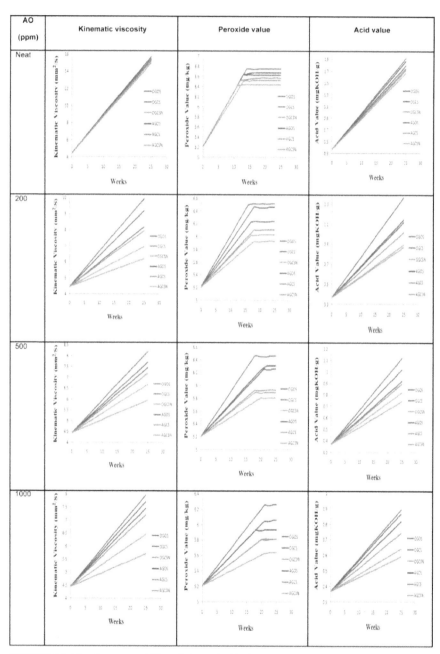

FIGURE 10.10 Storage stability of *Jatropha* biodiesel with TBHQ as antioxidant carried out for 25 weeks.

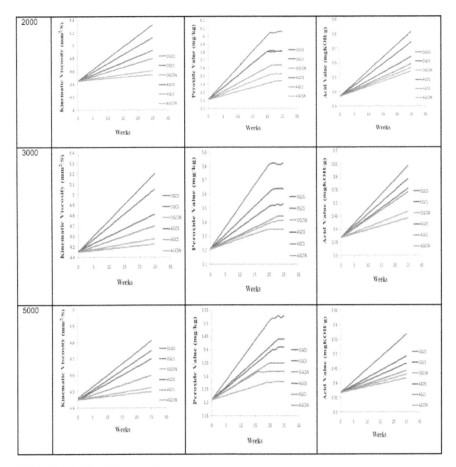

FIGURE 10.10   (*Continued*)

mm²/s, PV of 5.28 mg/kg and AV of 0.40 mg KOH/g. PBD with 5000 ppm TBHQ under similar conditions shows KV of 4.98 mm²/s, PV of 6.12 mg/kg and AV of 0.38 mg KOH/g. The storage stability increased when the concentration of TBHQ was increased.

10.3.1.9   *Jatropha* Biodiesel With PY as Additive

Storage stability of *Jatropha* biodiesel with PY as antioxidant carried out for 25 weeks is shown in Fig. 10.12.

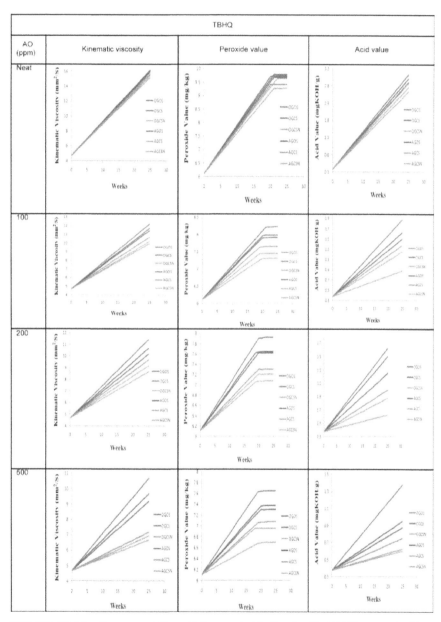

FIGURE 10.11 Storage stability of *Pongamia* biodiesel with TBHQ as antioxidant carried out for 25 weeks.

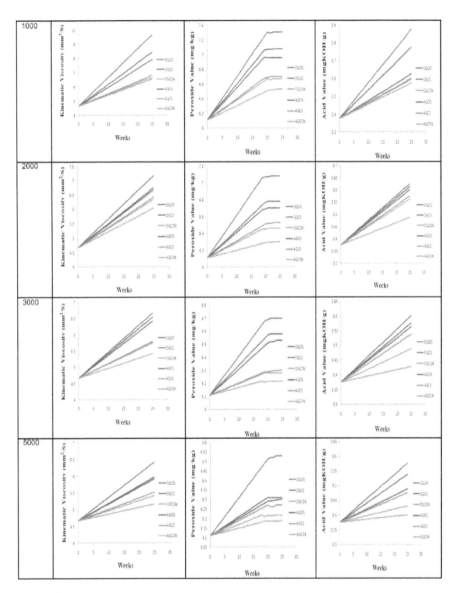

FIGURE 10.11 *(Continued)*

10.3.1.10 *Pongamia* Biodiesel With PY as Additive

Results of storage studies of *Pongamia* and *Jatropha* biodiesel using PY as the antioxidant were investigated under different storage conditions such as, (i)

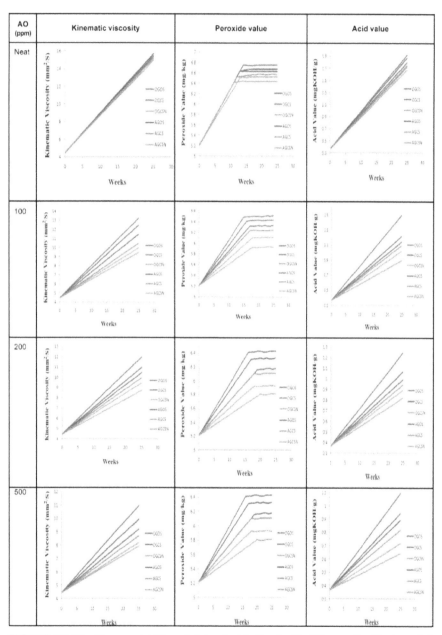

FIGURE 10.12   Storage stability of *Jatropha* biodiesel with PY as antioxidant carried out for 25 weeks.

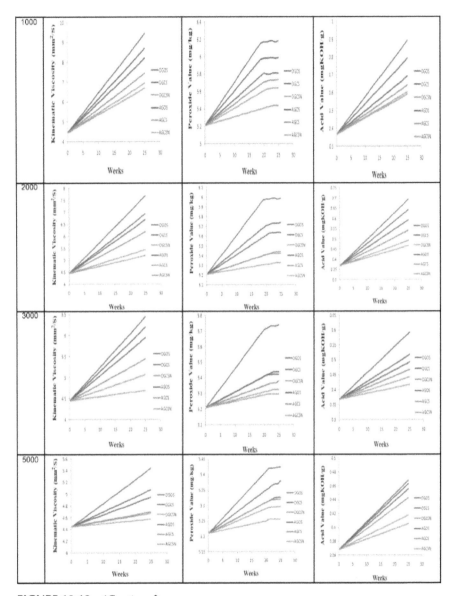

FIGURE 10.12   (*Continued*)

OGOS, (ii) OGCS, (iii) OGCSN, (iv) AGOS, (v) AGCS, and (vi) AGCSN.
Storage in amber glass bottle with nitrogen atmosphere was found to be the best
storage condition exhibiting minimum oxidation.

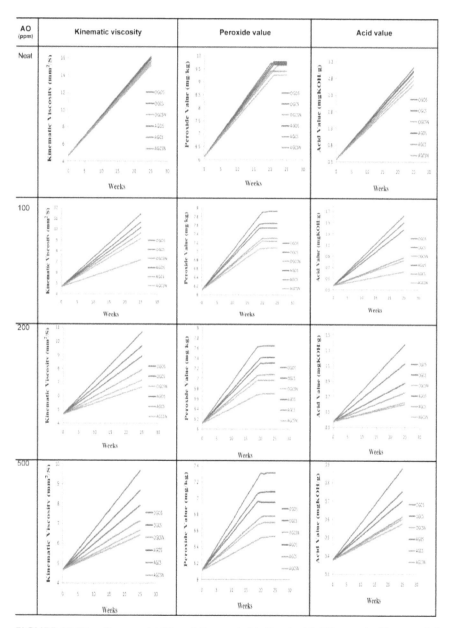

FIGURE 10.13   Storage stability of *Pongamia* biodiesel with PY as antioxidant carried out for 25 weeks.

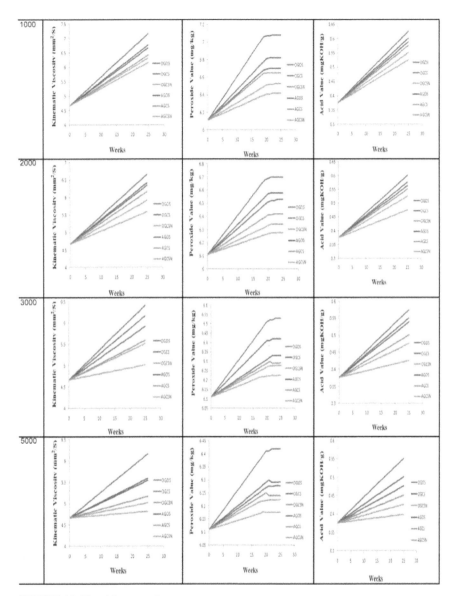

FIGURE 10.13 (*Continued*)

Figures 10.12 and 10.13 reveals that JBD with 5000 ppm PY, under nitrogen atmosphere stored in a amber glass bottle, exhibited a KV of 4.57 mm²/s, PV of 5.23 mg/kg and AV of 0.38 mg KOH/g. The storage studies reveal that PBD with 5000 ppm PY shows KV of 4.75 mm²/s, PV of 5.23 mg/kg and AV of 0.38 mg KOH/g. The storage stability increased when the concentration of antioxidants was increased.

10.3.2   OXIDATIVE STABILITY STUDIES OF PONGAMIA AND JATROPHA BIODIESEL USING COMMERCIALLY AVAILABLE ANTIOXIDANTS

Rancimat method, the most commonly employed method in the vegetable oil sector, was adopted for the determination of oxidation stability. A high content of unsaturated fatty acids, which is very sensitive to oxidative degradation, leads to decreased induction times. Thus, even the conditions of fuel storage directly affect the quality of product. Several studies showed that the quality of biodiesel over a longer period of storage strongly depends on the tank material as well as on contact to air and light. Increase in viscosity and acid values leads to decrease in induction periods [11, 16]. The oxidative stability of biodiesel was studied by the Rancimat method as per EN14112 methodology using commercially available antioxidants.

All the other antioxidants like BHA, BHT, GA, and TBHQ were also used in the studies with concentrations ranging from 100 ppm to 5000 ppm under various atmospheric conditions. The different storage conditions are ordinary glass bottle with open space, ordinary glass bottle with closed space, ordinary glass bottle with nitrogen atmosphere, amber colored bottle with open space, amber colored bottle with closed space and amber colored bottle with closed space containing nitrogen atmosphere.

When investigating phenolic antioxidants, it was found that their antioxidative capabilities bear a relationship to the number of phenol groups occupying 1, 2 or 1, 4 positions in an aromatic ring, as well as to the volume and electronic characteristics of the ring substituent present. Generally, the active hydroxyl group can provide protons that inhibit the formation of free radicals or interrupt the propagation of free radical and thus delay the rate of oxidation. The effectiveness of PY, TBHQ, GA, can be explained based on their molecular structure (Fig. 10.14). These additives possess two OH groups attached to the aromatic ring, while both BHT and BHA possess one OH group attached to the aromatic ring. Thus, based on their electro-negativities, TBHQ and PY offer more sites for the formation

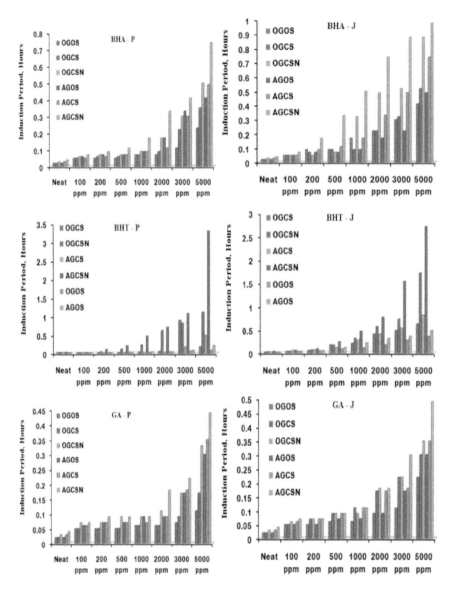

FIGURE 10.14 Oxidative stability results on *Pongamia* and *Jatropha* at different atmospheric conditions and different commercially available antioxidants.

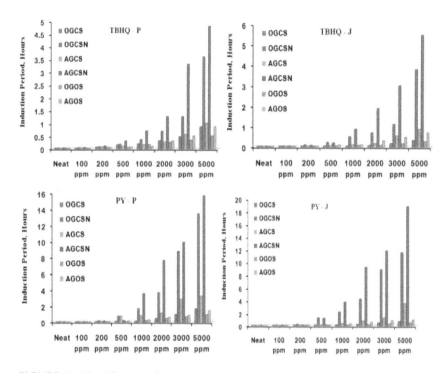

FIGURE 10.14   (*Continued*)

of a complex between free radical and antioxidant radical for the stabilization of the ester chain. Another contributing factor for the poor antioxidant performance of both BHT and BHA is their relatively low volatility, which under the operating conditions of the Rancimat method will lead to loss of additives during the early part of the test [18].

Figure 10.15 shows that the pyrogall was found to be the best antioxidant among those commercially available, with a induction period of 3.51 h at a concentration of 1000 ppm, 15.65 h at a concentration of 5000 ppm, and 19.21 hr at concentration of 5000 ppm for both *Pongamia* and *Jatropha* biodiesel.

10.4   CONCLUSION

Biodiesel samples stored in amber glass bottles with closed space containing nitrogen gave the best results as compared to the other five conditions

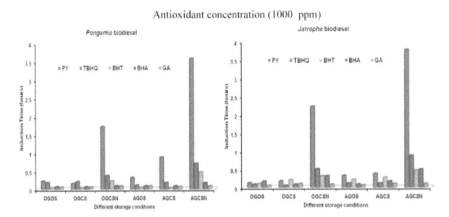

FIGURE 10.15   Oxidative stability studies on PBD and JBD at 1000 ppm commercially available antioxidants.

tested. Pyrogallol was found to be the best commercial antioxidant for both *Pongamia* and *Jatropha* biodiesels as it increased the induction time from 0.05 h to 4 hr, at 1000 ppm. Kinematic viscosity and acid values are good indicators of storage stability of biodiesels. BHT, TBHQ, PY, GA, and BHA, at concentrations of 1000 ppm, improved the storage stability of biodiesel.

KEYWORDS

- **Antioxidant**
- **Biodiesel**
- **Oxidation stability**
- **Pongamia oil**
- **Storage stability**

REFERENCES

1. Oyaizu. (2008). In vitro antioxidants estimation. *Food Res. Int. 41*, 1–15.
2. Dunn, R. O. (2005). Oxidative Stability of Soybean Oil Fatty Acid Methyl Esters by Oil Stability Index (OSI) *J. Am. Oil Chem. Soc. 82*, 381–387.

3. Lacoste, F., Lagardere, L. (2003). Quality parameters evolution during biodiesel oxidation using Rancimat test. *Eur. J. Lipid Sci. Technol. 105,* 49–55.
4. Schober, S., Mittelbach, M. (2004). The impact of antioxidants on biodiesel oxidation stability. *Eur. J. Lipid Sci. Technol. 106,* 382–389.
5. Mittelbach, M., Schober, S. (2003). Storage stability of biodiesel. *J. Am. Oil Chem. Soc. 80,* 817–823.
6. Schober, S., Mittelbach, M. (2004). The impact of antioxidants on biodiesel oxidation stability. *Eur. J. Lipid Sci. Technol. 106,* 382–389.
7. Dunn, O. (2005). Effect of antioxidants on the oxidative stability of methyl soyate (biodiesel) *Fuel Process. Technol. 86,* 1071–1085.
8. Ryu, J. H., Kim, T. J., Lee, T. Y., Lee, I. B. (2010). A study on modeling and simulation of capacitive deionization process for wastewater treatment. J. Taiwan Instit of Chem. Engg. *41 (4),* 506–511.
9. Monyem, A., Canakci, M., Gerpen, Jv. (2000). Investigation of biodiesel thermal stability under simulated in-use conditions. Appl. Eng. Agric. *16,* 373−8.
10. Bouaid, A., Martinez, M., Aracil, (2009). Production of biodiesel from bioethanol and Brassica carinata oil: oxidation stability study. J. *Biores. Technol. 100,* 2234–39.
11. McCormick, R. L., Westbrook, S. R. (2010). Storage Stability of Biodiesel and Biodiesel Blends. *Energy Fuels 24,* 690–98.
12. Evans, R. ABTS free radical scavenging modified method. 1997.
13. Cui, Y. (2005). Antioxidant effect of Inonotus obliquus *J. Ethno pharm. 96,* 79–85.
14. Selvakumar, K., Madhan, R., Srinivasan, G., Baskar, V. (2011). Antioxidant Assays in Pharmacological Research. *As. J. Pharm. Tech.* Vol. 1: Issue 4, 99–103.
15. Mittelbach, M., Gangl, S. (2001). Long storage stability of biodiesel made from rapeseed and used frying oil. *J. Am. Oil Chem. Soc. 78,* 573–77.
16. Bondioli, P., Gasparoli, A., Lanzani, A., Fedeli, E., Veronese, S., Sala, M. J. (1995). Storage stability of biodiesel. *Am. Oil Chem. Soc. 72,* 699–02.
17. Simkovsky, N. M., Ecker, A. (1999).Effect of Antioxidants on the. Oxidative Stability of Rapeseed Oil Methyl Esters. *Erdöl Erdgas Kohle 115,* 317–318.
18. Dunn, R. O. (2002). Effect of oxidation under accelerated conditions on fuel properties of methyl soyate (biodiesel). *J. Am. Oil Chem. Soc. 79,* 915–920.
19. Canakci, M., Monyem, A., Gerpen, J. V. (1999). Accelerated oxidation process in biodiesel. *Trans. ASAE 42,* 1565–1572.
20. Das, L. M., Bora, D. K., Pradhan, S., Naik, M. K., Naik, S. N. (2009). Long-term storage stability of biodiesel produced from Karanja oil. *Fuel 88,* 2315–2318.
21. Karavalakis, G., Hilari, D., Givalou, L., Karonis, D., Stournas, S. (2011). Storage stability and aging effect of biodiesel blends treated with different antioxidants. *Energy 36,* 369−374.
22. Knothe, G. (2006). Analysis of oxidized biodiesel by ^{1}H-NMR and effect of contact area with air. *Eur. J. Lipid Sci. Technol. 108,* 493−500.

CHAPTER 11

EFFECT OF FUNGAL BIOTIC STRESS ON PHYSIC NUT (*JATROPHA CURCAS* L.)

SEWETA SRIVASTAVA, and ASHA SINHA

Department of Mycology and Plant Pathology,
Institute of Agricultural Sciences, Banaras Hindu University, Varanasi
– 221 005, U.P., India; E-mail: shalu.bhu2008@gmail.com

CONTENTS

ABSTRACT

Jatropha curcas L. is a small tree that grows originally in areas near the equator. The oil plant *Jatropha curcas* L., a multipurpose drought resistant, perennial plant belonging to Euphorbiaceae family is gaining a lot of importance for the production of bio-diesel. Growth in the automotive industry, along with increases in population and gains in worldwide standards of living, has resulted in greater demand for energy sources such as biodiesel fuel. Seeds are regarded as highly effective means for transporting plant pathogens over long distances. Seed deterioration is defined as summation of all physical, physiological, biochemical changes occurring in a seed, which ultimately lead to its death. *Jatropha* seeds are constantly subjected to deterioration, which implies an irreversible degenerative change in the quality of seeds after it has reached its maximum quality of seeds after it has reached its maximum quality level. The fungi associated with seeds at the harvest stage and under storage bring about several undesirable changes and degradation of seed constituents, thus making the seed unfit for oil extraction, export purpose, consumption or sowing. These storage mycoflora affect the quality of seeds and as well as the diesel extracted by affecting their biochemical and physicochemical properties which is very essential. Biodiesel, an environmental friendly diesel fuel similar to petro-diesel in combustion properties, has received considerable attention in the recent past worldwide. There is therefore, need to explore alternative nonedible oil for use in production of biodiesel.

11.1 INTRODUCTION

Energy is the chief mover of economic growth, and plays a vital role in sustaining the modern economy and society. Utilization of waste is also way to help the nation by environmentally as well as economically. Biodiesel is the name for a variety of ester-based fuels (fatty ester) generally defined as monoalkyl ester made from renewable biological resources such as vegetable oils, recycled waste vegetable oil and animal fats. This renewable source is as efficient as petroleum diesel in powering unmodified diesel engine. Today's diesel engines require a clean burning, stable fuel operating under a variety of conditions. Using biodiesel not only helps maintaining our environment; it also helps in keeping the people around us healthy [1].

Biofuels are considered in part, a solution to such issues as sustainable development, energy security and a reduction of greenhouse gas emissions. Biodiesel, an environmental friendly diesel fuel similar to petro-diesel in combustion properties, has received considerable attention in the recent past worldwide. Our future economic growth considerably depends on the long-term accessibility of energy from the sources that are easily available, safe and affordable [2]. New sources of energy like biofuels may play a significant role in meeting the energy demands [3]. Biomass sources have turned out to be more effective in the recent days because of the insufficiency of conventional fossil fuels, their price hike and increased emissions of pollutants generated during combustion. The petroleum-based fuel reserves are concentrated in only some parts of the world and these resources are depleting day-by-day [4].

Biofuels can replace fossil fuels, such as petrol or diesel, either totally or partially in a blend. The history of biofuels goes back to the first part of last century. The inventor of the diesel engine, Rudolf Diesel, already stated in 1912 that "the diesel engine can be fed with vegetable oils and would help considerably in the development of agriculture of the countries which use it" and that "the use of vegetable oils for engine fuels may seem insignificant today; but, such oils may become in course of time as important as petroleum and the coal tar products of the present time" [5].

The seeds of physic nut are a good source of oil, which can be used as a diesel substitute [6]. Depending on the variety, the decorticated seeds contain 40–60% of oil [7–12], which is used for many purposes such as lighting, as a lubricant, for making soap [13] and most importantly as bio-diesel. The seeds can be burned directly, without extracting the oil, to provide an alternative to kerosene lamps [14]. Seed yield usually ranges from 1–5 t ha^{-1} [15–17]. About 30% of the seed weight is pure plant oil [15]. Mycorrhiza (a symbiotic association between a fungus and the roots of a plant) can increase the biomass and seed production with 30% after seven months [18]. If a holistic approach is taken to use *Jatropha* fruit, it will give three times the energy of bio-diesel alone [19].

Seeds are regarded as highly effective means for transporting plant pathogens over long distances [20]. Seed-borne mycoflora have been found to effect the growth and productivity of crop plants. A seed-borne pathogen present externally or internally or associated with the seed as contaminant, may cause abortion, seed rot, seed necrosis, reduction or elimination of germination capacity as well as seedling damage. This results in development of diseases at later stages of plant growth by systemic or local infection [21, 22].

Seed deterioration is defined by Delouche [23] as summation of all physical, physiological, biochemical changes occurring in a seed, which ultimately lead to its death. He also characterized seed deterioration as inexorable, irreversible, inevitable and minimal at the time of physiological maturity and variable among the seed kinds, varieties and seed lots of same variety and among individual seeds. Many researchers have opinion that seed deterioration is a progressive deleterious process, which has far reaching consequences [24, 25].

Jatropha seeds are constantly subjected to deterioration, which implies an irreversible degenerative change in the quality of seeds after it has reached its maximum quality of seeds after it has reached its maximum quality level [26, 27]. The fungi associated with seeds at the harvest stage and under storage bring about several undesirable changes and degradation of seed constituents, thus making the seed unfit for oil extraction, export purpose, consumption or sowing. Chelkowski [28] reported that in many cases, fungi infecting seeds are seed-borne pathogens. They play an important role in the transmission of numerous pathogenic fungal species to seedlings as well as to the soil.

Storage fungi are those that grow on stored products. Most of them are able to grow without free water, and on media with a high osmotic pressure. Most of the storage floras are the species of *Aspergillus* and *Penicillium,* which are active at relative humidities ranging from 70 to 90%. It has been established that storage fungi usually do not invade before harvest [29, 30] but they may be found on the seed in very low percentages often below 1 percent, nevertheless providing for the presence of inoculum of storage fungi [31–33]. They may be present not only as contamination but as dormant mycelium within the tissues of pericarp or seed coat [34].

Improper storage makes the oilseeds vulnerable to storage fungi, which deteriorates the stored oilseeds both qualitatively and quantitatively. They bring about the variety of biochemical changes in suitable conditions. It was found that *Alternaria dianthicola, Curvularia lunata, Fusarium oxysporum, Fusarium equiseti, Macrophomina phaseolina* and *Rhizopus stolonifer* causes decrease in reducing sugars of oilseeds. *Alternaria dianthicola, Curvularia pellescens, Macrophomina phaseolina, Penicillium digitatum* and *Penicillium chrysogenum* hampered the fat content of oilseeds. *Curvularia lunata, Curvularia pellescens, Fusarium oxysporum, Macrophomina phaseolina, Rhizopus stolonifer* and *Penicillium digitatum* increased the fiber content in oilseeds [35]. Therefore, keeping in the view of above facts, the present investigation aims to achieve the following objective, that is, Effect of seed-borne fungi on the deterioration of the seeds and extracted oil quality.

11.2 MAJOR EFFECTS OF SEED MYCOFLORA

11.2.1 EFFECT OF STORAGE ON MOISTURE CONTENT AND PERCENT GERMINATION OF JATROPHA CURCAS SEEDS

Parreno-de Guzman and Aquino were reported [36] that the storage behavior of *Jatropha curcas* seed is one of the main constraints. The slight moisture content increase against long-term storage period might be due to the hydrolysis or breakdown of lipid, carbohydrate and proteins and release of water molecules. The release of free fatty acids from lipid degradation by fungi is well known and hydrolysis of starch and proteins for utilization by the microorganisms also, has been studied from the earlier findings [29]. Storage under high relative humidity and temperature (85% and 30 °C) was observed to promote *Aspergillus* spp. growth that resulted in the loss of germination in seed within six months [37].

Seed storage behavior of Euphorbiaceae is generally orthodox [38]. Orthodox seed storage behavior means "Mature whole seeds not only survive considerable desiccation (to at least 5% moisture content) but their longevity in air-dry storage is increased in a predictable way by reduction inseed storage moisture content and temperature (e.g., to those values employed in long-term seed stores)" [39]. Physic nut also has orthodox seeds. Two- or six-month-old seeds received for the provenance trials described above, were stored in unsealed plastic bags at ambient temperatures (approximately 20 °C) for 5 months and germinated on average by 62% (ranging from 19 to 79%) after having been seeded in soil. When stored for 7 years in plastic bags (not sealed) at a temperature of approximately16 °C, the seeds still showed an average germinating capacity of 47% (ranging from 0 to 82%) when tested with the "between paper" method [40, 41]. When the seeds were analyzed for their chemical composition after 3 years of storage, they had moisture content of 6.2% (average of all provenances).

The fungi *Aspergillus flavus, Aspergillus niger, Pencillium* spp. *Fusarium* spp., and *Rhizopus* spp. were commonly found fungi causing considerable reduction in viability of stored seeds [42]. The safe moisture content for storing physic nut seeds was 7–9% [43]. The moisture contents increased after one month of storage, and became relatively constant up to 6 months of storage. The range of moisture contents (7.9–8.4%) was safe for storage of physic nut seeds. Lipid contents, viabilities and vigors decreased with the increase of storage duration, while free fatty acids and lipase activities increased. Under uncontrolled conditions, physic nut seeds packed in plastic material can be stored up to one month for seeds to be planted, while it can be stored up to five months for producing oil [26].

Seed germination percentage declined slowly with the longer storage periods when dry seeds stored under room temperature. The results indicated the slow deterioration of seeds with time [44]. However, Pasabutr and Suthiponpaiboon found that [45] physic nut seed stored under room temperature for 8 months gave germination up to 80% but decreased to 42% after 9 months storage. Therefore, it may be of high advantage to store seeds in a low temperature as low as 20 °C where high respiration rate could be reduced, thus it may be possible to prevent the rapid deterioration of seeds since oil seeds obviously loss its vigor more rapidly with time when store in a high environmental temperature. The germination of 80–90% has been obtained for seeds collected during October to December. The seed had been dried in shade, and stored dry indoors. The seeds retained germinating ability for >2 years. *Jatropha* planting material is mainly raised through seedlings currently [46].

11.2.2   EFFECT OF SEED MYCOFLORA ON BIOCHEMICAL PROPERTIES OF JATROPHA CURCAS SEEDS

Fungi association brings certain biochemical changes in seeds during storage by decreasing reducing sugars and oil content [47].

11.2.2.1   Effect on Protein Contents

Gradual loss of protein content was found in the deteriorated seeds [48]. Significant decrease in protein, oil content and total sugars and increase in free fatty acids and reducing sugars in deteriorated seeds [49] and carbohydrates increased with decrease in protein content in deteriorated seeds [50]. There was also reported that 1.5- to 3-fold decrease in protein level was observed in moderately infected and highly infected seeds compared to healthy seeds [51].

11.2.2.2   Effect on Amino Acid Contents

Concentrations of the isolated free amino acid fractions significantly decreased with seed size except for phenylalanine, which significantly increased, with seed size [52]. Significant changes also occurred in the free amino acid fractions across storage periods up to nine months in duration. These results are the first to document a change in amino acids during the storage of peanuts. Deswal and Sheoran reported that [53] permeability of membrane increases with increase in storage period and leads to loss of electrolytes, sugars, amino acids and phenols. Further decline in seeds and seedling vigor parameters was also related to decrease in both quantity and quality of food reserves like sugars, amino acids, oil content, protein, etc. in different crops [54, 55] but the total free amino acids increased with aging of seeds [56].

11.2.2.3   Effect on Carbohydrate Contents

Carbohydrates increase the severity of the infection and that they may serve as easily metabolized carbon substrates for the pathogen [57–59].

The gradual loss of carbohydrate content of seeds due to *A. flavus* and *A. niger* during storage [48]. The sharp decline in the levels of soluble carbohydrates during the desiccation of seeds [60]. Sucrose and raffinose level declined in stored seeds of maize although the monosacharides, glucose fructose and galactose diminished faster [61]. There was 1.5 to 3 fold decreases in protein, carbohydrate and phenol levels and 2–3 fold increase in reducing sugar were observed in moderately infected and highly infected seeds compared to healthy seeds [51] therefore indicate the degradation of protein and carbohydrates and accumulation of reducing sugar due to degradation of carbohydrates, by the pathogen.

11.2.2.4   Effect on Phenol Contents

The antimicrobial property of phenolic compounds and especially those of the flavonol group is well documented [62, 63]. The first stage of defense mechanism involves a rapid accumulation of phenols at the infection site, which restricts or slows the growth of the pathogens [64]. The changes in the level of total phenolic compounds in chickpea as influence by *Aspergillus flavus, A. niger, Fusarium maniliforme* and *Pencillium oxalicum* infection [65]. The four days after treatment, infected pods always contained the highest amount of phenolics followed by the wounded, and finally by intact pod [66]. The results showed that the increase in phenolic content could be correlated to the aging, the wounding and the infection of the pods. This could account for a higher synthesis of disease-related molecules like phenols and hydroxyproline-rich glycoproteins, which generally accumulate in fungus infected plants [67]. The increase amount of phenol might be responsible for resistance response in plant [68].

11.2.2.5   Effect on Lipid Contents

Storage duration gave very significant differences on lipid contents of physic nut seeds. The lipid contents decreased with the increase of storage duration [26]. The fatty acid contents of physic nut seeds were dominated by unsaturated fatty acids, that is, oleic acid followed by linoleic acid will cause hydrolyzes process [27]. This process caused the decrease of lipid and increased free fatty acid contents, either in the seeds or in the oil itself. Sukesh and Chandrashekar reported that [56] lipid peroxidation (refers to the oxidative degradation of lipids) is considered to be the primary cause for seed deterioration. The occurrence of fungi will accelerate the degradation of lipid during storage. The decrease of lipid content was probably due to the activity of lipolytic fungi, which grew dominantly and can live for long period. This condition caused the production of free fatty acid and rancidity. Lipid of seeds can be degraded by lipase into free fatty acid and glycerol, especially if the moisture content of seeds was high. Free fatty acid was an index of deteriorated seeds containing lipid during storage. *Aspergillus, Cladosporium* and *Penicillium* were capable to degrade lipid compound and have high lipolytic activity [69]. These fungal species were often isolated from seeds containing high lipid. There has been the possibility of lipid breakdown and release of free fatty acids (FFA) may occur during the growth of fungi in stored Jatropha seeds and other oil seeds [70, 71].

11.2.3 EFFECT OF SEED MYCOFLORA ON THE OIL QUALITY AND QUANTITY

Freshly harvested seeds are more likely to be free from fungal infection and will express maximum seed oil [37]. Similar results were observed by Gupta and Rao [72] that on keeping the *Jatropha* seeds for long time, oil content of the seeds declines. Cost of bio-diesel is sensitive to storage time of seeds as well as raw oil. This also supports to Oladimeji and Kolapo who observed decrease in oil content of sunflower after a period of storage. The preheating the seeds for 4 min by microwave before the extraction improved the oil recovery up to 2% [73]. Ultrasonication and microwave did not affect the oil quality in terms of free fatty acid content [74].

Basha and Pancholy reported a decrease in oil, iodine value, soluble carbohydrates and protein contents in seed infested with *Aspergillus* spp. [75]. Reduction in seed oil content at the end of the storage period may be attributed to the effect of the storage conditions, the seed moisture content and the fungi associated with the seed [37, 76]. Decrease in oil content was positively correlated with increase in lipase activity on account of fungal infection [77–79].

11.2.4 EFFECT OF FUNGI ON PHYSICOCHEMICAL PROPERTIES OF JATROPHA CURCAS OIL

11.2.4.1 Effect on Acid Value

Acid value increases during storage [80]. Higher acid value and free fatty acids (%) were reported in *J. curcas* [81]. Free fatty acid was an index of deteriorated seeds containing lipid during storage [69]. Free fatty acid contents increased with the increase of storage duration [27]. The free fatty acid content of the biodiesel should be less than 1 percent. It was observed that lesser the free fatty acid in oil, better is the bio diesel recovery. Higher free fatty acid oil can also be used but the bio diesel recovery will depend upon oil type and the amount of NaOH used [82]. FFA content of raw *Jatropha* oil increases with time. Higher FFA results in loss of more oil as precipitate while neutralizing. It is therefore recommended that raw oil should not be stored for a long time [72]. Drying at 80 °C gave the highest oil yield of 47.06% but also the highest acid value. The high acid value indicated high free fatty acid, which caused high abrasion on metal. Drying at 40 °C gave less oil yield at 36.83% but the lowest acid value, which indicated a better usage quality [83]. According to Sauer, the decrease of seed quality was followed by the increase of free fatty acid value. The level of free fatty acid value gave an indication concerning the decrease of seed quality. The level of free fatty acid depends on fungal species infecting the seeds [84].

11.2.4.2 Effect on Iodine Value

The iodine value should be less than 115 as per biodiesel standards [82]. Seed sources with higher iodine value may not be suitable for the production of bio-diesel [85]. *Jatropha* fatty acid methyl ester, with higher cetane number is preferred for use as biodiesel. The specifications of bio-diesel are such that it can be mixed with any diesel. Cetane number of the bio-diesel is in the range of 48–60 [82]. The acid value, peroxide value and viscosity, increased while the iodine value decreased with increasing storage time of the *brassica carinata* biodiesel [86]. Iodine values decreased as extraction temperature increased [87]. Gulla and Waghray reported that iodine value decreased gradually during storage in the oil blends studied [88]. The iodine value of the biodiesel decreased during storage period. It indicates that the oxidation of the biodiesel which results in reduction in double bonds [89]. The iodine value of *Jatropha* oil is higher than other oil reported. High iodine value of *Jatropha* is caused by high content of unsaturation fatty acid such as oleic acid and linoleic acid. The high iodine value and oxidative stability shows that the seed oil upholds the good qualities of semidrying oil purposes [75]. A decrease in oil, iodine value, soluble carbohydrates and protein contents in seed infested with *Aspergillus* spp. [75].

11.2.4.3   Effect on Saponification Value

Saponification value is an index of the average size of fatty acid present, which depends upon the molecular weight and percent concentration of fatty acids components in the oil. Saponification value will be higher if the oil contains more of saturated fatty acids (C14:0, C16:0, C18:0) as it determines the length of carbon chain and increase cloud point, Cetane number and improve stability of bio-diesel [82]. Similar results were earlier documented in *J. curcas* by several workers [81, 90]. Oil infested from *Fusarium* spp. shows increased saponification value but decreased iodine value [91]. The acid value, iodine value and saponification value increase during storage. The reasons for these increases may be attributed to the absorption of moisture from the surroundings, oxidation, heat, light and metal. When these occur there is the likelihood of microorganisms affecting the oils, which in turn may lead to spoilage. The low free fatty acid values suggest that the samples are good edible oils. High acid values are usually indicative of damage or high moisture, which enables the enzyme lipase to convert the triglycerides to free fatty acids. The higher levels obtained there after indicated deterioration. The iodine value is often the most useful figure for identifying oil or placing it into a particular group. It is an index of unsaturation which shows the molecular weight for their fatty acids in the two investigated oils to be mainly nondrying [92].

11.2.4.4 Effect on Cetane Value

Cetane number determines the ignition performance of transport fuel [93]. Conventional diesel must have a Cetane number of at least 40. This accords biodiesel better cold start properties, minimized formation of white smoke (emissions), less engine noise hence engine durability and reduced fuel consumption [94]. Cetane values of biodiesel are slightly higher than synthetic diesel, which is favorable for combustion [95]. The Cetane Number of the biodiesel was considerably increased and well within the ASTM specified limit, which indicates the better combustion quality of the fuel. The higher Cetane index of biodiesel compared to petrodiesel was indicated that it will be the high potential for engine performance [96].

11.2.4.5 Effect on Viscosity

The viscosity of the pure plant oil is more or less constant for a given kind of oil, but may increase with aging of the pure plant oil. The viscosity is dependent on the temperature of the oil. For pure plant oil from *Jatropha*, the viscosity at room temperature is much higher than for rapeseed [97]. *Jatropha* oil has higher viscosity than conventional diesel fuel and biodiesel. High viscosity leads to poorer atomization of the fuel spray and less accurate operation of the fuel injectors [96]. Similarly, Sundar Raj and Sahayaraj reported that the viscosity of *Jatropha* oil is higher than palm, coconut and sunflower oil [98]. High viscosity of the seed oil is not suitable if it is used directly as engine fuel, often results in operational problems, such as, carbon deposits, oil ring sticking, and thickening and gelling of lubricating oil as a result of contamination by the oils. Different methods such as preheating, blending, ultrasonically assisted methanol transesterification and supercritical methanol transesterification can reduce the viscosity and make them suitable for engine applications [99].

11.2.4.6   Effect on Refractive Index

The standard refractive index of *J. curcas* seed oil was 1.465 [99] and also reported the refractive index of the transesterified *Jatropha* oil was found to be 1.45 [100]. Fungal infested oil emitted moldy odor and the refractive index increased [91].

11.3   CONCLUSION

Deteriorated *Jatropha* seeds show decrement in their protein content, amino-acid content, carbohydrate content and lipid content. Phenol content of deteriorated *Jatropha* seeds were increased which might be responsible for resistance response against pathogen. These isolated seed mycoflora also affect the quality of *Jatropha* oil by effecting its physicochemical properties *viz.*, color, acid-value, iodine-value, saponification-value, cetane value, refractive index and viscosity. Oil content of *Jatropha curcas* L. was declined due to infestation of seed mycoflora and it is also the main cause of deterioration of seeds during storage.

Improper storage makes the oilseeds vulnerable to storage fungi, which deteriorate the stored oilseeds both qualitatively and quantitatively. They bring about the variety of biochemical changes in the suitable conditions. *Jatropha* seeds are constantly subjected to deterioration, which implies an irreversible degenerative change in the quality of seeds after it has reached its maximum quality level. The fungi associated with seeds at the harvest stage and under storage bring about several undesirable changes and degradation of seed constituents, thus making the seed unfit for oil extraction, export purpose, consumption or sowing. In many cases, fungi infecting seeds are seed-borne pathogens. They play an important role in the transmission of numerous pathogenic fungal species to seedlings as well as to the soil.

11.4   FUTURE PROSPECTS

Current global trends in energy supply and consumption are patently unsustainable: environmentally, economically and socially. That indicates a global need to secure the supply of reliable and affordable energy and to effect a rapid transformation to a low-carbon efficient and environmentally benign system of energy supply. It is also clear that current energy fossil fuel consumption trend will have severe consequences on natural ecosystems and social communities. Switching to renewable energy will therefore reduce global warming and curb actual trends [101]. These are the reasons why a search for alternatives to fossil fuels, such as renewable energy from solar, wind, water, biomass and nuclear, has been provoked. Biofuels are considered in part; a solution to such issues in sustainable development, energy security and a reduction of green house gas emission. Biodiesel, an environmental friendly diesel fuel similar to petro-diesel in combustion properties, has received considerable attention in the recent past worldwide. There is therefore, need to explore alter nonedible oil for use in production of biodiesel.

KEYWORDS

- **Biodiesel**
- **Ester-based fuels**
- *Jatropha curcas* **L.**
- **Petro-diesel combustion properties**

REFERENCES

1. Muche, A., Sahu, O. (2014). *Jatropha curcas* Oil for the Production of Biodiesel. *Internat. J. Biosci. Bioinform. 1 (1),* 001–007.
2. Liaquat, A. M., Kalam, M. A., Masjuki, H. H., Jayed, M. H. (2010). Potential emission reduction in road transport sector using biofuels in developing countries. *Atmos. Environ. 44,* 3869–3877.
3. Boey, P. L., Ganesan, S., Maniam, G. P., Khairuddean, M., Lim, S. L. (2012). A new catalyst system in transesterification of palm olein: Tolerance of water and free fatty acids. *Energy Convers. Manag. 56,* 46–52.
4. Anitescu, G., Bruno, T. J. (2011). Fluid properties needed in supercritical transesterification of triglyceride feedstocks to biodiesel fuels for efficient and clean combustion—A review. J. Supercrit. *Fluids, 63,*133–149.
5. Wood, J. (2008). 'An amazingly timely quote by Rudolph diesel.' http://dieselnews. wordpress.com/2008/05/15/an-amazingly-timely-quote-by-rudolph-diesel-the-inventor-of-the-diesel-engine/.
6. Kumar, A., Sharma, S. (2008). An evaluation of multipurpose oil seed crop for industrial uses (*Jatropha curcas* L.): a review. *Ind. Crops Prod. 28,* 1–10.

7. Liberalino, A. A., Bambirra, E. A., Moraes-Santos, T., Vieira, E. C. (1988). *Jatropha curcas* L. seeds: chemical analysis and toxicity. *Arq. Biol. Technol. 31,* 539–550.

8. Gandhi, V. M., Cherian, K. M., Mulky, M. J. (1995). Toxicological studies on ratanjyot oil. *Food Chem. Toxicol. 33,* 39–42.

9. Sharma, G. D., Gupta, S. N., Khabiruddin, M. (1997). Cultivation of *J. curcas* as a future source of hydrocarbons and other industrial products. In Biofuels and industrial products from *Jatropha curcas*; Gubitz, G. M., Mittelbach, M., Trabi, M., Ed., pp. 19–21.

10. Wink, M., Koschmieder, C., Sauerwein, M., Sporer, F. (1997). Phorbol esters of *J. curcas* – biological activities and potential applications. In Biofuels and industrial products from *Jatropha curcas*; Gubitz, G. M., Mittelbach, M., Trabi, M., Ed., pp. 160–166.

11. Makkar, H. P. S., Becker, K., Sporer, F., Wink, M. (1997). Studies on nutritive potential and toxic constituents of different provenanaces of *Jatropha curcas*. *J. Agric. Food Chem. 45,* 3152–3157.

12. Openshaw, K. (2000). A review of Jatropha curcas: an oil plant of unfulfilled promise. *Biom. Bioener. 19,* 1–15.

13. Rivera-Lorca, J. A., Ku-Vera, J. C. (1997). Chemical composition of three different varieties of *J. curcas* from Mexico. In Biofuels and industrial products from *Jatropha curcas*; Gubitz, G. M., Mittelbach, M., Trabi, M., (Ed.), pp. 47–52.

14. Slavin, T. 2008. When oil grows on trees: India's new oil bonanza could revitalize its wastelands – or starve its poor. *Green Futures: The Sustainable Solutions Magazine.* http://www.forumforthefuture.org/greenfutures/articles/whenoilgrowsontrees.

15. Jongschaap, R., Corré, W., Bindraban, P., Brandenburg, W. (2007). 'Claims and facts on *Jatropha curcas* L.' Report 158, Plant Research International, Droevendaalsesteeg 1, PO Box 16, 6700AA Wageningen, The Netherlands.

16. Abou Kheira, A., Atta, N. (2008). 'Response of *Jatropha curcas* L. to water deficit: Yield, water uses efficiency and oilseed characteristics.' Biom. Bioener. 1–8.

17. Heller, J. Physic Nut. Jatropha curcas L. Promoting the Conservation and Use of Underused and Neglected Crops. Institute of Plant Genetics and Crop Plant Research, Gatersleben/International Plant Genetic Resources Institute, Rome, 1996.

18. Achten, W., Verchot, L., Franken, Y., Mathijs, E., Singh, V., Aerts, R., Muys, B. (2008). 'Jatropha bio-diesel production and use.' *Biom. Bioener. 32,* 1063–1084.

19. Singh, R. N., Vyas, D. K., Srivastava, N. S. L., Narra, M. (2008). SPRERI experience on holistic approach to use all parts of Jatropha curcas fruit for energy. *Renew. Ener.* 33, 1868–1873.

20. Agarwal, V. K., Sinclair, J. B. (1996). *Principles of Pathology.* 2nd Edn., CRC Press, Inc., Boca Raton, Fl. pp: 539.

21. Bateman, G. L., Kwasna, H. (1999). Effects of number of winter wheat crops grown successively on fungal communities on wheat roots. *Appl. Soil Ecol. 13,* 271–282.

22. Khanzada, K. A., Rajput, M. A., Shah, G. S., Lodhi, A. M., Mehboob, F. (2002). Effect of seed dressing fungicides for the control of seed borne mycoflora of Wheat. *Asian J. Pl. Sci. 1 (4),* 441–444.

23. Delouche, J. C. (1973).Precepts of seed storage (Revised). *South Canada Proceedings*, Mississippi State University, pp. 97–122.

24. Ellis, R. H., Roberts, E. H. (1981). The quantification of aging and survival in ortho-dox seeds. *Seed Sci. Technol. 9,* 373–409.
25. Ghosh, B., Adhikari, J., Banerjee, N. C. (1981). Changes in some metabolites in rice seeds during aging. *Seed Sci. Technol. 9,* 469–473.
26. Worang, R. L., Dharmaputra, O. S., Syarief, R., Miftahudin, (2008). The quality of physic nut (*Jatropha curcas*) seeds packed in plastic material during storage. *Biotropia. 15 (1),* 25–36.
27. Dharmaputra, O. S., Worang, R. L., Syarief, R., Miftahudin, (2009). The quality of physic nut (*Jatropha curcas*) seeds Effected by water activity and duration of storage. *Microbiol. 3 (3),* 139–145.
28. Chelkowski, J. (1991). Fungal Pathogens Influencing Cereal seed Quality at Harvest. In: Cereal Grains; Mycotoxins, Fungi and Quality in Drying and Storage, Chelkowski, J. (Ed.). Elsevier Publisher, Amsterdam, pp. 53–66.
29. Christensen, C. M., Kaufman, H. H. (1969). Influence of moisture content, tempera-ture and time of storage upon invasion of rough rice by storage fungi. *Phytopathl. 59,* 145–148.
30. Christensen, C. M. (1971). Evaluating conditions and storability of sunflower seeds. *J. Stored Prod. Res. 7,* 163–169.
31. Tuite, J. (1959). Low incidence of storage molds in freshly harvested seed of soft red winter wheat. *Pl. Dis. Reptr. 43,* 470.
32. Tuite, J. (1960). The natural occurrence of tobacco ring spot virus. *Phytopathol. 50,* 296–298.
33. Qasem, S. A., Christensen, C. M. (1958). Influence of moisture content, temperature and time on the deterioration of stored corn by fungi. *Phytopathol. 48,* 544–549.
34. Warnock, D. W., Preece, T. F. (1971). Location and extent of fungal mycelium in grains of barley. *Trans. Br. Mycol. Soc. 56,* 267–273.
35. Kakde, R. B., Chavan, A. M. (2011). Deteriorative changes in oilseeds due to storage fungi and efficacy of botanicals. *Curr. Botany. 2 (1),* 17–22.
36. Parreño-de Guzman, L. E., Aquino, A. L. (2009). Seed characteristics and storage behavior of physic nut (*Jatropha curcas* L.). *Philippine J. Crop Sci. 34 (1),* 13–21.
37. Neergaard, P. (1977). Seed Pathology, Vol. I, The Macmillan Press Ltd London and Basingstoke, UK. pp. 285–288. *61,* 53, 294.
38. Ellis, R. H., Hong, T. D., Roberts, E. H. Handbooks for Genebanks No. 3. Handbook of Seed Technology for Genebanks. Vol. II. Compendium of Specific Germination Information and Test Recommendations. International Board for Plant Genetic Re-sources, Rome, 1985.
39. Hong, T. D., Linington, S., Ellis, R. H. Seed Storage Behaviour: a Compendium. IPGRI, Rome, 1996.
40. Heller, J. Investigation of the genetic potential and improvement of cultivation and propagation practices of physic nut (*Jatropha curcas* L.). Final report submitted to Deutsche Gesellschaft für Technische Zusammenarbeit (GTZ) GmbH (German Agency for Technical Cooperation), Eschborn, 1991.
41. Heller, J. Untersuchungen über genotypische Eigenschaften und Vermehrungsund Anbauverfahren bei der Purgiernuß (*Jatropha curcas* L.) [Studies on genotypic char-acteristics and propagation and cultivation methods for physic nuts (*Jatrophacurcas* L.)]. Dr. Kovac, Hamburg, 1992.

42. Narayanaswamy, S., Shambulingappa, K. G. (1994). Diversity in seed borne fungal pathogens in groundnut. *J. Oilseed Res. 11 (2)*, 294–299.
43. [Dirjenbun] Direktorat Jenderal Perkebunan. Pedoman Mutu Benih Jarak Pagar Sistem dan Prosedur Pembangunan Sumber Benih dan Peredaran Benih Jarak Pagar. Jakarta: Dirjenbun, 2006.
44. Ratree, S. (2004). A preliminary study on physic nut (*Jatropha curcas* L.) in Thailand. *Pak. J. Biologic. Sci. 7 (9)*, 1620–1623.
45. Pasabutr, R., Suthiponpaiboon, S. (2001). Outstanding features of Physic nut oil on diesel engines. Kaset *Kaona J. 14*, 60–66.
46. Deng, Z. J., Cheng, H. Y., Song, S. Q. (2005). Studies on *Jatropha curcas* seed. *Acta Botanica Yunnanica. 27 (6)*, 605–612.
47. Mathur, S. K., Sinha, S. (1978). Certain biochemical changes in bajra (*Pennisetum typhoides*) seeds during storage. *Seed Res. 6 (2)*, 172–180.
48. Saxena, N., Karan, D. (1991). Effect of seed borne – fungi on protein and carbohydrate contents of sesame and sunflower seeds. *Indian Phytopathol. 44*, 134–136.
49. Arulnandhy, V., Senanayake, D. A. (1991). Changes in viability, vigor and chemical composition of soybean seeds stored under humid tropical conditions. *Legume Res. 14 (3)*, 135–144.
50. Verma, S. S., Tomer, R. P. S., Verma, U. (2003). Loss of viability and vigor in Indian mustard seeds stored under Ambient conditions. *Seed Res. 31 (1)*, 98–101.
51. Begum, M., Lokesh, S., Ravishankar, R. V., Shailaja, M. D., Kumar, T. V., Shetty, H. S. Evaluation of certain storage conditions for okra (*Abelmoschus esculentus* (L.) Moench) seeds against potential fungal pathogens. *Internat. J. Agri. Biol. 7 (4)*, 550–554.
52. Pattee, H. E., Young, C. T., Giesbrecht, F. G. (1981). Free Amino Acids in Peanuts as Effected by Seed Size and Storage Time. *Peanut Sci. 8 (2)*, 113–116.
53. Deswal, D. P., Sheoran, I. S. (1993). A simple method of seed leakage measurement applicable to single seed of any size. *Seed Sci. Technol. 21*, 179–185.
54. Agrawal, P. K. (1980). Relative storability of seeds of ten species under ambient conditions. *Seed Res. 8*, 94–99.
55. Gupta, A., Aneja, K. R. (2004). Seed deterioration in soybean varieties during storage: Physiological Attributes. *Seed Res. 32 (1)*, 26–32.
56. Sukesh; Chandrashekar, K. R. (2011). Biochemical Changes During the Storage of Seeds of *Hopea ponga* (Dennst.) Mabberly: An Endemic Species of Western Ghats. *Res. J. Seed Sci. 4*, 106–116.
57. Hwang, B. K. (1983). Contents of sugars, fruit acids, amino acids and phenolic compounds of apple fruits in relation to their susceptibility to *Botryosphaeria ribis*. *Phytopathologische Zeitschrift. 108*, 1–11.
58. Patil, S. H., Hedge, R. K., Anahosur, K. H. (1985). Role of sugars and phenols in charcoal rot resistance of sorghum. *PhytopathologischeZeitschrift. 113*, 30–35.
59. Jeun, Y. C., Hwang, B. K. (1991). Carbohydrate, amino acid, phenolic and mineral nutrient contents of pepper plants in relation to agerelated resistance to *Phytophthora capsici*. *J. Phytopathol. 131*, 40–52.
60. Saha, P. K., Bhattacharya, A., Ganguly, S. N. (1992). Problems with regard to the loss of seed viability of *Shorea robusta* Gaertn. F. *Indian For. 118*, 70–76.
61. Bernal-Lugu, I., Leopold, A. C. (1992). Changes in soluble carbohydrates during seed storage. *Pl. Physiol. 30*, 195–236.

62. Smith, D. A., Banks, S. W. (1986). Molecular communication in interactions between plants and microbial pathogens. *Ann. Rev. Pl. Physiol. 41*, 339–367.

63. Hahlbrock, K., Scheel, D. (1989). Physiology and molecular biology of phenylpropanoid metabolism. *Ann. Rev. Pl. Physiol. Mol. Biol. 40*, 347–369.

64. Martern, U., Kneusel, R. E. (1988). Phenolic compounds in plant disease resistance. *Phytoparasitica. 16,* 153–170.

65. Dwivedi, S. N. (1990). Changes in the concentration of total phenolic compounds in gram seeds as influenced by fungal invasion during storage. *Indian Phytopathol. 43,* 96–98.

66. Ndoumou, D. O., Ndzomo, G. T., Djocgoue, P. F. (1996). Changes in carbohydrate, amino acid and phenol contents in cocoa pods from three clones after infection with *Phytophthora megakarya* Bra and Grif. *Ann. Bot. 77,* 153–158.

67. Esquerre-Tucaye, M. T., Lamport, D. T. A. (1979). Cell-surfaces in plant-microorganisminteractions. I- A structural investigation of cell-wallhydroxyproline-rich glycoproteins which accumulate infungus infected plants. *Pl. Physiol. 64*, 314–319.

68. Mishra, V. K., Biswas, S. K., Rajik, M. (2011). Biochemical Mechanism of Resistance to *Alternaria* Blight by Different Varieties of Wheat. *Internat. J. Pl. Pathol. 2,* 72–80.

69. Pomeranz, Y. (1992). Biological, Functional, and Nutritive Changes of Cereal Grains and Their Product. St. Paul: American of Cereal Chemist Inc. pp. 55–141.

70. Hanny, J. B. (2008). "Biodiesel production from crude *Jatropha curcas* L. seed oil with high content of free fatty acids." *Bioresour. Technol.* 1716–1721.

71. Bothast, R. J. Fungal deterioration and related phenomena in cereals, legumes and oilseeds. Northern Regional Research Center Agricultural Research Service U. S. Department of Agriculture Peoria, IL 61604, 2009.

72. Gupta, R. C., Rao, D. G. (2008). Effect of storage time on yield and free fatty acid (FFA) content of raw *Jatropha* oil. XXXII National Systems Conference, NSC, pp. 440–443.

73. Oladimeji, G. R., Kolapo, A. L. (2008). Evaluation of proximate Changes and Microbiology of stored defatted residues of some selected Nigerian oil seeds. *African J. Agricultural Res. 3 (2),* 126 – 129.

74. Sayyar, S., Abidin, Z. Z., Yunus, R., Muhammad, A. (2011). Solid liquid extraction of *Jatropha* seeds by microwave pretreatment and ultrasound assisted methods. *J. Appl. Sci. 11 (13),* 2444–2447.

75. Basha, S. M., Pancholy, S. K. (1986). Qualitative and quantitative changes in the protein composition of peanut (*Arachis hypogaea* L.) seed following infestation with *Aspergillus* spp. differing in aflatoxin production. *J. Agri. Food Chemistry* (*USA*). *34,* 638–643.

76. Simic, B., Popovic, R., Sudaric, A., Rozman, V., Kalinovic, I., Cosic J. (2007). Influence of storage condition on seed oil content of maize, soybean and sunflower. *CCS Agriculturae Conspectus Scienticus. 72 (3),* 211–213.

77. Saraswat, R. P., Mathur, S. K. (1985). Lipase production by seed borne fungi of linseed. *Seed Res. 13 (2),* 123–124.

78. Saxena, N., Karan, D., (1998). Mohana, Enzymatic studies of seed borne fungi of safflower seeds and their role in seed deterioration during storage. *Seed Tech News. 28,* 25–26.

79. Taung, U., McDonald, M. B. (1995). Changes in esterase activity associated with peanut (*Arachis hypogaea* L.) seed deterioration. *Seed Sci. Technol. 23,* 101–111.

80. Sadasivam, S., Manickam, A. Biochemicals Methods. (3rd ed.) New Age International (P) Limited, New Delhi, 2008.

81. Bhasabutra, R., Sutiponpeibun, S. (1982). *Jatropha curcas* oil as a substitute for diesel engine oil. *Renewable Energy Rev. J. 4 (2),* 56–67.

82. Anonymous. Report of the Committee on Development of Biofuel, Planning Commission. Government of India, New Delhi, 2003.

83. Sirisomboona, P., Kitchaiya, P. (2009). Physical properties of *Jatropha curcas* L. kernels after heat treatments. *Biosyst. Engin. 102,* 244–250.

84. Sauer, D. B. (1988). Effects of fungal deterioration on grain: nutritional value, toxicity, germination. *Int. J. Food Microbiol. 7,* 267–275.

85. Parthiban, K. T., Selvan, P., Paramathma, M., Umesh Kanna, S., Kumar, P., Subbulakshmi, V., Vennila, S. (2011). Physicochemical characterization of seed oil from *Jatropha curcas* L. genetic resources. *J. Ecol. and the Natural Environment, 3 (5),* 163–167.

86. Abderrahim, B., Mercedes M., Jose, A. (2009). Production of biodiesel from bio-ethanol and *Brassica carinata* oil: Oxidation stability study. *Biores. Technol. 100 (7),* 2234–2239.

87. Asoiro, F. U., Akubuo, C. O. (2011). Effect of Temperature on Oil Extraction of *Jatropha curcas* L. Kernel. *The Pacific J. Sci. Technol. 12 (2),* 456–463.

88. Gulla, S., Waghray, K. (2011). Effect of Storage on Physicochemical Characteristics and Fatty Acid Composition of Selected Oil Blends. *J. L. S. 3 (1),* 35–46.

89. Kapilan, N., Sekar, C. B. (2011). Production and evaluation of storage stability of Honge biodiesel. *Annals of Faculty Engineering Hunedoara – Internat. J. Engin.* IX (3), 267–269.

90. Bringi, N. V. Non-traditional oilseeds and oils of India. New Delhi: Oxford and IBH Publishing Co. Pvt. Ltd. 1987.

91. Ashraf, S. S., Basu Chaudhary, K. C. (1986). Effect of Seed-Borne *Fusarium* Species on the PhysicoChemical Properties of Rapeseed Oil. *J. Phytopathol. 117 (2),* 107–112.

92. Abulude, F. O., Ogunkoya, M. O., Ogunleye, R. F. (2007). Storage properties of oils of two Nigerian oil seeds *Jatropha curcas* (Physic Nut) and *Helianthus annus* (Sunflower). *Am. J. Food Technol. 2,* 207–211.

93. Foidl, N., Foidl, G., Sanchez, M., Mittelbach, M., Hackel, S. (1996). *Jatropha curcas* L. as a source for the production of biofuel in Nicaragua. *Biores. Technol. 58,* 77–82.

94. Russell, T. (1989). *Additives Influencing Diesel Fuel Performance,* John Wiley and Sons: Chichoster; pp. 65–104.

95. Singh, R. K., Padhi, S. K. (2009). Characterization of *Jatropha* oil for the preparation of biodiesel. *Nature Product Radiance. 8 (2),* 127–132.

96. Kyne, T. T., Oo, M. M. (2009). Production of Biodiesel from Jatropha Oil (*Jatropha curcas*) in Pilot Plant. *Proceedings of World Acad. Sci. Engin. Technol. 38,* 481–487.

97. Arrakis, J. de J., Ingenia, T. A. (2007). *Jatropha* oil quality related to use in diesel engines and refining methods. *Technical Note.* pp. 1–11.

98. Sundar Raj, F. R. M., Sahayaraj, J. W. (2010). A comparative study over alternative fuel (biodiesel) for environmental friendly emission. *IEEE,* pp. 80–86.

99. Akintayo, E. T. (2004). Characteristics and composition of *Parkia biglobbossa* and *Jatropha curcas* oils and cakes. *Biores. Technol. 92,* 307–310.

100. Payawan Jr., L. M., Damasco, J. A., Sy Piecco, K. W. E. (2010). Transesterification of Oil Extract from Locally Cultivated *Jatropha curcas* using a Heterogeneous Base Catalyst and Determination of its Properties as a Viable Biodiesel. *Philippine J. Sci. 139 (1),* 105–116.

101. International Energy Agency. *World Energy Outlook,* IEA, Paris, 2006.

PART 5

CATALYSIS FOR BIOFUELS

CHAPTER 12

A CHEMIST'S PERSPECTIVE ON BIOENERGY—OPPORTUNITIES AND CHALLENGES

SANNAPANENI JANARDAN,[1] B. B. PAVANKUMAR,[1]
SHIVENDU RANJAN,[2] NANDITA DASGUPTA,[2] MELVIN SAMUEL,[2]
CHIDAMBARAM RAMALINGAM,[2] AKELLA SIVARAMAKRISHNA,[1]
and KARI VIJAYAKRISHNA[1]

[1]*Department of Chemistry, School of Advanced Sciences,
VIT University, Vellore, Tamil Nadu, India*

[2]*School of Bio Sciences and Technology, VIT University, Vellore, Tamil
Nadu, India*

CONTENTS

12.1   INTRODUCTION

Since the crude oil resources are finite, there is however an increasing demand for ecofriendly, renewable and sustainable fuels, particularly for the transport purpose. At present, the entire world is looking for alternative, efficient and sustainable energy sources such as solar, wind, hydro, biomass and geothermal energy. The main drawback of these energy sources is that they need to be converted to energy carriers such as electricity, hydrogen or biofuels primarily before considering them to the transport purposes. Among all the energy sources, hydrogen is the ideal fuel for an energy efficient fuel cell with water as a by-product. However, there are several problems associated with conversion and storage of hydrogen gas although the efficiency of catalytic conversion of biomass to hydrogen was substantially improved recently [1, 2]. All these aspects are driving the scientists for the production of liquid biofuels from the biomass.

The main components in biomass are cellulose, a polymer of glucose, lignin (a highly cross-linked polymer built up of syringyl and sinapyl units) and hemi-cellulose (an oligomer of both C_6-glucose and C_5-xylose sugars), which are formed from CO_2 and H_2O using sunlight as energy source. It is significant to note that the energy content per mass unit increases with decreasing oxygen content of the building blocks. An overview of the components typically present in biomass is given in Fig. 12.1 [3]. Biofuels are already on the market in the form of ethanol for gasoline engines and fatty acid methyl esters (FAMEs) for diesel engines. Biomass can also be converted into various types of fuels such as hydrogen, synthetic natural gas (SNG), dimethyl ethanol (DME), and Fischer–Tropsch (FT) diesel [4, 5].

Higher oxygen content in biomass (carbohydrates) is the major difference between biomass and other fossil fuels like gasoline or diesel, which can bring the changes in the parameters as shown in Table 12.1. In view of this, the present techniques are focused on the elimination of oxygen from biomass (Fig. 12.2).

Elimination of molecular oxygen from cellulosic biomass to form hydrocarbons is exactly opposite to highly endothermic combustion of hydrocarbons. The removal of oxygen in the form of CO_2 or H_2O must have a combustion enthalpy equal to zero. The formation of carbon dioxide as the oxygen carrying molecule from a C6– hydrocarbon would result in $[C_3H_{12}]$, a hypothetical molecule which can split into either three methane (CH_4) units or propane with two hydrogen molecules $(C_3H_8 + 2H_2)$ (Fig. 12.2). It is very important to convert the biomass in to higher alkanes as methane, charcoal or synthesis gas cannot be blended with liquid transport fuels. Otherwise the coupling of methane (biogas) to form higher alkanes in the presence of a suitable catalyst would be a key discovery [6].

Elimination of all oxygen from carbohydrates in biomass thus automatically leads to a conversion route with synthesis gas as intermediate from which higher alkanes for diesel can be derived by the Fischer Tropsch reaction [7].

$$nCH_4 + \tfrac{1}{2} O_2 \rightarrow [-CH_2-]_n + nH_2O \qquad (1)$$

This chapter mainly focuses on possible routes from biomass to bioenergy systems as seen from a chemist's perspective.

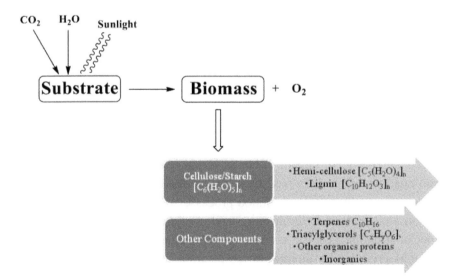

FIGURE 12.1   Components of Biomass.

TABLE 12.1 Difference Between Carbohydrate, Gasoline and Diesel

	Carbohydrate	Gasoline	Gasoil/Diesel
Chain length	[5–6]n	5–10	12–20
O/C molar ratio	1	0	0
H/C molar ratio	2	1–2	~2
Phase behavior	Solid	Liquid	Liquid
Polarity	Polar	Non polar	Non polar

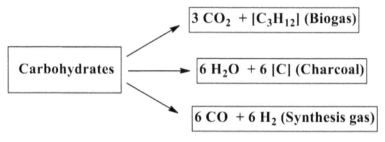

FIGURE 12.2 Elimination of oxygen species from carbohydrates.

12.2   METHODS

12.2.1   THERMOCHEMICAL APPROACHES

In general, the thermo-chemical conversion processes have two basic approaches. The first is the gasification of biomass and its conversion to hydrocarbons. The second approach is to liquefy biomass directly by high-temperature pyrolysis, high-pressure liquefaction, ultrapyrolysis, or super-critical extraction. These processes convert the waste biomass into energy rich useful products. Choice of conversion process depends upon the type and quantity of biomass feedstock, the desired form of the energy. Different thermo-chemical conversion processes include combustion, gasification, liquefaction, hydrogenation and pyrolysis. Pyrolysis has received special attention as it can convert biomass directly into solid, liquid and gaseous products by thermal decomposition of biomass in the absence of oxygen. Pyrolysis offers efficient utilization of particular importance for agriculture countries with vastly available biomass by-products. The efficiency of thermochemical methods strongly depends on temperature, holding time, rate of heating and nature of catalysts. Various available conventional processes have been discussed in brief (Fig. 12.3).

12.2.1.1   Combustion

The biomass is directly burnt with air to convert chemical energy stored in biomass into heat, mechanical power, or electricity, etc., for stoves, furnaces, boilers, steam turbines, turbo-generators. Though it is possible to burn any type of biomass but it is necessary that biomass should not have the moisture content for combustion. One of the major drawbacks of combustion process is that it requires some pretreatment like drying, chopping, grinding, etc., which in turn is associated with financial costs and energy expenditure [8, 9]. Combustion of biomass produces hot gases at temperatures around 800–1000°C. Net bio-energy conversion efficiencies for biomass combustion power plants range from 20% to 40%. [10].

12.2.1.2   Gasification

This process involves conversion of biomass into a syngas by the partial oxidation of biomass at high temperatures as summarized in Fig. 12.4. At high temperature solid biomass is converted into combustible gas mixtures (known as synthesis gas or syngas) through simultaneous occurrence of exothermic oxidation and endothermic pyrolysis under limited oxygen supply [10].

The advantage of gasification is that it is not constrained to a particular plant-based feedstock, and thus any lignocellulosic biomass can be considered appropriate. However, the amount of water in the biomass and impurities in the gases produced can be problematic in downstream processes, where a clean gas feed is required. Gasification is normally carried out at temperatures above 1000 K, but recently is has been demonstrated that H_2 and CO can be produced through the aqueous phase reforming of glycerol at lower temperatures (<620 K). The ratio of CO/H_2 can be modified by the water gas shift reaction ($CO+H_2O \rightarrow CO_2+H_2$) [11].

Fixed bed and fluidized bed gasification are more common. Different gasification agents can be applied, as air, oxygen or steam. The use of

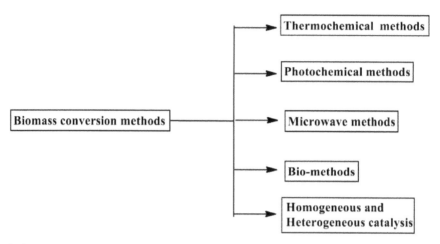

FIGURE 12.3 Broad classification of biomass treatment methods.

air as a gasifying agent is most common in industry but yields low heating value gas [12]. A high H_2/CO ratio is required for producing liquid fuels through Fischer–Tropsch synthesis and also benefits the production of H_2 for use in fuel cells. However, high capital costs and complex system design have hindered the applications of steam and oxygen gasification at a large industrial scale [13]. Earlier Fischer–Tropsch (FT) synthesis from atmospheric biomass gasification with a shift reactor was reported. In 2013, Nadia et al. highlighted the improved thermal process energy efficiencies by 10.7% without the shift process; although the liquid fuel energy production is 5% less [14].

In another approach by Kwangsu et al., to long-term operation of the biomass-to-liquid (BTL) process was conducted with a focus on the production of bio-syngas that satisfies the purity standards for the Fischer–Tropsch process. The integrated BTL system consists of a bubbling fluidized bed (BFB) gasifier, gas cleaning unit, syngas compression unit, acid gas removing unit, and an FT reactor. Since the raw syngas from the gasifier contains different types of contaminants, such as particulates, condensable tars, and acid gases, which can cause various mechanical problems or deactivate the FT catalyst, the syngas was purified by passing through cyclones, a gravitational dust collector, a two-stage wet scrubber (packing-type), and a methanol absorption

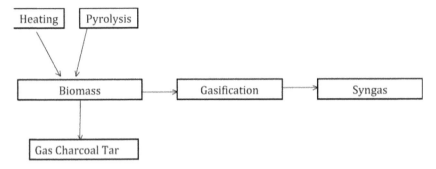

FIGURE 12.4   Schematic diagram for biomass gasification.

tower. The integrated system was operated for 500 h over several runs, and stable operating conditions for each component were achieved. The cleaned syngas contained no sulfur compounds and satisfied the requirements for the FT process [15].

In 2014, Berrueco et al. [16] researched on the pressurized gasification of torrefied woody biomass in a lab scale fluidized bed. This work reported experimental results concerning the influence of torrefaction level and pressure on product yields and composition during fluidized bed O_2/steam gasification of two different raw biomasses. The results show an increase in gas yield with pressure and torrefaction level for both types of biomass considered [16]. Danica et al., have evaluated costs for biofuel production in four biofuel polygeneration plants and the plants were integrated with district heating production. The revenues from by-produced electricity and heat were included as negative costs. The production costs depend on energy market and district heating conditions [17].

12.2.1.3   Liquefaction

Liquefaction consists of the catalytic thermal decomposition of large molecules to unstable shorter species that polymerize again into a bio-oil. Biomass is mixed with water and basic catalysts like sodium carbonate, and the process is carried out at lower temperatures than pyrolysis (525–725 K) but higher pressures (5–20 atm.) and longer residence times. These factors combine to make liquefaction a more expensive process; however, the liquid product obtained contains less oxygen (12–14%) than the bio oil produced by pyrolysis and typically requires less extensive processing [18, 19].

Recently liquefaction of many microorganisms and use of hydro-thermal liquefaction have been attempted to increase the efficiency also many byproducts/coproducts have been collected. α-glucan has been isolated as a coproduct of biofuel by hydrothermal liquefaction of *Chlo-rella sorokiniana* biomass, which will be beneficial for industries [20]. The comparative analysis has been done on land, water, and nutrient consumption between algae to biofuel conversion pathways. The study was based on spatial model of algae growth and resource availability for coterminous United States. Also, hydrothermal liquefaction com-pared to lipid extraction, with nutrient recycling technologies. And they have concluded with that for a given production target, hydrothermal liquefaction consumes far less land, water and phosphorous than lipid extraction. Also, nitrogen consumption is significantly increased over lipid extraction, but further recovery is still possible [21]. Also, Bahar et al., has givena cost effective approach for biofuel production, which is an alternative for synthetic fuels carried out by liquefaction of kenaf (*Hibiscus cannabinus* L.) biomass [22].

12.2.1.4   Pyrolysis

Pyrolysis, like gasification, is an advanced thermal treatment that converts a material into a syngas but at lower temperatures and in the absence of oxygen. Despite the calorific value of a gas derived from pyrolysis being higher than that of gasification, the volume of gas produced is usually much lower due to the lack of the oxygen carrier [23, 24].

Pyrolysis is the thermal decomposition of materials in the absence of oxygen or when significantly less oxygen is present than required for complete combustion. Pyrolysis also can be described as the direct thermal decomposition of the organic matrix that could obtain solid, liquid and gas products. Temperature is the most important factor for the product distribution of pyrolysis, most interesting range for the production of the pyrolysis products is between 625 and 775 K [25].

Pyrolysis process is an attractive, low cost strategy for the production of upgradeable liquid feedstocks (bio-oils) from lignocellulose. It does not require biomass pretreatment and fractionation, and offer potential for lignin utilization. However, bio oils require extensive hydrodeoxygenation before use in internal combustion engines or selective upgrading to larger hydrocarbons [26].

Recently microwave assisted pyrolysis is the emerging field to increase the efficiency of the biofuel production. Yinhas given a brief on 2nd-generation biofuels from biomass residues and waste feedstock. He presented a state-of-the-art review of microwave-assisted pyrolysis of biomass. First, conventional fast pyrolysis and microwave dielectric heating is briefly introduced. Then microwave-assisted pyrolysis process is thoroughly discussed stepwise from biomass pretreatment to bio-oil collection. The existing efforts are summarized in a table, providing a handy overview of the activities (e.g., feedstock and pretreatment, reactor/pyrolysis conditions) and findings (e.g., pyrolysis products) of various investigations [27]. Shoujie et al. first studied of integrated microwave torrefaction and pyrolysis of corn stover. They concluded that microwave torrefied pyrolysis favored phenols and hydrocarbons production. Organic acids were significantly reduced in bio-oils also up to 26.7 area % hydrocarbons in the bio-oil were produced [28].

In slow pyrolysis process, the biomass is pyrolyzed at slow heating rates (5–7 K/min) to lead to less liquid and gaseous product distribution and more of char production. Slow pyrolysis of sugarcane bagasse gives 23–28% yield of charcoal and fixed bed pyrolysis of Euphorbia rigida, sunflower presses bagasse and hazelnut shell were carried out [29–31]. These studies indicate that the efficiency increases with a rise in temperature in the presence of N_2 atmosphere.

As the slow pyrolysis produces more char, the higher yield of desirable liquid product can be obtained by fast pyrolysis with rapid heating of biomass (300°C/min). Fluidized bed reactors as it offers high heating rates, rapid de-volatilization, easy control, easy product collection, etc., can be used to carry out the fast pyrolysis [32]. The next option is flash pyrolysis process with special reactor configuration in which the reaction of small size samples (105–250 mm) completes in few seconds at very high temperature. Flash hydro-pyrolysis was also done in hydrogen atmosphere to get the saturated products [33, 34].

Other wet processes that produce bio-oils are hydrothermal gasification and hydrothermal liquefaction. The conversion of biomass in a wet environment at high pressure yields partly oxygenated hydrocarbons. Since the development of efficient wet processes is currently a growing area, the interest in liquefaction is low because the reactors and fuel-feeding systems are more complex and more expensive than pyrolysis processes.

12.2.2   MICROWAVE APPROACH FOR BIOFUEL PRODUCTION

Bio-diesel, an ideal alternative to fossil fuels, is very imperative for the sustainable development of mankind. Although the calorific value of the oils obtained from biological sources is as good as diesel fuel, the main drawback of these oils to use as fuel source is of low volatility and high viscosity, prohibits its direct application as fuel for diesel engines. To overcome these problems the oils obtained from natural sources (plant, bacteria, fungi, and oleaginous algae) undergo transesterification in which the triglycerides (vegetable oils) react with alcohol in presence of catalyst to produce bio-diesel and glycerin [35]. Promising options are there for converting the vegetable oils into prospective precursors for the preparation of activated carbon, and biodiesel. Among the reported techniques, microwave heating has been reported to be more energy efficient than other conventional methods for the transesterification reaction [36]. Figure 12.5 shows the biodiesel production process [37].

The influence of microwave on chemical reactions could be due to the molecular attrition caused by increase in the medium temperature and reaction rates. In the biodiesel production process, a satisfactory conversion can be achieved in shorter time duration under microwave irradiation, when compared with that obtained under conventional heating [38, 39]. The reaction for the transesterification process can undergo either by acid or base catalyzed. The mechanism for base catalyzed trans-esterification is shown in Scheme 12.1. The microwave efficiency is increased by the high polarity and high dielectric loss of methanol. Indeed, the dielectric constant of methanol at 25°C and 2.45 GHz is 32.7 compared with that of water at 78.5, which is high, and that of oil is only between 3 and 4. In general, conversions are high at very short reaction times in comparison with conventionally heated systems for homogenous base-catalyzed transesterification reactions under microwave irradiation in batch conditions.

The first pretreatment step required is to reduce the acidity of the oil obtained from waste cooking oils and from other biological sources. Both physical (i.e., drying, filtration, or distillation) and chemical pretreatment were used. Chemical treatment is carried out with an acid-catalyzed esterification reaction. The fatty acids in the oil react with methanol producing methyl esters and water. This is an immiscible liquid–liquid reaction, which is mass-transfer limited. The esterification reaction the tri, di and mono glycerides remains in the oil layer and the remaining waste was left in water layer. Following the esterification reaction and after a water removal step, the pretreated oil is then transformed using a base-catalyzed transesterification. Like the esterification reaction, the transesterification reaction is mass-transfer limited at the beginning due to the immiscibility of triglycerides and methanol, as well as at the end of the reaction because most of the catalyst is in the glycerol phase.

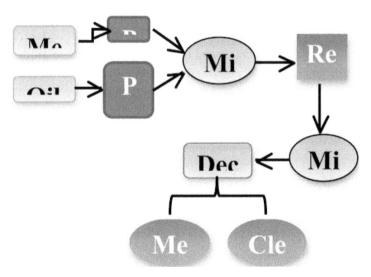

FIGURE 12.5   Biodiesel Production block diagram.

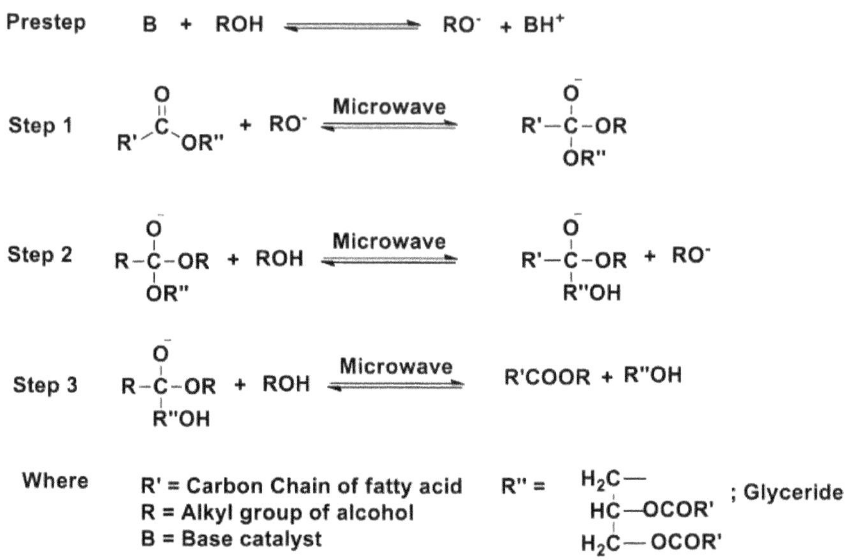

SCHEME 12.1   Step wise base catalyzed transesterification under microwave condition.

12.2.3 PHOTO-ELECTROCHEMICAL REACTIONS

In this method, photoelectrochemical (PEC) cell is used to degrade the organic waste by photocatalysis reaction to produce the biofuels under ambient conditions. However, not all organic substances result in the same energy outputs in terms of either electricity or molecular hydrogen production. In addition, water can be split in a PEC cell to produce fuel. A good amount of study is devoted on methanol, ethanol or glycerol as a model fuel due to their formation as bio-related compounds and their presence in biomass (Table 12.3). Generally the reaction is carried out in absence of oxygen and water. The reaction can be summarized as follows:

$$C_xH_yO_z + (2x - z)\,H_2O \rightarrow xCO_2 + (2x - z + (y/z))\,H_2 \qquad (2)$$

In another study, a wide variety of biological wastes is used in a PEC to liberate hydrogen that can be used as an energy source, which have been very well reviewed [60].

TABLE 12.2 Parameters Required For High Yields Conversion of Oils From Various Biological Sources to Biodiesel

S. No.	Source	Solvent/Catalyst	Watts/ temp	Time	Conver- sion, %	Ref.
1.	Cryptococcus curvatus (yeast strain)	Methanol/KOH	900 W	5 min	63.88	[40]
2.	Soy oil	Ionic liquid 4-allyl-4-methylmorpholin-4-ium bromine/ KOH	70°C	6 min	99.4	[36]
3.	Nannochlo-ropsis sp., CCMP1776	KOH	800W	9 min	71.5	[37]
4.	Pongamia pin-nata	NaOH: KOH (6:1)	60°C	10 min	96	[35]
5.	Oil palm empty fruit bunch (EFBAC)	KOH	600W	7 min	73.78	[41]
6.	Waste cooking oil	Cation ion exchange Resin particles (CERP)/ PES catalytic membrane	360W	90 min	97.4	[41]
7.	Chlorella pyre-noidosa (algae)	Chloroforl:methanol: sulfuric acid (1:1:0.1)	60°C	30 min	94	[42]
8.	Jatropha curcas L. seed	5 N KOH in ethanol	100 W	35 min	97.29	[43]
9.	Palm oil	Eggshells by simple calci-nation in air as catalyst	900 W	4 min	96.7	[44]
10.	Dry microalgae powder (Far East Bio-Tec Co., Taiwan)	Methanol/Li4SiO4	68°C	4 h	76.2	[45]
11.	Jatropha curcas	Methanol/NaOH	80 W	24 h	96.5	[46]
12.	Jatropha oil	KOH impregnated CaO	120 W	60 min	96.5	[47]
13.	Waste cooking oil	Sodium methoxide	750 W	3 m	97.7	[48]

TABLE 12.2 *(Continued)*

S. No.	Source	Solvent/Catalyst	Watts/ temp	Time	Conver- sion, %	Ref.
14.	Waste cooking oil	Aminophosphonic acid resin D418	81°C	420 min	90	[49]
15.	Waste frying Palm oil	Methanol/NaOH	800W	0.5 min	97	[50]
16.	Rapeseed oil	NaOH: MeOH:Oil (0.03: 1.27: 2.3)	60°C	1 min	97	[51]
17.	Soya bean oil	Nanopowder calcium oxide	65°C	60 min	96.6	[52]
18.	Sunflower oil	Carborundum	400 W	9 min	-	[53]
19.	Amazon flint kaolin	0.2 M KOH	400W	15 min	96.5	[54]
20.	Waste frying oil	$NaOCH_3$:Oil:Methanol (1.5:1:6)	60°C	9 min	97.74	[55]
21.	Palm oil	CH3ONa:Oil:Methanol (0.75:1:6)	750 W	6 min	99.5	[56]
22.	Pongamia pin- nata	Methanol–oil ratio 33.4 (w/w) % KOH catalyst concentra- tion 1.33 (w/w) %	180 W	2 min	89.9	[57]
23.	Carthamus tinctorius L	Methanol:Oil:NaOH (10:1:1)	60°C	6 min	98.4	[58]
24.	Nagchampa oil	KOH	88°C	210 min	-	[59]

12.2.4 BIO-APPROACHES

12.2.4.1 Enzyme Based Catalysis

Biofuels/bioethanol can be prepared by enzymatic conversion of bio-waste products. It has been observed that enzymes such as cellulase, hemicellu-lases, xylulase, amylase, lignicellulase, etc., can be used to act on differ-ent substrate and undergo chemical reaction to yield biofuel. Hereby we describe two of the commonly used enzymes – amylolytic enzymes and

TABLE 12.3 Performance of Various Cells (Photovoltaic, PEC and Gratzel) For Biomass Conversion

Type of Cell	Efficiency, %
Crystalline silicon	24
Multicrystalline silicon	18
Amorphous silicon	7
CuInSe$_2$	19
Dye-sensitized nanomaterial	10–11
Bipolar AlGaAs/Si	19–20
Organic solar cells	2–3

cellulosic enzymes that have been commercialized for biofuel production [61, 62].

12.2.4.1.1 Amylolytic Enzyme

Amylases have received a great deal of attention because of their significance especially in biotechnology. Amylases belong to a class of enzymes that has been extensively studied and is commercialized for different purposes. It can be obtained from different sources including plants, animals and microorganisms [63]. Amylases are starch hydrolyzing enzymes. They are mainly being employed in the starch processing industries for the hydrolysis of starch into simple sugar constituents [64–66]. Amylases are being commercially used in different industrial sectors. Amylase constitutes a class of industrial enzymes having approximately 25% of the enzyme market world-wide [67]. The main use of enzymes includes hydrolysis of starch to yield glucose syrup, amylase-rich flour and in the formation of dextrin during baking in food industries. Furthermore, in the textile industry, amylases are used for removal of starch sizing and as additives in detergents. In biotech industries, amylases are of the most important enzymes used [68]. Currently two types of amylases, glucoamylase and alpha-glucosidase are important for starch hydrolysis. Glucoamylase attacks – 1,4-bonds releasing D-glucose molecules [69]. This enzyme also attacks 1,6 bonds at branching points in the amylopectin molecule but much more slowly than 1,4 linkage [70]. Alpha-glucosidase catalyzes the splitting of alpha-D-glucozyl residues from the nonreducing end of substrates to release alpha-glucose [71]. Though, amylases are applicable for different sectors the production cost hinder its full utilization. The cost of procurement by developing countries can be even higher as a result of importation.

Thus, research has been focused to find alternate methods of amylase production. Microbial sources provide a great source for cheap and easy production of amylase. Many Bacillus species and thermostable Actinomycetes like *Actinomycetes thermomonospora* and *Actinomycetes thermoactinomyces* are versatile producers of amylase [72]. It has been observed that Bacillus species can produce a large range of extracellular enzymes such as amylases and proteases are of industrial importance [73, 74]. Cheap and readily available agricultural waste such as Potato peels, which presently constitutes a menace to solid waste management, may be a rich source of amylolytic bacteria [75].

Amylases have also been commercialized for enzymatic conversion of starch into bioethanol. For this process, starch has to be solubilized and then submitted to two enzymatic steps in order to obtain fermentable sugars [76, 77]. Since ethanol is the most used biofuel and starch provides a cheap and readily available raw material, amylases comprise some of the most reported groups of enzymes for the production of biofuels. Generally, these enzymes are of microbial origin: fungal and bacterial origins. Such enzymes have been used at industrial scale for decades, using conventional processes (dry grinding and wet milling) for the production of bioethanol, and recently nonconventional process is developed named cold starch hydrolysis (or granular starch hydrolysis) [78].

In the conventional process, starch is converted into ethanol by microorganisms that produce amylases extracellularly or externally adding enzymes such as glucoamylase and α-amylase, followed by fermentation. This process is known as saccharification wherein sugar is converted into ethanol using an ethanol-fermenting microorganism such as yeast *Saccharomyces cerevisiae* [79]. In order to obtain a new yeast strain that can directly produce ethanol from starch without the need for a separate saccharifying process, protoplast fusion was performed between the amylolytic yeast *Saccharomyces fibuligera* and *S. cerevisiae* [80]. Among bacteria, α-amylase obtained from thermoresistant bacteria like *Bacillus licheniformis* or from engineered strains of *Escherichia coli* or *Bacillus subtilis* is used during the first step of hydrolysis of starch suspensions [81]. In recent years, the many research projects have been performed for designing of microbial strains with more ethanol production capacity and enhancement of process ethanol efficiency. In addition, *Saccharomyces cerevisiae* has more specific characters biotechnologically, genetically, and physiologically than other microorganisms and considers the best microbial strain for bioethanol production. One of the best methods for bioethanol efficiency increment is the isolation of yeast mutant cells that resistant to high concentration of ethanol and capable to produce more bioethanol. Based on recent reports, the suitable yeast mutant strains have been isolated for more bioethanol productivity [82]. These mutant strains could produce 7% (W/V) bioethanol and tolerate up to 12% (V/V) exogenous ethanol but could not grow in the presence of other alcoholic compounds such as 2-propanol and 1-butanol. In addition, *Saccharomyces cerevisiae* is a safe microorganism and classify to GRAS (Generally Recognized As Safe) group. There are accepted biosafety rules in developed countries such as Japan about integration of other safe yeast genes to *Saccharomyces cerevisiae* genome and designing of recombinant *Saccharomyces* spp. by auxotrophic markers [83]. Details about the use of microbes for agro-industrials application is given in Table 12.4.

12.2.4.1.2   Cellulosic/Lignocellulosic Enzyme

Biofuels can also be generated using wood, grasses, or the inedible parts of plants. Cellulosic and lignocellulosic enzymes can act on these substrates to yield cellulosic ethanol. Lignocellulose is a structural material that comprises much of the mass of plants. Lignocellulose is composed mainly of cellulose, hemicellulose and lignin. Lignocelluloses can act on a variety of cheap and readily available substrates such as corn stover, *Panicum virgatum* (switch grass), *Miscanthus* grass species, wood chips and the by-products of lawn and tree maintenance which can be enzymatically converted for ethanol production. Lignocellulose usage has the advantage that abundant and diverse raw materials can be used compared to sources such as corn and cane sugars, but requires a greater amount of processing

TABLE 12.4   A Brief Summary on Biofuel Formation by Various Microorganisms

S. No.	Microorganism	Feedstock	Fermentative Process	Optimum temperature (°C)	Enzyme	Process time	Ref.
1.	Aspergillus awamori MTCC 7383	Soluble starch	Submerged fermentation	31	α-amylase or glucoamylase	96 h	[84]
2.	Aspergillus awamori NRRL 3112	Wheat flour		30	Glucoamylase	120 h	[85]
3.	Aspergillus kawachii IFO 4308	Wheat bran		30	α-amylase	48 h	[86]
4.	Aspergillus flavus A 1.1	Soluble starch		45	Glucoamylase	96 h	[87]
5.	Aspergillus oryzae PTCC5164	Soluble starch		35	α-amylase	48 h	[88]
6.	Bacillus circulans GRS 313	Soybean meal and wheat bran		40	α-amylase	48 h	[89]
7.	Rhizobium sp.	Peach palm flour		28	α-amylase	96 h	[90]
8.	Rhizopus microsporus var. Rhizopodiformis	Soluble starch and cassava flour		45	α-amylase	72 h	[91]
9.	Aspergillus niger 198	Soluble starch and potato starch		30	α-amylase	48 h	[92]
10.	Citrobacter sp.	Sorghum flour and soybean meal		28	α-amylase	60 h	[93]

11.	*Aspergillus niger* CCT 3312	Rice straw and bran	Solid-state fermentation	30	Glucoamylase	72 h	[94]
12.	*Aspergillus oryzae*	Wheat bran		30	Glucoamylase	120 h	[95]
13.	*Aspergillus sp. HA-2*	Wheat bran		28	Glucoamylase	96 h	[96]
14.	*Penicillium janthinellum NCIM 4960*	Wheat bran		35	α-amylase	96 h	[97]
15.	*Penicillium sp. X-1*	Wheat bran		30	Glucoamylase	36 h	[98]
16.	*Thermomucor indicae-seudaticae*	Wheat bran		40	Glucoamylase	96 h	[99]
17.	*Thermomyces lanuginosus ATCC 58160*	Wheat bean		50	α-amylase	120 h	[100]
18.	*Aplysia kurodai*	Sea lettuce		37	α-Amylase; α-Glucosidase	28 h	[101]

to make the sugar monomers available to the microorganisms typically used to produce ethanol by fermentation [102].

There are mainly two ways for cellulosic ethanol production, which includes cellulolysis and gasification. In cellulolysis processes, lignocellulosic materials are hydrolyzed and pretreated with enzymes to break complex cellulose into simple sugars such as glucose followed by fermentation and distillation. While in gasification process, the lignocellulosic raw materials are first converted to gaseous carbon monoxide and hydrogen and then converted to ethanol by fermentation or chemical catalysis. The ethanol obtained is further purified by distillation process [61, 103]. Thus, ethanol production by cellulosic enzyme can be summarized into four steps, pretreatment, cellulolysis, microbial fermentation and gasification.

The first step to ethanol production by cellulose is the "pretreatment" phase. In this step, lignocellulosic material such as wood or straw is made amenable to hydrolysis. However, most of the pretreatment processes are ineffective when the substrates used have with high lignin content, such as forest biomass or feedstocks. For this purpose, organosolv and SPORL (sulfite pretreatment to overcome recalcitrance of lignocellulose) are the only two processes that can achieve over 90% cellulose conversion for forest biomass, especially those of softwood species. SPORL is a better treatment process in terms of energy efficiency (sugar production per unit energy consumption in pretreatment) and robustness for pretreatment of forest biomass since very low fermentation inhibitors are produced. Organosolv pulping is particularly effective for hardwoods and offers easy recovery of a hydrophobic lignin product by dilution and precipitation [104].

The next step is the cellulose hydrolysis or "cellulolysis" which involves the breakdown of the molecules into sugars. Since, the cellulose molecules are composed of long chains of sugar molecules; it needs to be hydrolyzed to free sugars before it is fermented for alcohol production. Cellulolysis can be done either by chemical reaction using acids, or by enzymatic reaction. Chemical reaction using acids: In this process, acid is used for hydrolysis. Dilute acid may be used under high heat and high pressure, or more concentrated acid can be used at lower temperatures and atmospheric pressure. A significant obstacle to the dilute acid process is that the hydrolysis is so harsh that toxic degradation products are produced that can interfere with fermentation. The pretreated decrystalized cellulosic mixture of acid and sugars reacts in the presence of water to complete individual sugar molecules (hydrolysis). The product thus obtained, is then neutralized and yeast is added for fermentation.

Enzymatic hydrolysis: This process is similar to the reaction occurring in the stomachs of ruminants such as cattle and sheep, where the enzymes are produced by microbes. The reaction occurs at body temperature and usually involves a mixture of enzymes produced by different microbes. In this process, cellulose chains are broken into glucose molecules by cellulase enzymes. Similarly, such enzyme mixture can be used for lignocellulosic materials for effective cellulose breakdown without the formation of byproducts that would otherwise inhibit enzyme activity. The enzymatic hydrolysis requires relatively mild condition (50 °C and pH 5). All major pretreatment methods, including dilute acid, require an enzymatic hydrolysis step to achieve high sugar yield for ethanol fermentation. It should be noted that lignin and other sugars has to be separated before going for the microbial fermentation.

The sugar solution is then allowed to ferment by microbes for conversion of cellulose for ethanol production. Generally, baker's yeast (*Saccharomyces cerevisiae*), has long been used in the brewery industry to produce ethanol from hexoses (six-carbon sugars). Due to the complex nature of the carbohydrates present in lignocellulosic biomass, a significant amount of xylose and arabinose (five-carbon sugars derived from the hemicellulose portion of the lignocellulose) is also present in the hydrolysate. For example, in the hydrolysate of corn stover, approximately 30% of the total fermentable sugars is xylose. As a result, the ability of the fermenting microorganisms to use the whole range of sugars available from the hydrolysate is vital to increase the economic competitiveness of cellulosic ethanol and potentially biobased proteins. In recent years, metabolic engineering for microorganisms used in fuel ethanol production has shown significant progress. Besides *Saccharomyces cerevisiae*, microorganisms such as *Zymomonas mobilis* and *Escherichia coli* have been targeted through metabolic engineering for cellulosic ethanol production [105]. Recently, engineered yeasts have been described efficiently fermenting xylose [105, 106], arabinose [107] and even both together [108]. Yeast cells are especially attractive for cellulosic ethanol processes because they have been used in biotechnology for hundreds of years, are tolerant to high ethanol and inhibitor concentrations and can grow at low pH values to reduce bacterial contamination.

The "gasification" process does not rely on chemical decomposition of the cellulose chain (cellulolysis). Instead of breaking the cellulose into sugar molecules, the carbon in the raw material is converted into synthesis gas, using what amounts to partial combustion. The carbon monoxide, carbon dioxide and hydrogen may then be fed into a special kind of fermenter. Instead of sugar fermentation with yeast, this process uses *Clostridium ljungdahlii* bacteria. A recent study has found a novel Clostridium bacterium that seems to be twice as efficient in making ethanol from carbon monoxide as the one mentioned above [109].

12.2.4.2   Microbial Approaches

Energy is most critical for life and industry development. It is the major global economy as far as concerned. Fossil fuels include about 88% of the total global energy consumption, in which natural gas, oils and coals are the major economy fuels. The other energies *viz.* nuclear energy and hydroelectricity contribute only 5–6%. So, there is an urgent need to look for an alternative energy source apart from fossil fuels, as these fossils are getting depleted or unsustainable due to limited resources [110–113]. The first generation of biofuels is the greenhouse gases that contribute to global warming in large scale and also creates greater impact on the environment and human life. In recent years, potential threat has occurred to the marine environment due to the fossil fuels-associated CO_2 emission, which is directed into the oceans, where about 20 billion tons were dumped resulting in pH shift towards more acidic, affecting the marine ecosystem [114, 115]. The second generation of biofuels is from the plant tissue that includes energy crops, agricultural residues, harvesting residues from forest and waste of wood. The above-mentioned problems are needed to be exploited and global strategies for energy security should be addressed. In this chapter we have discussed about the biological methods for bioenergy production and its applications.

12.2.4.2.1   Biological Method For Bioenergy Production

The biological method for production of bioenergy is being explored nowadays. The recent exploration of microalgae, which could generate biofuels that are more effective than petroleum, and also the usage of land and water is minimal. Several eukaryotic microalgae have ability to store TAG (tri-acylglycerol) and starch, the energy rich compounds, which could be used for production of biofuels such as biodiesel and ethanol [116]. Microalgae or cyanobacteria is being used for the conversion of CO_2 into alcohols and also estimated that the recombinant of microalgae can yield up to 5.5 g/L biofuel. These cultures are now being used in few startup companies in order to commercialize algal fuels. Nevertheless these practices are still under trial and not being implemented for large scale productions [117–120]. There are few bacteria reported for production of biofuels is given in Table 12.5.

12.2.4.2.2   Microalgal Bacterial Consortia

In this consortia model, microalgae provide oxygen and organic molecules. The combined use of algae and bacteria as a consortium model has developed an interest in the field of biotechnology. This consortia model is being used as an alternative method for the enhancement of bioenergy and also reduces the algal biomass consumption in bioenergy production. This process is integrated with CO_2 and pollutant removal [128, 129]. *A. brasilense* stimulation of *Chlorella sp.* growth, lipids and starch accumulation indicates a significant potential in biotechnological application of microbial–bacterial consortium model to bioenergy. *Pseudomonas sp* is known for its bioremediation potential and for biodegradation of toxic metals such nitrophenol, etc. Similarly, the *Pseudomonas* GM41 strain along with *Synechocystissp* PCC6803 increased the productivity of bioenergy, hereby helping to degrade toxic compounds commonly found in polluted water [130]. Another application

TABLE 12.5   Consolidated Bioprocessing Microbes and Their Production of Biofuels

Strains	Substrate	Pretreatment condition	Final product	Ref.
Yeast	Cellulose with yeast extract	Phosphoric acid	Ethanol	[121]
E. coli	Brich wood xylan	Acid hydolysis	Ethanol	[122]
Marine microbes	Sugarcane bagasse with chicken manure	Lime	Carboxylic acid	[123]
C. cellulolyticum	Saccharified sugar and beet pulp	NA	Isobutanol	[124]
Bacillus subtilis	Avicel	Phosphoric Acid	Lactate	[125]
Thermobifidiafusca	Switchgrass	NA	n-propanol	[126]
Caldicellulosiruptor saccharolyticus DSM 8903	Switchgrass	NA	Hydrogen gas	[127]

of microalgal–bacterial consortia is the development of microbial solar cells that comprises photoautotrophic microorganisms to harvest solar energy and release organic compounds that are used by electrochemically active microorganisms to generate electricity [131]. Genetic engineering technique is employed in microalgae to enhance their valuable compounds. Most of genetic engineering techniques are involved in the improvement of microalgae and identifying the metabolic pathway for biofuel production. Especially genetic engineering tools are applicable for finding out the complete genome sequence of microalgae species that reveals several pathways involved in their metabolic processes. The application of genetic engineering tool in microalgae is mainly for the improvement of biofuel production in microalgae.

12.2.4.2.3   Microbial Fuel Cell

Microbial fuel cell (MFC) is a recent technique considered for the treatment of organic wastes and it can be used for the recovery of bioenergy from wastes such as waste from the food industries, sludge from sewerage, etc. MFC mainly works on the microbial catabolic activities to generate bioenergy (electricity) from the organic waste material. Researches in this area have found that MFC is a better technique than the anaerobic biogas technology for bioenergy production. Earlier, the conversion of biogas to electricity results in a significant energy loss. MFC are applied to wastewater that contains mainly volatile fatty acids and high amount of CO_2, nitrogen, and sulfur [132, 133]. This technique is currently under discussion and not been implanted in the practical waste material treatment. This is mainly because of the insufficient technical support and the efficiency of bioenergy production is also less when compared to the methanogenic anaerobic digesters. So based on the above-mentioned challenges it has been considered difficult to construct in large scale. When it comes to the improvement of the MFC performance, the values of the organic-treatment and electricity-generation comes in to play. While considering MFC process as an alternative to wastewater-treatment, the organic-treatment and electricity-generation treatment has limitations and the quality of the treated water becomes more important than energy efficiency [134, 135]. This MFC requires less energy compared to other conventional process.

12.2.4.2.4   MFC Fundamentals

In the early 1990s it was reported that in some bacterial species self-mediated extracellular electron transfer occur at substantial rates. The bacteria *Shewanellaputrefaciens* and *Shewanellaoneidensis* are able to use electrodes for conserving the electrochemical energy called as electrode respiration. Reports reveal that a ferric iron-reducing bacterium *Shewanellaputrefaciens* grew on lactate by using an electrode as the sole electron acceptor without the addition of an artificial mediator required for their growth. Many bacterial species such as *Shewanella-oneidensis* [136], *Geobactersulfurreducens* [137, 138], *Pseudomonas aeruginosa* [139] and *Clostridium butyricum* [140] have been reported for its ability to perform electrode respiration. Among these few bacteria are able to self-produce mediator compounds. While other makes use of cell surface cytochromes. Species like *Shewanellaoneidensis* can use both mechanisms for extracellular electron transfer. These findings have facilitated the construction of better MFC's that do not require any additional artificial mediators.

12.2.4.2.5 *Microbial Nanowires For Bioenergy Application*

Nanowires or nanowire mimetic is the recent finding in bioenergy application the scientist and the researchers in the field believe that this method will improve the bioenergy strategies, which relies upon the extracellular electron exchange, such as microbial electrosynthesis. As far as reports are concerned, there are only two microorganisms, *Shewanellaoneidensis* and *Geobactersulfurreducens* studied for microbial nanowires formation. Microbial nanowires have been implicated in extracellular electron transfer in many organisms. The nanowires formed by *G. sulfurreducens*, a type IV pili, is reported to possess metal conducting property. Microbial nanowires plays a vital role in contributing to bioenergy. Anaerobic digestion is one of the well-known methods for obtaining/extracting energy from organic wastes in the form of methane [141]. The *Geobacter* sp. and its closely related microorganisms are used in anaerobic digesters. These microorganisms are likely to function as anaerobic fermentation/methanogenesis or syntrophs. *G. metallireducens* microorganisms growth/capacity in the presence of anaerobic fermentation is studied. *G. metallireducens* is similar to *G. sulfurreducen,* which requires a pili for long range extracellular electron transfer [142, 143]

The advantages are:
- Microalgae or cyanobacteria used in biofuel production consumes less energy.
- These bacteria have exponential/growth phase which could actually double the biomass production.
- The time taken for biomass production in exponential phase is very less, that is, 3.5 hr.
- It is being estimated that the Oleaginous microalgae, produces oil more productively *viz.* exceeding ten-fold more than that of the best oilseed crops.

- They require only less water than the terrestrial crops used for bioenergy production.
- These microorganisms can uptake nutrients such as CO_2, nitrogen and phosphorous from the industrial or municipal waste, thereby helping in waste disposal.
- They do not require additional nutrients, herbicides or pesticides for growth.
- They produce proteins and oils as by-products where few strains produce biohydrogen.

The disadvantages do arise with the biological method in the production of bioenergy due to technical barriers such as slow growth rate, less the amount of biomass production, less lipid content and ability to survive under extreme pH and temperature [111, 112]. So, in order, to overcome all these drawbacks conventional microalgae breeding technique is being adopted for efficient and economical bioenergy production.

12.2.5 CATALYTIC METHODS

12.2.5.1 Ionic Liquids in Bio-fuel Production

The biofuel can be considered as a potential alternative for the nonrenewable fossil fuel. From renewable feedstock biodiesel can be produced by simple and well-known transesterification. But this process needs lots of organic or inorganic solvents, which have environmental, constrains. To overcome these problems ionic liquids (ILs) has been widely applied as an alternative green solvent because of their unique properties. ILs which are organic salts with melting point below 100 °C are currently receiving significant global attention due to their unique physiochemical properties such as high thermal stability, chemical stability, high ionic conductivity, controllable hydrophobicity and low vapor pressure. Owing to their intrinsic properties like nonflammability, nonvolatility and fire resistance; ILs are often being known as "green solvents" replacing traditional organic solvents in different reactions [144]. Also, ILs acquired a wide range of applications in different research areas. The biodiesel is said to be carbon neutral due to the net CO_2 emission is zero. This means whatever the CO_2

emitted from the biodiesel vehicles will be used by plants for photosynthesis [145] as shown in Fig. 12.6.

Based on the type, the applied ILs in biofuel generation will have multiple roles such as solvent, lipid extractant, catalyst support and catalyst. These ILs can be recycled and reused that supports the economical prospective. IL mediated biodiesel synthesis is considered as homogeneous catalysis. The advantages of ILs mediated biodiesel process include higher catalyst activity, easy separation of the final product, recyclability of ILs, high stability, environmentally benign. These advantages make ILs as superior catalysts than conventional catalysts (liquid and solid) [146]. Bronsted acid ILs are more commonly used catalysts in biodiesel production [147] and they are preferred over basic ILs (Table 12.6). This is because the formation of soap in based ILs catalyzed reactions won't occur when Bronsted acid ILs as it involves in the transesterification of feedstock having high content of free fatty acids and prevent the problem related to the formation of soap. In general, elevated temperatures are required for the high biodiesel production [148].

TABLE 12.6   Ionic Liquid Used as Catalyst in Biodiesel Synthesis

Catalyst	Feedstock	Time (h)	Reaction Temp. (°C)	Alcohol oil molar ratio	Biodiesel yield (%)	Ref.
$[BMIM][CH_3SO_3]$-$FeCl_3$	Jatropha oil	5	120	2:1	99.7	[149]
$[NMP][CH_3SO_3]$	Oleic acid	5	70	2:1	96.5	[150]
$[C_3SO_3HMIM]$ $[HSO_4]$	Tung oil	6	150	17:1	97.7	[151]
IMC_2OH	Cotton-seed oil	5	60	12:1	98.6	[152]
$[BMIM][OH]$	Glycerol triolate	8	120	9:1	87.2	[153]
$ChCl.ZnCl_5.H_2SO_4$	Palm oil	4	65	15:1	92.0	[154]
$[Et_3NH][Cl]$-$AlCl_3$	Soybean oil	9	70	12:1	98.5	[155]

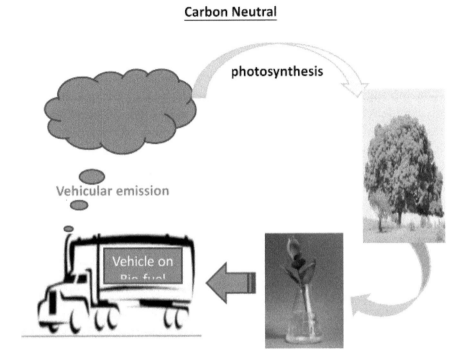

FIGURE 12.6. Carbon neutral cycle.

12.2.5.1.1 Ionic Liquids For Cellulose Dissolution and Pretreatment

Generally the raw material used for the production of biofuel is feedstocks, which are abundantly enriched with cellulose. So for the production of biofuel one can dissolve the cellulose into sugars and these sugars are subjected to dehydration process results HMF a key intermediate in biofuel production. For this pretreatment of cellulose in feed stocks generally conventional solvents are used but these solvents having the solubility problems. In order to overcoming this solubility problem, recent day's ILs are attaining greater attention as an ideal solvent for the dissolution of cellulose and pretreatment of biomass [156].

12.2.5.1.2 Acid Catalyzed Hydrolysis of Biomass in ILs

ILs in combination with acid catalyst is an efficient system for dis-solution of lignocellulosic materials under mild conditions that pro-duce wonderful conversions of key intermediates in biofuels and total reducing sugars like some of monosaccharides (e.g., glucose, fructose, glyceraldehyde and galactose) and also disaccharides (maltose and lac-tose). For instance 1-butyl-3-methyl imidazolium chloride with cata-lytic amount of HCl hydrolyzes biomasses like pine wood bagasse, rice straw, corn stalk at 100 °C for 60 min that results in 81% of TRS [157]. The same catalyst showed 100% conversion at mild temperatures in the case of carbohydrates [158]. In the same way 1-ethyl-3-imidazolium chloride with catalytic amount of HCl gives 90% of conversion of cel-lulose to monosaccharide at 105 °C with continuous addition of water. However, water also plays a key role in ILs mediated biomass dissolu-tion. In some cases it will increase the rate of reaction, and with some ILs it will decrease the rate of reaction by aggregating the cellulose moieties. For Instance 1-butyl-3 methylimidazolium chloride, needs high water, but 1,3-dimethyl imidazolium dimethyl phosphate works if it is anhydrous [159].

12.2.5.1.3 Metal Catalyzed Hydrolysis of Biomass in ILs

Transition metals also used as catalysts for biomass production [160–162]. Various metal chlorides ($CrCl_2$, $FeCl_2$, CuCl, $RuCl_3$, etc.) with ILs are efficient systems for dehydration of monosaccharides to HMF with good yields. N-heterocyclic carbenes can selectively hydrolyze the monosaccharide and the conversion will depend up on the catalyst load-ing and substrate used for the reaction [163]. Metal mediated microwave method in combination with ILs emerging as a cost effective and energy efficient system for HMF production [164–166]. In some of the metal catalyzed hydrolysis of biomass in ILs the HMF formation depends up on the hydrophobic nature of the imidazolium ring. Yetterbium based metal salts are preferred over chromium salts due its toxic nature towards nature [167].

12.2.5.1.4 *Functional Ionic Liquids Used in Biofuel Synthesis*

There are different types of ILs that played an important role in biofuel production (Fig. 12.7). In IL mediated biofuel synthesis it is observed that functionalized ionic liquids showing better results than normal ILs. These functionalized ILs can be used as alternative for the mineral acids like sulfuric acid and hydrochloric acid [168]. Based on the functionalities these ILs can be classified as (i) bifunctional ionic liquids (BILs), (ii) functional polymeric reusable ionic liquis (FPILs), (iii) dicationic acidic ILs (DAILs), and (iv) dicataionic basic ILs (DIBILs) [169–171]. Cr^{3+} and SO_4^{2-} ions derived from BFIL weaken the glycosidic bonds through binding with a glycosidic oxygen atom of cellulose that eases the dissolution to monosaccharide. Polymeric ionic liquids (FPILs) functionalized with SO_3H and different anionic counterparts show good catalytic activity that

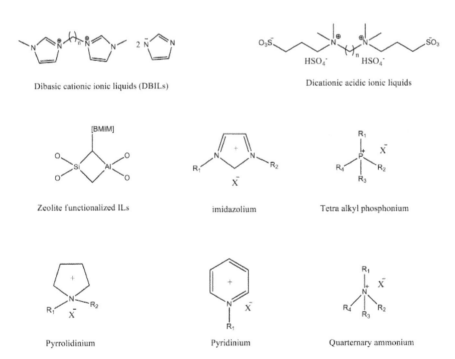

Dibasic cationic ionic liquids (DBILs)

Dicationic acidic ionic liquids

Zeolite functionalized ILs

imidazolium

Tetra alkyl phosphonium

Pyrrolidinium

Pyridinium

Quarternary ammonium

FIGURE 12.7   Ionic liquids used in biofuel synthesis.

involved in dissolution of cellulose to monosaccarides on further isomerization followed by dehydration to give HMF in good yields (96%).

12.2.5.2 Homogenous and Heterogeneous Catalysis For Transesterification

Among different existing methods for the conversion of biomass to biofuel, transesterification is the one of the best processes that produces biodiesel in pure quality [172]. In general, the triglycerides of biomass are treated with alcohol (methanol) and the applied catalyst catalyzes the transesterification to produce glycerol and the biodiesel as shown in Scheme 12.2. The quality of biodiesel strongly depends on various factors including temperature, reaction period, type of feedstock, amount of catalyst to alcohol ratio, etc. [173].

Both homogenous and heterogeneous catalysts were well studied in the transesterification process for production of biofuel. Homogenous catalysts require neutralization, separation, washing and recovery including salt waste disposal operations with serious economic and environmental penalties. There are different types of homogeneous catalysts used in the biofuel conversion and they can be classified as shown in Fig. 12.8. The homogeneous catalysts have some significant disadvantages about their reusability, neutralization, requirement of huge amount of water to leach out the homogenous catalyst, production of undesired by-products and a low-grade glycerol [174].

The general reaction mechanism of transesterification of triglycerides with acid and base catalysts is shown in Schemes 12.3 and 12.4, respectively. Although there are several types of homogenous base catalysts

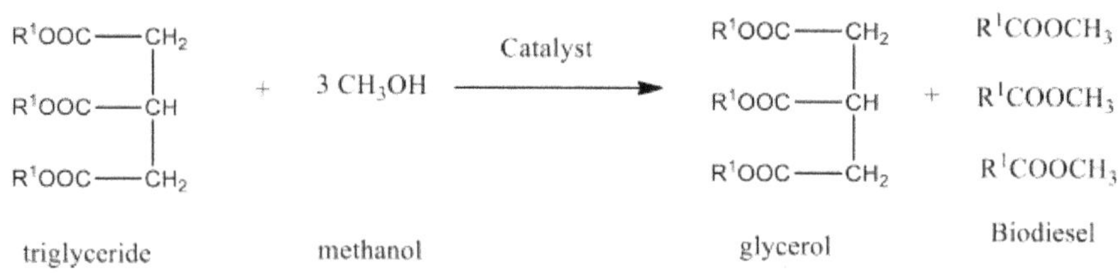

SCHEME 12.2 General reaction involved in transesterification of triglycerides with alcohol.

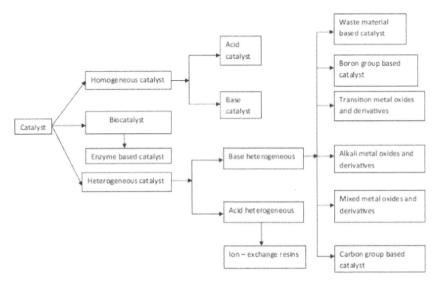

FIGURE 12.8   Classification of catalyst used for transesterification [175].

used in the biofuel production, the commonly used catalysts are NaOH, NaOMe KOH and KOMe. Among bases like NaOH, KOH, NaOMe, and KOMe, 1% NaOMe (w/v) with methanol/oil molar ratio 6 showed better conversion efficiency (98%) at 60 °C [176]. The best optimized conditions for transesterification reactions are found to be temperature between 55 to 70°C at 1–2 hr. Sodium methoxide is more effective than sodium hydroxide because of ionization of sodium methoxide to CH_3O^- and Na^+ and the ionization of sodium hydroxide leads to the formation of water that affects the rate of reaction. Even though sodium and potassium alkoxides show better conversions, NaOH and KOH are generally preferred because of their low cost [175]. Among these bases, NaOH is more favored because of ease of emulsification, ease of separation and economically more viable [177, 178].

It is significant to note that acid catalysts (homogenous and heterogeneous) more preferred over base catalysts for biofuel production. This is because all the feedstock used for biofuel production having water and free fatty acids, which decrease the activity of base catalyst.

Some of the benefits of heterogeneous catalysis over homogenous catalysis are separation of the products; recycling and reusability of the

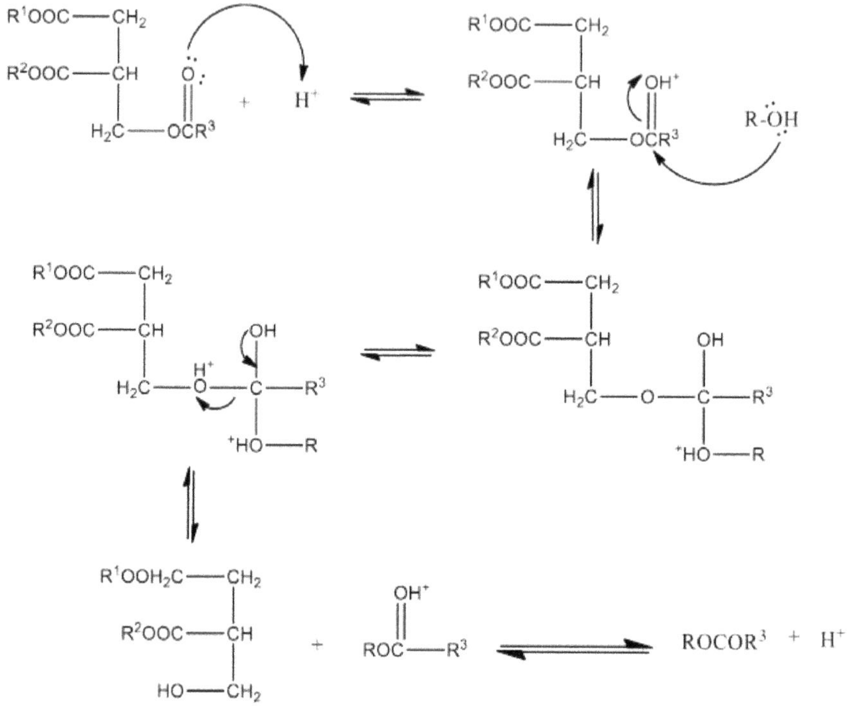

R^1OOC—CH_2

R^2OOC—CH $\overset{..}{O}:$ $+$ H^+ $\rightleftharpoons$

H_2C—$O\overset{\|}{C}R^3$

R^1OOC—CH_2

R^2OOC—CH OH^+

H_2C—$O\overset{\|}{C}R^3$ R-$\overset{..}{O}H$

R^1OOC—CH_2

R^2OOC—CH OH

H_2C—$\overset{H^+}{O}$—C—R^3

^+HO—R

$\rightleftharpoons$

R^1OOC—CH_2

R^2OOC—CH OH

H_2C—O—C—R^3

^+HO—R

R^1OOH_2C—CH_2

R^2OOC—CH $+$ ROC—R^3 $\overset{OH^+}{\|}$ $\rightleftharpoons$ $ROCR^3$ $+$ H^+

HO—CH_2

SCHEME 12.3 General reaction mechanism for acid catalyzed transesterification of triglycerides [174].

ROH $+$ B^+ $\rightleftharpoons$ RO^- $+$ BH^+

R^1OOC—CH_2

R^2OOC—CH $+$ $R\bar{O}H$ $\rightleftharpoons$

H_2C—$O\overset{\|}{C}$—R^3

R^1OOC—CH_2

R^2OOC—CH OR

H_2C—O—C—R^3

O^-

B $+$ R^2OOC—CH $\overset{R^1OOC—CH_2}{}$ $\overset{BH^+}{\rightleftharpoons}$ R^2OOC—CH $\overset{R^1OOC—CH_2}{}$ $+$ $ROOCR^3$

H_2C—OH H_2C—$\bar{O}$

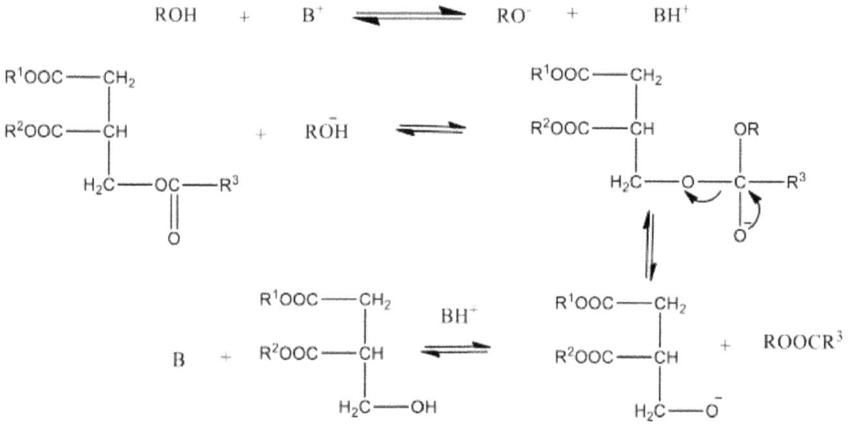

SCHEME 12.4 General reaction mechanism for base catalyzed transesterification of triglycerides [174].

catalyst; ultimately the product is free from any major catalytic impurities. Commonly used solid base catalysts are alkaline earth metal oxides, zeolite, KNO_3 loaded on Al_2O_3, KNO_3/Al_2O_3, BaO, SrO, CaO, MgO, etc. Among the solid base catalysts BaO takes minimum time while MgO takes the maximum time to complete the reaction.

12.2.5.2.1 Different Heterogeneous Catalysts For Biodiesel Production

The development of stable and cost-effective heterogeneous catalysts has been studied using different transesterification conversion technologies for biodiesel synthesis based on ultrasonication and microwave- assisted treatment, super critical fluid extraction, solvent engineering and synergetics of different mixed metal oxides. As a bulk commodity with low revenue, biodiesel production is strongly affected by the prices of oil feedstock. Hence, value-added derivatives are highly significant [179]. In esterification and transesterification processes, the triglycerides are converted to FAMEs and glycerol as a by-product. Additionally, the high yield production of biodiesel is influenced by the purity of the feedstock, amount of catalyst, temperature and oil to methanol ratio. In general, it is required to give a pretreatment to the feedstock to remove the FFAs and water. In fact, many researchers have reported the potential advantages of the use of mixed metal oxide catalyst over homogeneous catalysis. The heterogeneous requires 4–20% refining cost, which is comparatively less compared to homogeneous catalyst. However, many additional and challenging issues remain to be resolved before efficient solid catalyst can be developed, and many investigations are underway. The transesterification with the mixed metal oxides can be catalyzed either by either acid or base catalyzed but base catalyzed manifest much higher catalytic than acid catalysts, but are suitable for deriving biodiesel from the refined oils having low content of free fatty acids (FFA) less than 0.5% [180]. Xu et al. synthesized KF/Zn(Al)O catalyst and its activity was compares with that of Zn-Al hydrotalcite-like compounds Zn(Al)O, KF, KF/γ-Al_2O_3 and KF/Zn, they observed that the KF/Zn(Al) O catalyst shows highest activity among other mixed oxides. For the effective results of catalyst in transesterification reaction is optimized with the oil to methanol ratio 6:1, 3wt.% KF/Zn(Al)O and temperature 65°C for 3 hr yielding more than 95%, so the optimization is the key aspect for the effective catalytic activity. Materials to show enhanced catalytic activity they should have higher basicity, specific surface area, pore volume and size [181]. However, the optimization conditions like temperature, oil to methanol/ethanol ratio and catalyst in mole percentage are the prerequisites to obtain bio diesel in high yields. Among the current transesterification processes, the heterogeneous catalysis, that is, mixed metal oxides shows promising conversions than others.

12.2.5.2.2 Design of An Ideal Heterogeneous Catalyst

It is necessary for the researchers to find an ideal catalyst and also to know what are the factors affecting the stability, productivity, inactivity of a catalyst in transesterification process. This can lead to the designing of a catalyst in such a manner that it is showing highly stable and highly productive under different reaction conditions. There are some chemical and physical properties that can affect the stability and productivity of a heterogeneous catalyst. The chemical properties include: (i) acidic and basic properties of a catalyst, (ii) hydrophobic/hydrophilic character, (iii) calcination temperature, and (iv) leaching/reusability and physical properties include size, shape, pore structure, surface area and mechanical strength [182–184]. The basic catalysts, for example, KI, KF and KNO_3 doped with Al_2O_3 show good catalytic activity due to their basic sites involved in the formation of active species like K_2O and Aluminum alkoxides at low temperatures [66–68]. Active basic sites originated by the ionization of the Na and K from catalysts like Na/NaOH/γ-Al_2O_3 and K/KOH/γ-Al_2O_3 [185, 186] but the leachablity of these ions makes the process least stable towards the reaction conditions. Among the catalysts $Ca(NO_3)_2$/Al_2O_3, $LiNO_3$/Al_2O_3, $NaNO_3$/Al_2O_3 and $Mg(NO_3)_2$/Al_2O_3, calcium nitrate/aluminate is showing better catalytic activity and also stable towards transesterification than any other because of the formation of inactive forms like Na and Mg aluminum phases and leaching of Li ions form lithium nitrate aluminates [187]. Metal oxides are widely (CaO, SrO, BaO) used in the biofuel production, but they required high calcination temperatures (500–1050°C) and CaO undergo calcination process by nucelophic reaction and increases the basic properties of catalyst [188]. This could be due to the reaction of CaO with alcohol to form an protonated catalyst and alkoxide (RO⁻) ion which undergoes a nucleophilic attack on carbonyl group of the triglycerides then generates a tetrahedral intermediate [189] and then the ester is generated. It is imperative that in all metallic catalysts should be designed in such a way that the catalyst is having the property of making the carbonyl group is more electophilic during catalyst- substrate interaction and it is susceptible to the nucleohilic attack [190–192]. In the case of CaOMe the ionized methoxide ion is a stronger base and increases the rate of reaction. In case of SrO, the calcination temperature needed is 1200 °C for high yields of biofuel [193]. In the case of hydrotalcites (MgO/Al_2O_3) the temperatures required is above 180 °C for achieving 90% yields [194]. Further, the doping of small amount of Al^{3+} to pure MgO makes it more basic than in its pure form because of generation of Lewis acid-strong base sites [195] this clearly shows the dependence of catalytic activity on surface area and basic sites [196]. The catalytic activity of zeolites depends up on the ion exchange capacity, which is directly related to its basicity and size and porosity of the zeolite. So the bigger size zeolite like CsX is having least conversion efficiency than KX. Even water, FFA present in feedstock affects the rate of reaction. The presence of 1.0% to 4.0% of water decreases the yield of methyl ester drastically from 95.39% to 86.72% [197]. However, the catalytic activity can be increased by introducing hydrophobic nature to the catalyst.

(a) Influence of Hydrophobic Character of the Catalyst
The hydrophilic and hydrophobic characters of a catalyst exhibit important role in the catalyst design. The hydrophobic character of a catalyst requires the following properties: (i) it is necessary to absorb the oil on to its surface, (ii) the adsorption is not to deactivate the catalyst sites [198]. When it is strongly adsorbed and the by-products like glycerol and water are produced which are going to be involved in saponification process leading to low yields. Hydrophobicity of acidic and basic zeolites zeolite is controlled by the use of Si/Al ratio [199, 200]. Similarly hydrophobic nature is also achieved by sulfonation of incompletely carbonized starch [201].

(b) Influence of Pore Size of the Material

Porosity of a material plays an important role in activity of the catalyst and the pore size starts from 2 to 50 nm [202]. Even though mesoporous materials are thermally least stable and showing least reactivity than zeolite materials [203], but a combination of these mesoporous materials with different catalysts exhibit wonderful results. Sulfonated mesoporous carbon having strong acidic sites gives more stability and catalytic activity to the catalyst [204]. It is necessary to make the pore size more than 1.2 nm which allows the fatty acid molecule to undergo transesterification otherwise it blocks the pore and activity of the catalyst decreases [205]. For example, MCM-41 catalyst gives lower yields than polystyrene [206]. Mesoporous calcium methoxide (H_3PO_4/Al_2O_3) catalysts having a bigger pore size (100 and 1400 nm) on its surface area make the material more catalytic and high conversion rates when compared with the lower pore size materials. This is because of an accumulation of water and methanol at hydrophilic sites and hinders the activity of the catalyst.

(c) Influence of Surface Area

Generally an increase in the surface area leads to better catalytic activity [207, 208] and this property depends upon the calcination temperatures. The surface area of tungstated zirconia is decreased to 20% (58-m^2/g) from the actual (325-m^2/g) during the calcination temperature changed from 800 to 900°C [209]. Among catalysts like TiO_2, ZrO_2, TiO_2-SO_4^{2-} and ZrO_2-SO_4^{2-} surface area of TiO_2 and ZrO_2 increases by introducing sulfate, efficiency of the catalyst also increases [210]. For instance mixed metal carbonate precursor (CaO/ZnO) are showing better activity than the calcinated metal oxides, this is because of lose in the active sites in surface area at calcinated temperatures [211, 212].

12.3 CONCLUSIONS AND FUTURE PROSPECTS

There are many issues remain still unclear and require further research for the efficient bioenergy systems. The most studied examples were carried out with small-scale experiments using small batch mini-autoclaves. The product separation from thermochemical conversion of biomass is also a critical issue as it may affect the conversions of biocrude oil. The high heating rate may also influence the recovery of either catalysts or microorganisms, and there are not much details available on this topic. Experiments should be designed in such a way that they represent the real conditions as close as possible.

Conversion of biomass materials to energy is an environmentally friendly approach to meet the energy demands and this can address waste issues. Biomass, an alternative feedstock, can be converted into liquid and gaseous biofuels through thermochemical, photochemical, catalytic and bio-methods. The photocatalytic treatment of biomass using nanocrystalline titania photocatalysts has recently become more attractive. Also, hydrogen production through the photocatalysis of water as the cathodic reaction of an optically illuminate photochemical is effective in an aqueous electrolyte containing a biomass product (carbohydrates) thus producing liquid fuels and useful chemicals. Therefore, fuel cell system provides a novel way of energy conversion for the future research. The two main technologies presently used to convert biomass into energy are thermochemical and biochemical approaches. The selection of a conversion technology for biomass depends on the form in which the energy is required. Among them, pyrolysis and transesterification produce liquid fuels suitable for use as transportation fuels. Gasification/pyrolysis and anaerobic digestion is currently a cost-effective fuel.

For this purpose a defined set of technology breakthroughs will be required to develop the optimum utilization of algal biomass for the commercial production of biofuel. If these technological breakthroughs occur, biofuels based on algal biomass will play a role in the future energy systems. At the current stage of development, it is still too early to comments on any preferred routes of biofuels production from algal biomass. Finally, comprehensive life cycle assessment of algal biofuels illustrating environmental benefits and impacts can and should be a tool for guiding technology development as well as for policy decisions. Hydrothermal liquefaction of microalgae appears to be a very promising technology for biofuel production, but still in a very early stage of development. Nowadays, the development of bioenergy industries has gained importance for its sustainable energy production and deliverance all over the world. Even though fossil fuels, nuclear energy and hydroelectricity contribute to the bioenergy production, the availability of bioenergy is still in scarce level. There are various techniques being adopted for the production of bioenergy in sustainable environment considering the pollution as a major factor. Due to the limitations from the conventional technology, MFC is an attractive application in terms of bioenergy production as well wastewater treatment.

12.4   ACKNOWLEDGEMENTS

The authors are grateful to VIT University for providing the facilities and constant encouragement.

KEYWORDS

- Cellulose
- Crude oil resources
- Dimethyl ethanol
- Fatty acid methyl esters
- Fischer–tropsch diesel
- Hemi-cellulose
- Hydrogen
- Lignin
- Polymer of glucose
- Synthetic natural gas

REFERENCES

1. Freni, S., Cavallaro, S. (2003). Production of hydrogen for MC fuel cell by steam reforming of ethanol over MgO supported Ni and Co catalysts, Catal. Commun., *6,* 259–268; (1a) Frusteri, F., Freni, S., Spadaro, L., Chiodo, V., Bonura, G., Donato, S; Cavallaro, S. (2004). H2 production for MC fuel cell by steam reforming of ethanol over MgO supported Pd, Rh, Ni and Co catalysts, Catal. Commun., *5,* 611–615.
2. Huber, G. W., Shabaker, J. W., Dumesic, J. A. (2003). Raney Ni–Sn catalyst for H2 production from biomass-derived hydrocarbons, Science, *300,* 2075–2077.

3. Ragauskas, A. J., Williams, C. K., Davison, B. H., Britovsek, G., Cairney, J., Eckert, C. A., Frederick, W. J., Hallett, J. P., Leak, D. J., Liotta, C. L., Mielenz, J. R., Murphy, R., Templer, R., Tschaplinski, T. (2006). The path forward for biofuels and biomaterials, Science, *311,* 484–489.

4. Moon, J. H., Lee, J. W., Lee, U. D. (2011). Economic analysis of biomass power generation schemes under renewable energy initiative with Renewable Portfolio Standards (RPS) in Korea. Bioresour. Technol. *102,* 9550–9557.

5. Tijmensen, M. J. A., Faaij, A. P. C., Hamelinck, C. N., van Hardeveld, M. R. M. (2002). Exploration of the possibilities for production of Fischer Tropsch liquids and power via biomass gasification. Biomass Bioenergy. *23,* 129–152.

6. Liu, S., Tan, X., Li, K., Hughes, R. (2001). Methane coupling using catalytic membrane reactors, Catal. Rev. Sci. Eng. *43,* 147–198.

7. Petrus, L., Minke, A. (2006). Noordermeer. Biomass to biofuels, a chemical perspective, Green Chem. *8,* 861–867.

8. McKendry P. Energy production from biomass (Part 2): overview of biomass. Biores- our. Technol. (2002). *83,* 37–46.

9. Bridgwater, A. V., Peacocke, G. V. C. (2000). Fast pyrolysis process for biomass. Renew Sustain Energy Rev. *4,* 1–73.

10. Goyal, H. B., Seal, D., Saxena, R. C. (2008). Bio-fuels from thermochemical conversion of renewable resources: a review. Renew Sustain Energy Rev. *12,* 504–517.

11. Simonetti, D. A., Hansen, R. J., Kunkes, E. L., Soares, R. R., Dumesic, J. A. (2007). Coupling of glycerol processing with Fischer-Tropsch synthesis for production of liquid fuels. Green Chem. *9,* 1073–1083.

12. Kalinci, Y., Hepbasli, A., Dincer, I. (2012). Exergoeconomic analysis and performance assessment of hydrogen and power production using different gasification systems. Fuel. *102,* 187–198.

13. Swanson, R., Platon, A., Satrio. J., Brown, R. C. (2010). Techno-economic analysis of biomass-to-liquids production based on gasification. Fuel. *89,* 11–19.

14. Nadia, H. L., Akinwale, O. A., Johannes, H. K., Johann, F. G. (2013). Process efficiency of biofuel production via gasification and Fischer–Tropsch synthesis. Fuel. *109,* 484–492.

15. Kwangsu, K., Youngdoo, K., Changwon, Y., Jihong, M., Beomjong, K., Jeongwoo, L., Uendo, L., Seehoon, L., Jaeho, K., Wonhyun, E., Sangbong, L., Myungjin, K., Yunje, L. (2013). Long-term operation of biomass-to-liquid systems coupled to gasification and Fischer–Tropsch processes for biofuel production. Biores Technol. *127,* 391–399.

16. Berrueco, C., Recari, J., Matas, G. B., Del, A. G. (2014). Pressurized gasification of torrefied woody biomass in a lab scale fluidized bed. Energy. *70,* 68–78.

17. Danica, D. I., Erik, D., Louise, T., Göran, B. (2014). Integration of biofuel production into district heating- part I: an evaluation of biofuel production costs using four types of biofuel production plants as case studies. J. Cleaner Prod. *69,* 176–187.

18. Elliott, D. C. (2007). Historical Developments in Hydroprocessing Bio-oils. Energy Fuels *21,* 1792–1815.

19. Diego, L. B., Wolter, P., Frederik, R., Wim, B. (2013). Hydrothermal liquefaction (HTL) of microalgae for biofuel production: State-of-the-art review and future prospects. Biomass Bioenergy. *53,* 113–127.

20. Moumita, C., Armando, G. M., Caleb. N., Shulin, C. (2013). An α-glucan isolated as a coproduct of biofuel by hydrothermal liquefaction of *Chlorella sorokiniana* biomass. Algal Res. *2,* 230–236.
21. Erik, R. V., Richard, L. S., Mark, S. W., Andre, M. C. (2014). A national-scale comparison of resource and nutrient demands for algae-based biofuel production by lipid extraction and hydrothermal liquefaction. Biomass Bioenergy. *64,* 276–290.
22. Bahar, M., Arif, H., Sibel, I., Oktay, E. (2014). Biofuel production by liquefaction of kenaf (*Hibiscus cannabinus* L.). Biores Technol. *151,* 278–283.
23. Lupa C. J., Wylie, S. R., Shaw, A., Al-Shamma'a. A., Sweetman, A. J., Herbert, B. M. J. (2012). Experimental analysis of biomass pyrolysis using microwave-induced plasma. Fuel Process Technology. *97,* 79–84.
24. Carlos, J. D. N., Eduardo, B. S., Ricardo, A., Dolivar, C. N., Carlos, R. S. (2014). Production of biofuels from algal biomass by fast pyrolysis. Biofuels from Algae. *1,* 143–153.
25. Mohan, D., Pittman, C. U., Steele, P. H. (2006). Pyrolysis of wood/biomass for bio-oil: a critical review. Energy Fuels. *20,* 848–889.
26. David, M. A., Jesse, Q. B., James, A. D. (2010). Catalytic conversion of biomass to biofuels. Green Chem. *12,* 1493–1513.
27. Yin, C. (2012). Microwave-assisted pyrolysis of biomass for liquid biofuels production. Biores Technol. *120,* 273–284.
28. Shoujie, R., Hanwu, L., Lu, W., Gayatri, Y., Yupeng, L., James, J. The integrated process of microwave torrefaction and pyrolysis of corn stover for biofuel production. J. Anal. Appl. Pyrolysis. 2014; DOI: 10.1016/j.jaap.2014.04.008.
29. Putun, A. E., Ozcan, A., Gercel, H. F., Putun, E. (2001). Production of biocrudes from biomass in a fixed bed tubular reactor. Fuel. *80,* 1371–1378.
30. Ozbay, N., Putun, A. E., Uzun, B. V., Putun, E. (2001). Biocrude from biomass: pyrolysis of cotton seed cake. Renew Energy. *24,* 615–25.
31. Onay, O., Kockar, O. M. (2004). Fixed bed pyrolysis of rapeseed (Brassica napus L.). Biomass Bioenergy. *26,* 289–99.
32. Luo, Z., Wang, S., Liao, Y., Zhou, J., Gu, Y., Cen, K. (2004). Research on biomass fast pyrolysis for liquid fuel. Biomass Bioenergy. *26,* 455–62.
33. Gercel, H. F. (2002). Production and characterization of pyrolysis liquids from sunflower pressed bagasse. Bioresource Technol. *85,* 113–7.
34. Lede, J., Bouton, O. Flash pyrolysis of biomass submitted to a concentrated radiation. Application to the study of the primary steps of cellulose thermal decomposition. In: Division of fuel chemistry; reprints of symposia, vol. 44 (2), 217th ACS meeting, 21–25 March, Anaheim, USA, 1999.
35. Kumar, R., Kumar, R. G. (2011). Microwave assisted alkali-catalyzed transesterification of Pongamia pinnata seed oil for biodiesel production. Bioresour. Technol. *102,* 6617–662.
36. Lin, Y. C., Yang, P. M. (2013). Biodiesel production assisted by 4-allyl-4-methylmorpholin-4-ium bromine ionic liquid and a microwave heating system. Appl. Therm. Eng. *61,* 570–576.
37. Wali, W. A., Hassan, K. H. (2013). Real time monitoring and intelligent control for novel advanced microwave biodiesel reactor. Measurement, *46,* 823–839.

38. Yan, Y., Xiang Li, X. (2014). Biotechnological preparation of biodiesel and its high-valued derivatives: A review. Appl. Energy, *113*, 1614–1631.
39. Patil, P. D., Gnaneswar, V. G. (2011). Optimization of microwave-assisted transesterification of dry algal biomass using response surface methodology. Bioresour. Technol. *102*, 1399–1405.
40. Cui, Y., Liang, Y. (2014). Direct transesterification of wet Cryptococcus curvatus cells to biodiesel through use of microwave irradiation. Appl. Energy. *119*, 438–444.
41. Foo, K. Y., Hameed, B. H. Microwave-assisted preparation and adsorption performance of activated carbon from biodiesel industry solid reside: Influence of operational parameters. Bioresour. Technol.2012, 103, 398–404.
42. Cheng, J., Tao Yu, T. (2013). Using wet microalgae for direct biodiesel production via microwave irradiation. Bioresour. Technol. *131*, 531–535.
43. Jaliliannosrati. H; Amin, N. A. S. (2013). Microwave assisted biodiesel production from Jatropha curcas L. seed by two-step insitu process: Optimization using response surface methodology. Bioresour. Technol. *136*, 565–573.
44. Khemthong, P., Luadthong, C. (2012). Industrial eggshell wastes as the heterogeneous catalysts for microwave-assisted biodiesel production. Catal. Today. *190*, 112–116.
45. Dai, Y. M; Chen, K. T. (2014). Study of the microwave lipid extraction from microalgae for biodiesel production. Chem. Eng. J. *250*, 267–273.
46. Liaob, C. C., Chung, T. W. (2011). Analysis of parameters and interaction between parameters of the microwave-assisted continuous transesterification process of Jatropha oil using response surface methodology. Chem. Eng. Res. Des. *89*, 2575–2581.
47. Liao, C. C., Chung. T. W. (2013). Optimization of process conditions using response surfacemethodology for the microwave-assisted transesterificationof Jatropha oil with KOH impregnated CaO as catalyst. Chem. Eng. Res. Des. *91*, 2457–2464.
48. Chen K. S., Lin, Y. C. (2012). Improving biodiesel yields from waste cooking oil by using sodium methoxide and a microwave heating system. Energy. *38*, 151–156.
49. Liu, W., Yin, P. (2013).Microwave assisted esterification of free fatty acid over a heterogeneous catalyst for biodiesel production. Energy Convers. Manage. *76*, 1009–1014.
50. Lertsathapornsuka, V., Pairintra, R. (2008). Microwave assisted in continuous biodiesel production from waste frying palm oil and its performance in a 100 kW diesel generator. Fuel Process. Technol. *89*, 1330–1336.
51. Hernando, J; Leton. P. (2007). Biodiesel and FAME synthesis assisted by microwaves: Homogeneous batch and flow processes. Fuel. *86*, 1641–16.
52. Hsiao, M. C., Lin, C. C. (2011). Microwave irradiation-assisted transesterification of soybean oil to biodiesel catalyzed by nanopowder calcium oxide. Fuel *90*, 1963–1967.
53. Manco, I., Giordani, L. (2012). Microwave technology for the biodiesel production: Analytical assessments. Fuel *95*, 108–112.
54. Alex de Nazaré de Oliveira. (2013). Microwave-assisted preparation of a new esterification catalyst from wasted flint kaolin. Fuel. *103*, 626–631.
55. Azcan, N; Yilmaz, O. (2013). Microwave assisted transesterification of waste frying oil and concentrate methyl ester content of biodiesel by molecular distilation. Fuel. *104*, 614–619.

56. Lin, Y. C., Hsu, K. H. (2014). Rapid palm-biodiesel production assisted by a microwave system and sodium methoxide catalyst. Fuel. *115,* 306–311.

57. Kamath, V. H., Regupathi, I. (2011).Optimization of two step karanja biodiesel synthesis under microwave irradiation. Fuel Process. Technol. *92,* 100–105.

58. Duz, Z. M., Saydut, A. (2011).Alkali catalyzed transesterification of safflower seed oil assisted by microwave irradiation. Fuel Process. Technol. *92,* 308–313.

59. Gole, V. L., Gogate, P. R. (2013). Intensification of synthesis of biodiesel from nonedible oil using sequential combination of microwave and ultrasound. Fuel Process. Technol. *106,* 62–69.

60. Ibrahim, N., Kamarudin, S. K., Minggu, L. J. (2014). Biofuel from biomass via photo-electrochemical reactions: An overview. J. Power Sour. *259,* 33–42.

61. Trent, C. Y., Jyothi, K., Samuel, A., Miranda, M., Xiangling, L., Fan, L., Wensheng Q. (2014). Chapter 5: Biofuels and Bioproducts Produced through Microbial Conversion of Biomass. Bioenergy Research: Advances Applications. *1,* 71–93.

62. Nicolas, L. F., Antoine, M., Senta, B., Jean-Guy, B. (2014). Chapter 17—Use of Cellulases from Trichoderma reesei in the Twenty-First Century—Part I: Current Industrial Uses and Future Applications in the Production of Second Ethanol Generation. Biotechnology Biology of Trichoderma. *1,* 245–261.

63. Horikoshi, K. (1996). Alkaliphiles from an industrial point of view. FEMS Microbiol Rev. *18,* 259–270.

64. Mitchell, D. A., Lousane, B. K., In general principles of solid state fermentation, Monogram, Ed. By Doelle H. W., Ro13c. Publication of Oxford Univ., London. 1990.

65. Akpon, A. A., Adebiyi, A. B., Akinyanju, J. A. (1996). Partial purification and characterization of Alpha amylase from *Bacillus cereus* BC 19. J Agric Sci Technol. *2,* 152–157.

66. El-Saadany, R. M. A., Salem, F. A., El-Manawaty, H. K. (2006). A high yield of amylase from *Aspergillus niger*by effect of gamma irradiation. Starch. *36,* 64–66.

67. Sindhu, M. K., Singh, B. K., Prased, T. (1997). Changes in starch content of anhar seed due to fungal attack. In. Phytopathol. *34,* 269–271.

68. Aneja, K. R. (2003). Experiments in Microbiology Plant Pathology and Biotechnology, New Age International (P) Ltd., Publishers, New Delhi, Fourth Edition.

69. Guzman-Maldonato, H., Paredes-Lopes. O. (1995). Amylolytic enzymes and products derived from starch: a review. Crit Rev Food Sci Nut. *35,* 373–403.

70. James, J. A., Lee, B. H. (1997). Glucoamylases: Microbial sources, industrial application and molecular biology-A review. J Food Biochem. *21,* 1–52.

71. Pandey, A., Nigam, P., Soccol, C. R., Soccol, V. T., Singh. D., Mohan, R. (2002). Advances in microbiological amylases. Biotechnol Appl Biochem. *31,* 136–152.

72. Buzzini, P., Martini, A. (2002). Extracellular enzymatic activity profiles in yeast and yeast-like strains isolated from tropical environments. J. Applied Microbiol. *93,* 1020–1050.

73. Bajpai, P., Bajpai, P. K. (1999). High-temperature alkaline alpha-amylase from *Bacillus licheniformis* TCRDC-B13. J. Biotechnol. *33,* 72–78.

74. Demiorijan, D., Moris-Varas, F., Cassidy, C. (2001). Enzymes from extremophiles. J. Microbiol, *57,* 144–151.

75. Burhan, A., Nisa, U., Gokhan, C., Colak, Ö., Ashabil, A., Osmair, G. (2003). Enzymatic properties of a novel thermostable, thermophilic, alkaline and chelator resistant amylase from an alkaliphilic Bacillus sp. isolate ANT-6. Process Biochem. *38*, 1397–1403.

76. Sanchez, O. J., Cardona, C. A. (2008). Trends in biotechnological production of fuel ethanol from different feedstocks. Biores Technol. *99*, 5270–5295.

77. Moraes, L. M. P., Filho, S. A., Ulhoa, C. J. (1999). Purification and some properties of an α-amylase glucoamylase fusion protein from *Saccharomyces cerevisiae*. World J Microbiol Biotechnol. *15*, 561–564.

78. Aline, M. C., Leda, R. C., Denise, M. G. F. (2011). An overview on advances of amylases productionand their use in the production of bioethanol by conventional and nonconventional processes. Biomass Conv Bioref. *1*, 245–255.

79. Toksoy, O. E. (2006). Optimization of ethanol production from starch by an amylolytic nuclear petite *Saccharomyces cerevisiae* strain. Yeast. *23*, 849–856.

80. Chi, Z., Chi, Z., Liu, G., Wang, F., Ju, L., Zhang, T. (2009). *Saccharomycopsis fibuligera* and its applications in biotechnology. Biotechnol Adv. *27*, 423–43.

81. Saxena, R. K., Dutt, K., Agarwal, L., Nayyar, P. A. (2007). A highly thermostable and alkaline amylase from a *Bacillus sp*. PN5. Biores Technol. *98*, 260–265.

82. Mobini-Dehkordi, M; Nahvi. I., Zarkesh-Esfahani, H., Ghaedi, K., Tavassoli, M., Akada, R. (2008). Isolation of a novel mutant strain of Saccharomyces cerevisiae by ethyl methyl sulfonate mutagenesis approach as a high producer of bioethanol. J Biosci Bioengin. *105*, 403–408.

83. Mobini-Dehkordi, M., Nahvi, I., Zarkesh, H., Ghaedi, K., Akada, R. (2011). Characterization of an interesting novel mutant strain of commercial *Saccharomyces cerevisiae*. Iranian J. Biotechnol. *9*, 109–114.

84. Prakasham, R. S., Rao, C. S., Rao, R. S., Sarma, P. N. Enhancement of acid amylase production by an isolated *Aspergillus awamori*. J Appl Microbiol 2007; 102, 204–211.

85. Al-Turki, A. I., Khattab, A. A., Ihab, A. M. (2008). Improvement of glucoamylase production by *Aspergillus awamori* using microbial biotechnology techniques. Biotechnol. *7*, 456–462.

86. Nagamine, K., Murashima, K., Kato, T., Shimoi, H. (2003). Mode of α-amylase production by the shochu koji mold *Aspergillus kawachii*. Biosci Biotechnol Biochem. *67*, 2194–2202.

87. Gomes, E., Souza, S. R., Grandi, R. P., Silva, R. (2005). Production of thermostable glucoamylase by newly isolated *Aspergillus flavus* A 1.1 and *Thermomyces lanuginosus* A 13.37. Braz J Microbiol. *36*, 75–82.

88. Esfahanibolandbalaie, Z., Rostami, K., Mirdamadi, S. S. (2008). Some studies of α-amylase production using Aspergillus oryzae. Pak J Biol Sci. *11*, 2553–2559.

89. Gargi, D., Abhijit, M., Rintu, B., Maiti, B. R. (2001). Enhanced production of amylase by optimization of nutritional constituents using response surface methodology. Biochem. Eng. J. *7*, 227–231.

90. Oliveira, A. N., Oliveira, L. A., Andrade, J. S., Chagas Junior, A. F. (2007). Rhizobia amylase production using various starchy substances as carbon substrates. Braz J Microbiol. *38*, 208–216.

91. Peixoto, S. C., Jorge, J. A., Terenzi, H. F., Polizeli, M. L. T. M. (2003). Rhizopus microsporus var. rhizopodiformis: a thermotolerant fungus with potential for production of thermostable amylases. Int Microbiol. *6,* 269–273.

92. Shafique, S., Bajwa, R., Shafique, S. (2009). Screening of *Aspergillus niger* and *A. flavus* strains for extra cellular alpha-amylase activity. Pak J Bot. *41,* 897–905.

93. Orji, J. C., Nweke, C. O., Nwabueze, R. N., Nwanyanwu, C. E., Alisi, C. S., Etim-Osowo, E. N. (2008). Production and properties of α-amylase from Citrobacterspecies. Ambiente e Água Interdiscip. J Appl Sci. *4,* 45–57.

94. Costa, J. A. V., Eliane, C., Glênio, M., Lucielen, O. D. S., Mauricio, V., Telma, E. B. (2007). Simultaneous Amyloglucosidase and Exo-polygalacturonase Production by *Aspergillus niger* using Solid-state Fermentation. Brazilian Archives of Biology And Technology. *50,* 759–766.

95. Zambare, V. (2010). Solid state fermentation of *Aspergillus oryzae* for glucoamylase production on agro residues. Int J Life Sci. *4,* 16–25.

96. Anto, H., Trivedi, U. B., Patel, K. C. (2006). Glucoamylase production by solid-state fermentation using rice flake manufacturing waste products as substrate. Biores Technol. *97,* 1161–1166.

97. Sindhu, R., Suprabha, G. N., Shashidhar, S. (2009). Optimization of process parameters for the production of α-amylase from *Penicillium janthinellum* (NCIM 4960) under solid state fermentation. Afr J Microbiol Res. *3,* 498–503.

98. Haiyan, S., Xiangyang, G., Weiguo, Z. (2007). Production of novel raw-starch-digesting glucoamylase by *Penicillium* sp. X-1 under solid state fermentation and its use in direct hydrolysis of raw starch. World J Microbiol Biotechnol. *23,* 603–613.

99. Kumar, S., Satyanarayana, T. Statistical optimization of a thermostable and neutral glucoamylase production by a thermophilic mold *Thermomucor indicae-seudaticae* in solid-state fermentation. World J Microbiol Biotechnol. (2004). *20,* 895–902.

100. Kunamneni. A., Permaul. K., Singh, S. (2005). Amylase production in solid state fermentation by the thermophilic fungus *Thermomyces lanuginosus*. J Biosci Bioeng. *100,* 168–171.

101. Akihiko, T., Nami, N., Miki, O., Saori. M., Shuji. K., Masataka. S., Keizo. Y., Comprehensive enzymatic analysis of the amylolytic system in the digestive fluid of the sea hare, Aplysia kurodai: Unique properties of two α-amylases and two α-glucosidases. FEBS Open Bio 2014; DOI: 10.1016/j.fob.2014.06.002.

102. Edem, C. B., Moses. M. Chemical Pretreatment Methods for the Production of Cellulosic Ethanol: Technologies and Innovations. Int J Chem Eng 2013; DOI: 10.1155/2013/719607.

103. Zhu, J. Y., Pan, X. J., Wang, G. S., Gleisner, R. (2009). Sulfite pretreatment (SPORL) for Robust enzymatic saccharification of spruce and red pine. Biores Technology. *100,* 2411–2418.

104. Macfarlane, A. L., Farid, M. M. Chen JJJ. Organosolv delignification of willow Lambert Academic Press, 2010; ISBN 978-3-8383-9155-7.

105. Jeffries, T. W., Jin, Y. S. (2004). Metabolic engineering for improved fermentation of pentoses by yeasts. Appl Microbiol Biotechnol. *63,* 495–509.

106. Ohgren, K., Bengtsson, O., Gorwa-Grauslund, M. F., Galbe, M., Hahn-Hagerdal, B., Zacchi, G. (2006). Simultaneous saccharification and cofermentation of glucose

and xylose in steam-pretreated corn stover at high fiber content with Saccharomyces cerevisiae TMB3400. J Biotechnol. *126,* 488–498.

107. Becker, J., Boles, E. A modified *Saccharomyces cerevisiae* strain that consumes L-Arabinose and produces ethanol. Appl Environ Microbiol. (2003). *69,* 4144–4150.

108. Karhumaa, K., Wiedemann, B., Hahn-Hagerdal, B., Boles, E., Gorwa-Grauslund, M. F. (2006). Co-utilization of L-arabinose and D-xylose by laboratory and industrial *Saccharomyces cerevisiae* strains. Microb Cell Fact. DOI: 10.1186/1475-2859-5-18.

109. Srini, R., Rohit, P. D., Randy, S. L. (2002). Formation of ethanol from carbon monoxide via a new microbial catalyst. Biomass and Bioenergy. 23. 487–493.

110. British Petroleum, BP Statistical Review of World Energy; 2009.

111. Brennan, L., Wende, P. O. (2010). Biofuels from microalgae—a review of technologies for production, processing, and extractions of biofuels and coproducts. Renew. Sust. Energ. Rev. *14,* 557–577.

112. Khan, S. A., Rashm, R., Hussain, Z., Prasad, S., Banerjee, U. C. (2009). Prospects of biodiesel production from microalgae in India. Renew.Sust. Energ. Rev. *13,* 2361–2372.

113. Ugarte, D. G., Walsh, M. E., Shapouri, H., Slinsky, P. (2003). The economic impacts of bioenergy crop production in US agriculture, USDA Agricultural Economic Report. *41,* 816.

114. Banerjee, A., Sharma, R., Chisti, Y., Banerjee, U. C. (2002). Botryococcusbrauni: a renewable source of hydrocarbons and other chemicals. Crit. Rev.Biotechnol. *22,* 245–279.

115. Ormerod, W. G., Freund, P., Smith, A., Davison, J. (2002). Ocean storage of CO. IEAgreenhouse gas R&D program. UK, IE.

116. Radakovits, R., Jinkerson, R. E., Darzins, A., Posewitz, M. C. (2010). Genetic engineering of algae for enhanced biofuel production. Eukaryot. Cell. *9,* 486–501.

117. Wijffels, R. H., Barbosa, M. J. (2010). An outlook on microalgal biofuels. Science. *329,* 796–799.

118. Rasala, B. A. (2013). Genetic engineering to improve algal biofuels production. In Algae for Biofuels and Energy (Borowitzka, M.A and Moheimani, N. R., eds), 99–113.

119. Chisti, Y., Yan, J. (2011). Energy from algae: current status and future trends: algal biofuels: a status report. Appl. Energy. *88,* 3277–3279.

120. Petkov, G. (2012). A critical look at the microalgae biodiesel. Eur. J. Lipid Sci. Technol. *114,* 103–111.

121. Den Haan, R., Rose, S. H., Lynd, L. R., van Zyl, W. H. (2007). Hydrolysis and fermentation of amorphous cellulose by recombinant Saccharomyces cerevisiae. Metab. Eng. *9,* 87–94.

122. Shin, H. D., McClendon, S., Vo, T., Chen, R. R. (2010). Escherichia colibinary culture engineered for direct fermentation of hemicellulose to a biofuel. Appl. Environ. Microbiol. *76,* 8150–8159.

123. Fu, Z., Holtzapple, M. T. (2010). Consolidated bioprocessing of sugarcane bagasse and chicken manure to ammonium carboxylates by a mixed culture of marine microorganisms. Bioresour. Technol. *101,* 2825–2836.

124. Higashide, W., Li, Y., Yang, Y., Liao, J. C. (2011). Metabolic engineering ofClostridi-umcellulolyticum for isobutanol production from cellulose. Appl. Environ. Micro-biol. *77*, 2727–2733.

125. Zhang, X. Z., Sathitsuksanoh, N., Zhu, Z., Percival Zhang, Y. H. (2011). One-step production of lactate from cellulose as the sole carbon source without any other or-ganic nutrient by recombinant cellulolyticBacillussubtilis. Metab. Eng. *13*, 364–372.

126. Deng, Y., Fong, S. S. (2011). Metabolic engineering of Thermobifidafuscafor direct aerobic bioconversion of untreated lignocellulosic biomass to 1-propanol. Metab. Eng. *13*, 570–577.

127. Talluri, S., Raj, S. M., Christopher, L. P. (2013).Consolidated bioprocessing of un-treated switchgrass to hydrogen by the extreme thermophile Caldicellulosiruptorsac-charolyticus DSM 8903. Bioresour. Technol. *139*, 272–279.

128. Choix, F. J. (2012). Enhanced accumulation of starch and total carbohydrates in alginate-immobilized Chlorella spp. induced by Azospirillumbrasilense: II. Hetero-trophic conditions. Enzyme. Microb. Technol. *51*, 300–330.

129. de-Bashan, L. E. (2002). Removal of ammonium and phosphorus ions from synthetic wastewater by the microalgae Chlorella vulgaris co immobilized in alginate beads with the microalgae growthpromoting bacterium Azospirillumbrasilense. Water. Res. *36*, 2941–2948.

130. Abed, R. M. M. (2010). Interaction between cyanobacteria and aerobic heterotrophic bacteria in the degradation of hydrocarbons. Int. Biodeterior. Biodegrad. *64*, 58–64.

131. Strik, D. P. B. T. B. (2011). Microbial solar cells: applying photosynthetic and elec-trochemically active organisms. Trends Biotechnol. *29*, 41–49.

132. Rittmann, B. E. (2008). Opportunities for renewable bioenergy using microorgan-isms. Biotechnol. Bioeng. *100*, 203–212.

133. Kaspar, H. F., Wuhrmann, K. (1978). Product inhibition in sludge digestion. Microb. Ecol. *4*, 241–248.

134. Lettinga, G. (1995). Anaerobic digestion and wastewater treatment systems. Antonie Van Leeuwenhoek, *67*, 3–28.

135. McCarty, P. L., Smith, D. P. (1986). Anaerobic wastewater treatment. Environ. Sci. Technol. *20*, 200–1206.

136. Logan, B. E., Regan, J. M. (2006). Electricity-producing bacterial communities in microbial fuel cells. Trends Microbiol. *14*, 512–518.

137. Lovley, R. D. (2006). Microbial fuel cells: novel microbial physiologies and engi-neering approaches. Curr. Opin. Biotechnol. *17*, 327–332.

138. Bond, D. R., Lovley, D. R. (2003). Electricity production by Geobactersulfurredu-cens attached to electrodes. Appl. Environ.Microbiol. *69*, 1548–1555.

139. Rabaey, K., Boon, N., Hofte, M., Verstraete, W. (2005). Microbial phenazine pro-duction enhances electron transfer in biofuel cells. Environ. Sci. Technol. *39*, 3401–3408.

140. Park, H. S., Kim, B. H., Kim, H. S., Kim, H. J., Kim, G. T., Kim, M., Chang, I. S., Park, Y. K., Chang, H. I. (2001). A novel electrochemically active and Fe (III)-reducing bacterium phylogenetically related to Clostridium butyricumisolated from a microbial fuel cell. Anaerobe. *7*, 297–306.

141. McCarty, P. L., Bae, J., Kim, J. (2011). Domestic wastewater treatment as a net en-ergy producer – can this be achieved. Environ. Sci. Technol. *45*, 7100–7106.

142. Lovley, D. R. (2011). Live wires: direct extracellular electron exchange for bioenergy and the bioremediation of energy-related contamination. Energy. Environ. Sci. *4*, 4896–4906.

143. Smith, J. A., Lovley, D. R., Tremblay, P. L. (2013). Outer cell surface components essential for Fe (III) oxide reduction by Geobactermetallireducens. Appl. Environ. Microbiol. *79*, 901–907.

144. Mallakpour, S., Dinari, M. (2012). Ionic Liquids as Green Solvents: Progress and Prospects. In *Green Solvents II*, Mohammad, A., Inamuddin, D., Eds. Springer Netherlands, 1–32.

145. Atadashi, I. M., Aroua, M. K., Aziz, A. (2010). A High quality biodiesel and its diesel engine application: a review. Renew Sust Energ Rev. *14*, 1999–2008.

146. Wu, Q., Chen, H., Han, M., Wang, D., Wang, J. (2007). Transesterification of cottonseed oil catalyzed by Brønsted acidic ionic liquids. Industrial and Engineering Chemistry Research *46*, 7955–7960.

147. Zhang. L., Xian, M., He, Y., Li, L., Yang, J., Yu, S et al. (2009). A Bronsted acidic ionic liquid as an efficient and environmentally benign catalyst for biodiesel synthesis from free fatty acids and alcohols. Bioresour. Technol. *100*, 4368–4373.

148. Wang, C., CuiG., LuoX., XuY., LiH., Dai S. Highly efficient and reversible SO_2 capture by tunableazole-based ionic liquids through multiple-site chemical absorption. J. Am. Chem. Soc. *133*, 11916–11919.

149. Zhang. L., Xian, M., He, Y., Li, L., Yang, J., Yu S, et al. (2011). A Bronsted acidic ionic liquid as an efficient and environmentally benign catalyst for biodiesel synthesis from free fatty acids and alcohols. Bioresour. Technol. (2009). *100*, 4368–73.

150. Zhou, J., Lu, Y., Huang, B., Huo, Y., Zhang, K. (2011). Preparation of biodiesel from tung oil catalyzed by sulfonic-functional Bronested acidic ionic liquids. Adv. Mater. Res. 314–316, 1459–1462.

151. Liang, J. H., Ren, X. Q., Wang, J. T., Jinag, M., Li, Z. J. (2010). Preparation of biodiesel by transesterification from cottonseed oil using the basic dication ionic liquids as catalysts. Journal of Fuel Chemistry and Technology, *38*, 275–280.

152. Zhou, S., Liu, L., Wang, B., Xu, F., Sun, R. C. (2012). Biodiesel preparation from transester- ification of glycerol trioleate catalyzed by basic ionic liquids. Chin. Chem. Lett. *23*, 379–82.

153. Isahak, W. N. R. W., Ismail, M., Mohd Jahim, J., Salimon, J., Yarmo, M. A. (2011). Transester- ification of palm oil by using ionic liquids as a new potential catalyst. Trends Appl. Sci. Res. *6*, 1055–62.

154. Liang, X., Gong, G., Wu, H., Yang, J. (2009). Highly efficient procedure for the synthesis of biodiesel from soybean oil using chloroaluminate ionic liquid as catalyst. Fuel. *88*, 613–616.

155. Advances on biomass pretreatment using ionic liquids: An overview Haregewine Tadesse and Rafael Luque Energy Environ. Sci., (2011). *4*, 3913.

156. Charles, G., Cellulose Solution, US Patent 1 943 175, 1934.

157. Marzialetti, T., Olarte, M. B. V., Sievers, C., Hoskins, T. J. C., Agrawal, P. K., Jones, C. W. (2008). Dilute Acid Hydrolysis of Loblolly Pine: A Comprehensive Approach. Ind. Eng. Chem. Res. *47*, 7131–7140.

158. Mazza, M., Catana, D. A., Vaca-Garcia, C., Cecutti, C. (2009). Influence of water on the dissolution of cellulose in selected ionic liquids. Cellulose. *16*, 207–215.

159. Binder, J. B., Raines, R. T. (2009). Simple Chemical Transformation of Lignocellulosic Biomass into Furans for Fuels and Chemicals. J. Am. Chem. Soc. *131*, 1979–1985.

160. Stahlberg, T., Sorensen, M. G., Riisager, A. (2010). Direct conversion of glucose to 5-(hydroxymethyl) furfural in ionic liquids with lanthanide catalysts. Green Chem. *12*, 321–325.

161. Chun, J. A., Lee, J. W., Yi, Y. B., Hong, S. S., Chung, C. H. (2010). Catalytic production of hydroxymethylfurfural from sucrose using 1-methyl-3-octylimidazolium chloride ionic liquid. Korean J. Chem. Eng. *27*, 930–935.

162. Yong, G., Zhang, Y. G., Ying, J. Y. (2008). Efficient Catalytic System for the Selective Production of 5-Hydroxymethylfurfural from Glucose and Fructose. Angew. Chem., Int. Ed. *47*, 9345–9348.

163. Li, C. Z., Zhang, Z. H., Zhao, Z. B. K. Direct conversion of glucose and cellulose to 5-hydroxymethylfurfural in ionic liquid under microwave irradiation Tetrahedron Lett. (2009). *50*, 5403–5405.

164. Zhang, Z. H., Zhao, Z. B. K. (2010). Microwave-assisted conversion of lignocellulosic biomass into furans in ionic liquid Bioresour. Technol. *101*, 1111–1114.

165. Hu, S. Q., Zhang, Z. F., Song, J. L., Zhou, Y. X., Han, B. X. (2009). Efficient conversion of glucose into 5-hydroxymethylfurfural catalyzed by a common Lewis acid $SnCl_4$ in an ionic liquid. Green Chem. *11*, 1746–1749.

166. Stahlberg, T., Sorensen. M. G., Riisager, A. (2010). Direct conversion of glucose to 5-(hydroxymethyl) furfural in ionic liquids with lanthanide catalysts. Green Chem. *12*, 321–325.

167. Zhao, H. B., Holladay, J. E., Brown, H., Zhang, Z. C. (2007). Metal Chlorides in Ionic Liquid Solvents Convert Sugars to 5-Hydroxymethylfurfural.Science. *316*, 1597–1600.

168. Fan, M., Yang, J., Jiang, P., Zhang, P., Li, S. (2013). Synthesis of novel dicationic basic ionic liquids and its catalytic activities for biodiesel production. RSC Adv. *3*, 752.

169. Fang, D., Yang, J., Jiao, C. (2011). Dicationic Ionic Liquids as Environmentally Benign Catalysts for Biodiesel Synthesis. ACS Catal. *1*, 42–47.

170. Hu, L., Wua, Z., Xu, J., Sun, Y., Linc, L., Liu, S. (2014). Zeolite-promoted transformation of glucose into 5-hydroxymethylfurfural in ionic liquid. Chem. Eng. J. *244*, 137–144.

171. Guo, F., Fang, Z., Tian, X. F., Long, Y. D., Jiang, L. Q. (2011). One-step production of biodiesel from Jatropha oil with high-acid value in ionic liquids. Bioresour. Technol. *102*, 6469–6472.

172. Sinha, S., Agarwal, A., K, Garg S. (2008). Biodiesel development from rice bran oil: transesterification process optimization and fuel characterization. Energy Conver Manage, *49*, 1248–57.

173. Ejaz, M., Shahid, E., M., Jamal, Y. (2011). Production of biodiesel: A technical review Renewable and Sustainable Energy Reviews, *15*, 4732–4745.

174. Sivasamy, A., Cheah, K., Y., Fornasiero, P., Kemausuor, F., Zinoviev, S., Miertus S. (2009). Catalytic Applications in the Production of Biodiesel from Vegetable Oils. Chem. Sus. Chem. *2*, 278–300.

175. Singh, S. P., Singh, D. (2010). Biodiesel production through the use of different sources and characterization of oils and their esters as the substitute of diesel: a review. Renew Sustain Energy Rev, *14*, 200–216.
176. Rashid, U., Anwar, F., Moser, B., R., Ashraf, S. (2008). Production of sunflower oil methyl esters by optimized alkali-catalyzed methanolysis. Biomass Bioenergy, *32*, 1202–5.
177. Math, M., C., Irfan G. (2007). Optimization of resturant waste oil methyl ester yield. J Sci Ind Res, *66*, 772–226.
178. Saifuddin, N., Chua, K. H. (2004).Production of ethyl ester (biodiesel) from used frying oil: optimization of transesterification process using microwave irradiation. Malaysian J Chem, *6*, 77–82.
179. Yan, Y., Li, X. (2014).Biotechnological preparation of biodiesel and its high-valued derivatives: A review. Appl. Energy, *113*, 1614–1631.
180. Lee, J. S., Saka, S. (2010). Biodiesel production by heterogeneous catalysts and supercritical technologies. Bioresour. Technol, *101*, 7191–7200.
181. Xu, C., Sun, J. (2010). On the study of KF/Zn(Al)O catalyst for biodiesel production from vegetable oil. Appl. Catal., B, *99*, 111–117.
182. Xie, W., Huang, X. (2006). Synthesis of biodiesel from soybean oil using heterogeneous KF/ZnO catalyst. Cat. Lett. *107*, 53–59.
183. Xie, W., Li, H. (2006). Alumina-supported potassium iodide as a heterogeneous catalyst for biodiesel production from soybean oil. J. Mol. Catal. A: Chem. *255*, 1–9.
184. Vyas, A. P., Subrahmanyam, N., Patal, P. A. (2009). Production of biodiesel through transesterification of Jatropha oil using KNO_3/Al_2O_3 solid catalyst. Fuel *88*, 625–628.
185. Ma, H., Li, S., Wang, B., Wang, R., Tian, S. (2008). Transesterification of rapeseed oil for synthesizing biodiesel by $K/KOH/-Al_2O_3$ as heterogeneous base catalyst. J. Am. Oil Chem. Soc. *85*, 263–270.
186. Kim, H. J., Kang, B. S., Kim, M. J., Park, Y. M., Kim, D. K., Lee, J. S., Lee, K. Y. (2004). Transesterification of vegetable oil to biodiesel using heterogeneous base catalyst. Catal. Today 93–95, 315–320.
187. Benjapornkulaphong, S., Ngamcharussrivichai, C., Bunyakiat, K. (2009). Al_2O_3-supported alkali and alkali earth metal oxides for transesterification of palm kernel oil and coconut oil. Chem. Eng. J. *145*, 468–474.
188. Kouzu, M., Kasuno, T., Tajik, M., Sugimoto, Y., Yamanaka, S., Hidaka, J. (2008). Calcium oxide as a solid base catalyst for transesterification of soybean oil and its application to biodiesel production. Fuel *87*, 2798–2806.
189. Lotero, E., Goodwin, J. G., Bruce, D. A., Suwannakarn, K., Liu, Y., Lopez, D. E. (2006). The catalysis of biodiesel synthesis. J. Catal. *19*, 41–84.
190. Tateno, T., Sasaki, T. Process for producing fatty acid fuels comprizing fatty acids esters. US Patent 6818026, 2004.
191. Antunes, W. M., Veloso, C. D. O., Assumpc, C., Henriques, O. (2008). Transesterification of soybean oil with methanol catalyzed by basic solids. Catal. Today, 133–135, 548–554.
192. Leclercq, E., Finiels, A., Moreau, C. (2001). Transesterification of rapeseed oil in the presence of basic zeolites and related solid catalysts. J. Am. Oil Chem. Soc. *78*, 1161–1165.

193. Liu, X., He, H., Wang, Y., Zhu, S. (2007). Transesterification of soybean oil to bio-diesel using SrO as a solid base catalyst. Catal. Commun. *8*, 1107–1111.
194. Serio, M. D., Ledda, M., Cozzolino, M., Minutillo, G., Tesser, R., Santacesaria, E. (2006). Transesterification of soybean oil to biodiesel by using heterogeneous basic catalysts. Ind. Eng. Chem. Res. *45*, 3009–3014.
195. Cosimo, J. I. D., Diez, V. K., Xu, M., Iglesi, E., Apestegui, C. R. (1998). Structure and surface and catalytic properties of Mg–Al basic oxides. J. Catal. *178*, 499–510.
196. Albuquerque, M. C. G., Santamaria-Gonzalez, J., Merida-Robles, J. M., Moreno-Tost, R., Rodriguez-Castellon, E., Jimenez-Lopez, A., Azevedo, D. C. S., Cavalcante Jr., C. L., Maireles-Torres, P. (2008). MgM (M = Al and Ca) oxides as basic catalysts in transesterification processes. Appl. Catal. A *347*, 162–168.
197. Liu, X., Xiong, X., Liu, C., Liu, D., Wu, A., Hu, Q., Liu, C. (2010). Preparation of biodiesel by transesterification of rapeseed oil with methanol using solid base cata-lyst calcined $K_2CO_3/-Al_2O_3$. J. Am. Oil Chem. Soc, *87*, 817–823.
198. Melero, J. A., Grieken, R. V., Morales, G. (2006). Advances in thesynthesis and cata-lytic applications of organosulfonic-functionalized mesostructured materials. Chem. Rev. *106*, 3790–3812.
199. Kiss, A. A., Omota, F., Dimian, A. C., Rothenberg, G. (2006). The heterogeneous advantage: biodiesel by catalytic reactive distilation. Top. Catal. *40*, 141–150.
200. Canakci, M., Gerpen, J. V. (1999). Biodiesel production via acid catalysis. Trans. ASAE, *42*, 1203–1210.
201. Lou, W. Y., Zong, M. H., Duan, Z. Q. (2008). Efficient production of biodiesel from high free fatty acid-containing waste oils using various carbohydrate-derived solid acid catalysts. Bioresour. Technol, *99*, 8752–8758.
202. Islama, A., Taufiq-Yapb, Y. H., Chua, C., M., Chanc, E., S., Ravindraa P. (2013). Studies on design of heterogeneous catalysts for biodiesel production. Process Safety and Environ. Prot. *91*, 131–144.
203. Beck, J. S., Vartuli, J. C., Roth, W. J., Leonowicz, M. E., Kresge, C. T., Schmitt, K. D., Chu, C. T. W., Olson, D. H., Sheppard, E. W., McCullen, S. B. (1992). A new family of mesoporous molecular sieves prepared with liquid crystal templates. J. Am. Chem. Soc, *114*, 10834–10843.
204. Xiao, F. S. (2004). Hydrothermally stable and catalytically active ordered mesopo-rous materials assembled from preformed zeolites nanoclusters. Catal. Surv. Asia, *35*, 151–159.
205. Liu, X., Piao, X., Wang, Y., Zhu, S. (2008). Calcium ethoxide as a solid base catalyst for the transesterification of soybean oil to biodiesel. Energy Fuel, *22*, 1313–7.
206. Mbaraka, I. K., Shanks, B. H. (2006). Conversion of oils and fats using advanced mesoporous heterogeneous catalysts. J. Am. Oil Chem. Soc, *83*, 79–91.
207. Sercheli, R., Vargas, R. M., Schuchardt, U. Alkyguanidine-catalyzed heterogeneous transesterification of soybean oil. J. Am. Oil Chem. Soc, (1999). *76*, 1207–1210.
208. Buchmeiser, M. R. (2001). New synthetic ways for the preparation of high-perfor-mance liquid chromatography supports. J. Chromatogr. A, *918*, 233–266.
209. Kirkland, J. J., Truszkowski, F. A., Dilks Jr., C. H., Engel, G. S. (2000). High pH mobile phase effects on silica-based reversed-phase high-performance liquid chro-matographic columns. J. Chromatogr. A, *890*, 3–19.

210. Lopez, D. E., Suwannakarn, K., Bruce, D. A., Goodwin Jr., J. G. (2007). Esterification and transesterification on tungstated zirconia: effect of calcination temperature. J. Catal, *247,* 43–50.
211. He, C., Baoxiang, P., Dezheng, W., Jinfu, W. (2007). Biodiesel production by the transesterification of cottonseed oil by solid acid catalysts. Front. Chem. Eng. Chin, 1, 11–15.
212. Ngamcharussrivichai, C., Totarat, P., Bunyakiat, K. (2008). Ca and Zn mixed oxide as a heterogeneous base catalyst for transesterification of palm kernel oil. Appl. Catal, A, *341,* 77–85.

INDEX